Love the SAT Test Prep presents

SAT® Math Mastery

2nd Edition

Volume 1

By Christian Heath

Founder of Love the SAT Test Prep

www.LovetheSAT.com

For information regarding bulk purchases, reprints, & foreign rights, please email Help@LovetheSAT.com.

Also by Christian Heath

SAT Math Mastery, Vol.2: Advanced Algebra, Geometry and Statistics

Get higher SAT math scores - guaranteed!

The second volume of SAT Math prep adds another 19 critical math lessons that break the math test down into easy topics to master before test day. Over 325 more realistic SAT practice questions exclusive to this textbook. Comprehensive Pretest & Posttest diagnostics to quickly identify your weak spots.

Available on Amazon at https://amzn.to/2z6hMqe.

SAT & ACT Grammar Mastery, Ed.2

Get higher SAT & ACT grammar scores - guaranteed!

A revolutionary new grammar textbook for higher SAT & ACT scores. Master the seventeen rules of the SAT Writing and Language and ACT English sections in record time.

17 lessons break the grammar tests down into easy topics to master before test day. Over 320 realistic SAT & ACT practice questions exclusive to this textbook. Comprehensive Pretest & Posttest diagnostics to quickly identify your weak spots.

Available on Amazon at https://amzn.to/36LW9Nl.

Ultimate Time Management for Teens and Students

If there's one thing that unites every high school student, it's that they never have enough time or energy to get everything done. It's time for that to change.

This book contains an arsenal of tips, tricks, and strategies from a veteran SAT & ACT tutor and elite-college graduate that will work for every high school student at any point in their high school career.

Get better grades, have more fun, reduce your anxiety, enjoy life more, win more scholarships, and get into a better college!

Available on Amazon at https://amzn.to/2SxWPj8.

A Special Bonus for Readers of this Book

As one of the top SAT tutors in the world, I can help you increase your SAT score beyond the lessons in this math book.

I've got some special SAT stuff available for you at www.LovetheSAT.com/sat-math-mastery-bonus.

Follow the link above or enter it into your browser. This is *only* available for people who have this math book.

If you're trying to get a higher SAT score, I *know* these bonus materials will help you. Before you read any further, please follow the link and claim your score-raising bonuses!

A Note on the Two Volumes of this Book:

When I originally wrote this textbook, I planned for it to be a single giant book that completely covered every single math topic on the SAT test.

Unfortunately, I learned that the book-printing service I use could not physically print the size of book that I had written. My manuscript simply exceeded the physical page limit by almost 700 pages. There was no way to shorten the book down without losing a huge amount of essential content and practice questions.

As a result, I decided the best choice was to split the textbook into two volumes to fit the page requirements of the printer and simultaneously lower the price of both volumes to make them more affordable.

Volume 1 covers Algebra 1 with some basic Algebra 2 and related skills. These are foundational skills that make up over half of the SAT Math questions. The entirety of the SAT Math test is based upon this foundation.

Volume 2 covers Advanced Algebra 2, Geometry, and Statistics. These are more advanced and/or rarer topics that, when combined, make up the other half of the SAT Math test. These topics are just as essential to a high score as the topics covered in Volume 1. Volume 2 is available on Amazon at https://amzn.to/2z6hMge.

It's best to think of these two volumes as a single textbook, the way it was originally written. I will frequently make cross-references between the two volumes that you will miss out on unless you have both texts.

I strongly urge you to get both volumes of this book from the very beginning of your SAT Math studies. Together, they form the ideal single SAT Math book that I envisioned when I started the project. The only reason it's split into two volumes is because it was not physically possible to print it all as a single book.

Both volumes of the book are available on Amazon.com. If you only have one of the volumes, I highly recommend that you order the other volume right away so you have access to the complete "original" textbook, the way I always intended it.

Recommended Lesson Study Plan

The recommended study plan follows the chapters of the two volumes in order, working from front of Volume 1 to back of the Volume 2.

If you follow this study plan, each lesson will build on previous lessons. It is the best plan to follow for most students in most situations.

Volume 1

1. Basic Algebra 1
2. Advanced Algebra 1
3. Absolute Value
4. Algebra 1 Word Problems
5. d=rt
6. Averages with Algebra
7. Ratios & Proportions
8. Unit Conversions
9. Linear Equations (Algebraic)
10. Linear Equations (Words, Charts & Tables)
11. Probabilities
12. Charts & Tables
13. Percents
14. Exponents & Roots
15. Exponential Growth & Decay
16. Basic Algebra 2
17. The Quadratic Formula
18. Advanced Algebra 2
19. Algebra 2 (Parabolas)

Volume 2

20. Polynomial Long Division
21. Equation of a Circle
22. Imaginary & Complex Numbers
23. Conjugate Fractions
24. Functions
25. Systems of Equations
26. Graphs (Other)
27. Angles
28. Pythagorean Theorem
29. Special Right Triangles
30. Similar Triangles
31. Soh Cah Toa
32. Additional Trigonometry Topics
33. Circles, Arcs, Sectors & Radians
34. Area & Volume
35. Geometry into Algebra
36. Combined Shapes
37. Basic Statistics
38. Statistics Conclusions

Topics By Frequency of Appearance on the SAT Math Test

This list ranks the topics by the *frequency* of their appearance on the entire SAT Math test. This list gives the topics out-of-order and you may find yourself lacking critical basic skills if you try to study the lessons in this order. I've given the approximate percentage that each topic appears on an average SAT Math test.

If you are not sure what study plan to use, then follow the "Recommended Study Plan" on the previous page.

1. (10.7%) Algebra 1 Word Problems
2. (8.6%) Linear Equations (Words, Charts & Tables)
3. (7.9%) Systems of Equations
4. (6.9%) Basic Algebra 1
5. (4.8%) Advanced Algebra 1
6. (4.7%) Linear Equations (Algebraic)
7. (4.3%) Graphs (Other)
8. (4%) Percents
9. (3.8%) Basic Algebra 2
10. (3.8%) Charts & Tables
11. (3.6%) Functions
12. (3.3%) Algebra 2 (Parabolas)
13. (2.8%) Exponential Growth & Decay
14. (2.7%) Basic Statistics
15. (2.4%) Ratios & Proportions
16. (2.2%) Statistics Conclusions
17. (2.2%) Advanced Algebra 2
18. (2%) Geometry into Algebra
19. (1.9%) Angles
20. (1.6%) Circles, Arcs, Sectors & Radians
21. (1.4%) Exponents & Roots
22. (1.4%) d=rt
23. (1.3%) Unit Conversions
24. (1.2%) Equation of a Circle
25. (1.2%) The Quadratic Formula
26. (1.2%) Probabilities
27. (1.1%) Similar Triangles
28. (0.9%) Additional Trigonometry Topics
29. (0.9%) Averages with Algebra
30. (0.9%) Absolute Value
31. (0.7%) Conjugate Fractions
32. (0.7%) Imaginary & Complex Numbers
33. (0.6%) Combined Shapes
34. (0.6%) Area & Volume
35. (0.5%) Polynomial Long Division
36. (0.4%) Soh Cah Toa
37. (0.4%) Special Right Triangles
38. (0.1%) Pythagorean Theorem

Topics By Importance to No-Calculator Section

This list ranks the topics by their importance to the No-Calculator Section of the SAT Math test. There are seven topics in this book that do not seem to appear on the No-Calculator Section. I've given the approximate percentage that each topic appears on an average SAT No-Calculator Math test.

This list may be useful if you are scoring lower on the No-Calculator Section than on the Calculator Section.

1. (12.5%) Systems of Equations
2. (9.5%) Basic Algebra 1
3. (8%) Linear Equations (Words, Charts & Tables)
4. (7.5%) Algebra 1 Word Problems
5. (6.5%) Basic Algebra 2
6. (6.5%) Advanced Algebra 1
7. (5.5%) Linear Equations (Algebraic)
8. (4.5%) Advanced Algebra 2
9. (4.5%) Functions
10. (3.5%) Exponents & Roots
11. (3%) Angles
12. (3%) Algebra 2 (Parabolas)
13. (2.8%) Circles, Arcs, Sectors & Radians
14. (2.5%) Similar Triangles
15. (2.5%) The Quadratic Formula
16. (2%) Imaginary & Complex Numbers
17. (2%) Conjugate Fractions
18. (2%) Geometry into Algebra
19. (2%) Graphs (Other)
20. (1.5%) Polynomial Long Division
21. (1.5%) Percents
22. (1%) Soh Cah Toa
23. (1%) Absolute Value
24. (1%) Additional Trigonometry Topics
25. (1%) Equation of a Circle
26. (1%) Exponential Growth & Decay
27. (0.5%) Area & Volume
28. (0.5%) Ratios & Proportions
29. (0.3%) Special Right Triangles
30. (0.3%) Combined Shapes
31. (0.3%) Averages with Algebra

Topics By Importance to Calculator Section

This list ranks the topics by their importance to the Calculator Section of the SAT Math test. There are three topics on this book that do not seem to appear on the Calculator Section. I've given the approximate percentage that each topic appears on an average SAT Calculator Math test.

This list may be useful if you are scoring lower on the Calculator Section than on the No-Calculator Section.

1. (12.4%) Algebra 1 Word Problems
2. (8.9%) Linear Equations (Words, Charts & Tables)
3. (5.8%) Charts & Tables
4. (5.5%) Systems of Equations
5. (5.5%) Basic Algebra 1
6. (5.5%) Graphs (Other)
7. (5.3%) Percents
8. (4.2%) Linear Equations (Algebraic)
9. (4.1%) Basic Statistics
10. (3.9%) Advanced Algebra 1
11. (3.7%) Exponential Growth & Decay
12. (3.4%) Algebra 2 (Parabolas)
13. (3.4%) Ratios & Proportions
14. (3.3%) Statistics Conclusions
15. (3.2%) Functions
16. (2.4%) Basic Algebra 2
17. (2.1%) d=rt
18. (2%) Geometry into Algebra
19. (2%) Unit Conversions
20. (1.8%) Probabilities
21. (1.3%) Angles
22. (1.3%) Equation of a Circle
23. (1.3%) Averages with Algebra
24. (1.2%) Combined Shapes
25. (1.1%) Advanced Algebra 2
26. (1.1%) Circles, Arcs, Sectors & Radians
27. (0.8%) Absolute Value
28. (0.8%) Additional Trigonometry Topics
29. (0.7%) Area & Volume
30. (0.5%) The Quadratic Formula
31. (0.5%) Special Right Triangles
32. (0.4%) Similar Triangles
33. (0.3%) Exponents & Roots
34. (0.1%) Soh Cah Toa
35. (0.1%) Pythagorean Theorem

Formulas and Equations to Memorize (Volume 1)

Note: This is a partial list of some important formulas, equations, and concepts that are commonly seen on the SAT Math test that we cover in Volume 1 of *SAT Math Mastery*. Remember that Volume 2 (available on Amazon at https://amzn.to/2z6hMge) covers 19 additional SAT Math topics that also have their own formulas, diagrams, and concepts to memorize.

You should have all of these committed to memory before test day.

Volume 1

Distance, Speed, and Time: $d = rt$ (Lesson 5)

Averages: $\text{Avg} = \frac{\text{sum}}{\#}$ (Lesson 6)

Linear Equation: $y = mx + b$ (Lessons 9 and 10)

Probability: $\frac{\text{desired}}{\text{total}}$ (Lesson 11)

Percentage *of*: $\frac{\text{desired}}{\text{total}} \times 100\%$ (Lesson 13)

Percent Change: $\frac{\text{change}}{\text{original}} \times 100\%$ (Lesson 13)

Fractional & Negative Exponents: for example, $2x^{-\frac{5}{7}} = \frac{2}{\sqrt[7]{x^5}}$ (Lesson 14)

Exponential Growth & Decay: $(\text{Starting Value})(\text{Percent Multiplier})^{\text{number of changes}}$ (Lesson 15)

Quadratic Equation: $ax^2 + bx + c = 0$ (Lessons 16 and 17)

The Quadratic Formula: $X = \frac{-b \pm \sqrt{b^2 - 4ac}}{2a}$ (Lesson 17)

Table of Contents

About Me (and Why I Can Help You Succeed!)

Before we get into the book, I want you to understand why I'm going to be able to help you improve your SAT Math score so much.

In high school, my best SAT score was a 1590 out of 1600 - nearly perfect, but off by just one wrong question on the Verbal section. Later in my tutoring career I came back and scored perfect 1600s on official SAT tests.

Now I'm the Founder and Lead Instructor of Love the SAT Test Prep. I've been teaching SAT and ACT Prep full-time for over 10 years. And I run the top-rated SAT & ACT Prep center in Austin, Texas - a university town filled with great tutors and smart students, so the competition is stiff.

This is my newest book, and it's my best one yet (my previous favorite was *SAT & ACT Grammar Mastery*, which is still helping students get higher Verbal scores and is available on Amazon here: https://amzn.to/2vOcO6L).

Here are the highlights of my SAT / ACT experience:

- 10+ Years of experience teaching SAT & ACT Prep.
- Perfect Score of 1600 on the SAT Test (including a perfect 800 on the Math section).
- Over 1500 students taught.
- Over 12,000 hours of SAT & ACT teaching experience.
- Top-Rated SAT & ACT Tutor in Austin, Texas by customer reviews.

Most of my physical students are in Texas, but I've worked with on Skype with students from dozens of countries around the world. I've also been invited three times to teach SAT Prep to top students in China.

I've written over 200 free blog articles focused on SAT / ACT Testing & College Readiness.

Basically, the last 10 years of my life have been devoted to finding the best ways of helping students around the world improve their SAT & ACT scores.

It is deeply meaningful to me to help so many students improve their results - and by extension, their college & scholarship opportunities - but most importantly, I want to make a positive impact on your *whole life*. That's the power of education. That's the power of higher SAT Math scores.

This is a mission for me and, not to brag (well, maybe a little bit) - it's one I've gotten pretty good at.

You can learn more about me at www.lovethesat.com/about-founder-Christian-Heath.

And now it's time for me to help *you*!

How to Use This Book

I have carefully planned & designed the two volumes of this book to be useful to self-studying students, SAT tutors, and teachers. To get the most out of them, I recommend you follow the study plan below.

Study Plan:

1. Study the **Prelessons** either before or after you take the Pretest - or whenever review is needed.
2. Take the **Pretests** & identify your personal list of math topics that need further attention. You can either take both pretests immediately or save the second one for later. Start with the first one, though.
3. Study each **lesson** from the list of questions you miss on the Pretests. Work from the front of the book to the back. Prioritize fundamental prerequisite lessons before moving onto more advanced topics.
4. After you study each lesson, work the **Practice Problems** for that lesson. Check and correct them. The answers and explanations are at the end of each lesson.
5. Finish studying & completing the practice problems for all the lessons on your personal list of topics.
6. Take the **Posttests** at the end of the book. You can either take both posttests immediately or save the second one for later. Again, check your answers to identify topics that need further attention.
7. For any questions you miss from the Posttest, review the lessons carefully. You should work the Practice Problems again on a fresh sheet of paper.
8. Follow the instructions in **What to Do Next** at the end of the book.
9. Get the second volume of *SAT Math Mastery* (available on Amazon at https://amzn.to/2z6hMge) and study the next set of SAT Math lessons.
10. Return to this book whenever you want or need to review specific SAT Math topics.

More Details

The Pretest at the beginning of the book is a diagnostic test. It's split into two halves. You can take the first half by itself and save the second half for later, or take both halves together. Either way, be sure to take the first half of the Pretest first, because the second half is slightly more difficult.

After you take the **Pretests**, check your answers. Notice that each question is connected to a specific math topic covered in the book. Make a list of topics you missed. Prioritize studying these lessons.

The **Prelessons** are highly recommended supplemental content - particularly the lesson on **Careless Mistakes**. These are overarching lessons that have an impact on every other lesson in the book. They include lessons on technical content like **Fractions** as well as mindset, time management & SAT strategy.

The bulk of the book is focused on the first 19 of the **38 Math Lessons** themselves. This is where most of your time with this book should be spent. (The next 19 lessons are covered in Volume 2 of *SAT Math Mastery*).

From the list of topics you missed from the Pretest, study each lesson one at a time. I recommend simply starting with the earlier lessons and working front-to-back through the book, because many of the later lessons are based on previous topics.

Each lesson shows the Prerequisite lessons that you should master before studying it. I've also given the approximate percentages that each topic appears on the SAT Math test to help you decide what's worth your time if you're in a hurry.

Near the beginning of each lesson is a "Quick Reference" where the key points of the lesson are summarized. This is not a substitute for studying the lesson in-depth. However, it is a useful tool to quickly review key points - for example, in the final days before you take the SAT, or whenever you just need a quick refresher.

The lessons thoroughly teach everything you need to understand about each Math topic on the SAT. Some lessons are shorter and some are significantly longer - it just depends on the topic. Each lesson also explains the two Pretest questions in detail.

The lessons are filled with content. Many contain practice examples that you should complete as you move through the lesson. Study the lessons and answer any practice examples inside the lesson. I also recommend taking notes to keep yourself engaged and improve your retention of what you study.

At the end of each lesson there is a set of **Practice Problems**. Once you've thoroughly studied the lesson, move onto the Practice Problems. I recommend completing all of of the Practice Problems before moving onto the next lesson.

Check and correct your Practice Problems. The answers and explanations are at the end of each lesson. Be sure to review any questions you get wrong (or can't figure out). Generally, you want to make the most of each lesson and understand all the practice questions before moving on to the next topic.

The **Posttests** at the end of the book are a pair of final diagnostic tests. They are similar to the Pretests, but many of the questions are more difficult. I recommend taking the Posttests only after you've studied all the topics you missed on the Pretest.

Again, the Posttest is divided into two sections. You can take them both at once, or just do the first part and save the second half for later review.

Either way, check and correct your Posttest. Make a list of all the topics you miss on this final diagnostic test. You should return to each of these lessons and study them again in depth. I also recommend re-doing the Practice Problem from each of these lessons on a fresh set of paper.

After you finish everything I've listed above, follow the steps I describe in **What to do Next** at the end of the book. Also make sure you've got your copy of the second volume of *SAT Math Mastery* (available on Amazon at https://amzn.to/2z6hMge). There is an entire second book of SAT Math topics to study after this one.

Understand: this project will take time and patience. Your rewards on the SAT Math test will directly relate to the investment of time and energy you commit to this book.

Therefore, be patient with yourself. I know how frustrating it can be to feel stuck on a topic. But once a math topic "clicks" for you, you'll realize it was never even hard in the first place.

Now go forth and conquer!

Prelesson 1: Overview of the SAT & SAT Math Test

Before we get started into this book, I'd like to give a broad overview of the SAT Test and its Math sections from my point of view as a SAT & ACT tutor with perfect scores the test and over a decade of experience working with high school students and their parents.

As you probably know, the SAT test is a college admissions test considered alongside your GPA and extracurriculars when applying to colleges. It's used as a benchmark, a way of comparing the academic levels of students from different high schools. It is one of the longest, most challenging, and most important tests that students take in their high school career. Your SAT scores will typically play an important role in your applications to most U.S. colleges, universities, and even scholarships.

The SAT test breaks into four timed sections and an optional Essay section - in order, Critical Reading, Writing & Language (i.e. "Grammar"), Math without Calculator, and Math with Calculator. The optional timed Essay section, if you take it, comes after these four sections.

SAT scores are reported out of a possible total of 1600 points (a perfect score). This breaks down into a "Verbal" score out of 800 points and a "Math" score, also out of 800 points.

The SAT Math test itself is composed of two timed sections - a shorter "No Calculator" section and a longer "Calculator" section. Each of these two sections also divides into a longer Multiple-Choice section and a shorter Free Response section (more on this later).

The No Calculator section is 20 questions long with a time limit of 25 minutes. There are 15 multiple-choice questions and 5 free response questions in the No-Calculator section. The Calculator section is 38 questions long with a time limit of 55 minutes. There are 30 multiple-choice questions and 8 free response questions in the Calculator section.

There is no penalty for guessing on the SAT test. Wrong answers are not counted against you. As a result, you should always answer every question even if you don't have time to solve it before time is called. With luck, you'll get a few of them right for some extra points.

Each SAT Math section starts with more basic questions and moves to more advanced questions. In other words, it is in order of "ascending difficulty." The difficulty resets at the end of the multiple-choice portion, so the first Free Response question will be easier again. Then the Free Response questions follow an ascending difficulty pattern once more.

Note: this "ascending difficulty" pattern is exclusive to the SAT Math sections. The verbal sections of the SAT are *not* arranged in ascending difficulty.

The SAT Math test includes topics from basic arithmetic, Algebra 1, Geometry, and Algebra 2, including a small amount of basic Trigonometry. There are also a few Statistics questions. You do *not* need to have taken any classes on Statistics to succeed. However, the lessons learned in your high school classes on Algebra 1, Geometry, and Algebra 2 will play a major role in your success.

In its current format, the SAT Math test places a huge emphasis on Algebra 1 and 2. The test is nearly 60% Algebra 1 and 25% Algebra 2. Geometry is less significant, at only about 10% of the test, and some basic Statistics concepts round things out with another 5% of questions.

Since most students take Algebra 2 in Junior year, the SAT test is primarily intended for Juniors and Seniors. Still, the test is sometimes taken by Sophomores or even Freshman - but usually just for early practice at this point, not for "high scores."

Most students will be fully prepared for the SAT test by about halfway through Junior Year. When I say "fully prepared," I mean that they will have covered everything in school that they need to understand the concepts on the SAT test.

That does *not* mean that school classes alone are enough to get a great SAT score. My entire career is based on the reality that a huge number of students and parents need to invest additional time, energy, and yes, money, into more thorough preparation for the SAT test.

There is a lot to study. We should all view SAT prep as a long-term project. When possible, I prefer for students to start serious SAT prep in the summer before Junior year. However, many students start later than this. Don't worry - I've seen hundreds of students get great SAT scores even if they didn't start preparing until Senior year. Nevertheless, it's *safer* to start as soon as possible. As with any big, important project, it's typically better to give yourself more time to work.

Statistically, most students will get their highest scores near the end of Junior Year or beginning of Senior Year. Although many families wish they could just get a high score early and be done with SAT testing, the reality is most students need to be patient, keep studying, and test several times - usually leading up to a high score near the beginning of Senior year.

It's common to take the SAT test multiple times. For example, it's pretty standard to take a PSAT (a simplified version of the SAT test) at the beginning of Sophomore year, then another PSAT at the beginning of Junior year, followed by an official SAT in the Spring of Junior year, another official SAT near the end of Junior year, and a final SAT near the beginning of Senior year. Some students even add an extra one or two official tests to this schedule. However, it's uncommon and generally unnecessary to take more than two PSATs and four official SATs in total.

Some colleges follow a "Superscoring" admissions policy. This is where they will combine your highest section scores from multiple SAT test dates. For example, if you get a high score in your SAT Verbal section in March, and a high score in your SAT Math score in October, a "superscore" college will combine your Verbal and Math scores from the two test dates for a higher total score than you got on either individual test day.

Not all colleges will superscore the SAT. If they don't, then you'll simply need to get a high total score on a single test day. It's important for you to do your own research to discover which of your top colleges follow superscore policies and which do not.

In this book, we'll focus on improving your skills and scores on the SAT Math test. And if you work hard, your focus and dedication will be rewarded with higher Math scores that make a substantial difference to your college and scholarship opportunities.

Prelesson 2: Free Response Questions

The Free Response questions, also known as "Grid-Ins" or "Student-Produced Response Questions," are found at the end of both the No-Calculator and Calculator sections of the SAT Math test.

It's important to review the rules of the Free Response questions before test day. You don't want to waste time learning the rules while the timer runs down on test day, and you don't want to be penalized for getting the correct answer but entering it incorrectly.

In contrast to the multiple-choice questions, which each have four answer-choice options, these Free Response questions have no answer choices and instead require you to enter your answer by hand into four blank spaces. There is room for a maximum of four characters per answer - up to four numbers, decimal points, and "slash marks" for fractions. There are no dollar signs or percent signs in the final answers to these questions.

If your final answer doesn't fill all four spaces, then it doesn't matter if you leave blank spaces to the left or right. You will not lose points either way.

There are no negative number answers on any Free Response questions (in fact, there isn't even a negative sign available to use on the bubble response sheet).

You can enter answers as either a decimal or an equivalent fraction. If the answer is a repeating decimal (such as $\frac{2}{3}$, which equals $.666...$ repeating), then you can either "truncate" your answer as $.666$ or round up to $.667$. Either is acceptable. *However*, if you enter this answer as $.66$ or $.6$, you will be counted as wrong.

Mixed Numbers, such as $2\frac{1}{2}$, cannot be entered directly into the Free Response spaces. You must enter this answer as either $\frac{5}{2}$ or 2.5. If you attempt to enter $21/5$, the answer will be read as "$\frac{21}{5}$" and counted wrong.

On the other hand, fractions do *not* have to be reduced to their simplest form, as long as you have room to enter your answer. For example, you could enter your answer as $\frac{4}{12}$ without reducing the fraction to $\frac{1}{3}$. Both answers would be counted as correct.

Some Free Response questions will have more than one possible answer (usually phrased within the question as "one *possible* value" that solves the question). In this case, you are only required to enter any *one* of the possible answers. You only need one of the possible answers to get all the points - don't waste time finding the other possibilities.

Many students report feeling very nervous about the Free Response questions, because there are no multiple-choice answers to check your results against. It's also common to run out of time to work on them, because they're at the end of the math sections.

However, the difficulty levels and the math topics tested in the Free Response are identical to those in the multiple-choice questions. The Free Response questions are actually *not* any more difficult than the multiple-choice questions - they just might "feel" like they are.

In my experience, it's mostly a psychological difference that causes students to feel uncertain or anxious without the safety net of multiple choice options. In other words, the difference is just in your head!

Also, without the safety net of multiple-choice answers, reading the exact wording of the question becomes even more important than ever. Many questions have very precise requirements for the format of your final answer.

For example, imagine that the Free Response question says to solve for x and "round your final answer to the nearest tenth."

If the exact value of x is 6.27 and you enter 6.27 as your final answer, you would be counted wrong and receive no credit. In this imaginary question, you were required to round to the nearest tenth, so the only acceptable final answer would be 6.3.

There is no partial credit for such rounding errors, even if you did all the other work correctly. Making even a couple of mistakes like this one can be very punishing to your SAT Math score. On the Free Response questions, you either enter your answer exactly as required, or you get the question entirely wrong.

Because of the importance of Free Response questions to your overall SAT Math score, I've made sure to include a good mix of Free Response questions throughout this book.

Be careful on these questions! To help you train, I've also included some common SAT-style traps for students who rush or misread the exact wording of these questions...

Prelesson 3: SAT Math vs ACT Math

Many families I consult want advice on the differences between the SAT and the ACT tests. Again, I've been teaching both tests for over a decade, and I also have a perfect score on both tests, so I think that I can provide you with some useful insights.

Without going overly in-depth, I'm going to give some key points that compare and contrast the SAT and the ACT tests, leading up to a particular focus on the differences between their Math sections.

You may already know that the ACT is an alternative test to the SAT. It fills the same role as a comprehensive college admissions test.

First off, I think the tests are more similar than different. They are both long, timed tests on Reading, Grammar, and Math, with an optional Essay section. The ACT does also include a Science section that's not on the SAT test, but this section is more about reading charts and graphs than actual "science knowledge."

Therefore, I usually ballpark the two tests at about "80% similarity." This is a natural consequence of them both being used for the same purpose: broadly evaluating the academic skills of high school students as part of the college admissions process.

Still, there are some key differences between the SAT and ACT tests.

One of the most noticeable differences is the timing. The ACT test is notorious for being very fast-paced and having a brutally short time limit. In contrast, the SAT is a bit more generous with the time you get for each section. Of course, most students would *prefer* to have more time to work on the SAT, but it's the ACT that really puts your high-speed work to the test.

Another noticeable difference is difficulty. If I had to sum things up in one sentence, I'd say that the individual questions throughout the ACT tend to be *slightly* easier than the equivalent questions on the SAT. The SAT Reading test, in particular, features harder reading passages, more advanced vocabulary, and more challenging questions than the ACT Reading.

The Grammar sections of the two tests are nearly identical - so similar that I was able to write a single textbook, *SAT & ACT Grammar Mastery*, that covers all the Grammar topics for both tests. It's available on Amazon, and I'd highly recommend getting it if you want to improve your verbal scores on either the SAT or ACT.

So overall - speaking in a *very* broad generalization - I think the ACT is *slightly* easier on a question-to-question basis, but you have to work much *faster*. The SAT is perhaps *slightly* harder on a question-to-question basis, especially in the Reading section, but you have noticeably more time to work on each question.

Focusing more specifically on the Math sections, the ACT test gives 60 questions in 60 minutes, compared to the two SAT Math sections that total a combined 58 questions in 80 minutes. You can immediately see that the pace of the ACT is faster - it has more questions to answer in less time.

The ACT Math also tends to involve more *variety* of questions. It tests a wider variety of math topics and the mix is more randomized. In contrast, the SAT places a huge emphasis on Algebra - especially Algebra 1 - so I feel that the SAT Math test actually requires mastery of *fewer* math topics than the ACT Math does. If you're preparing for the SAT test, this is good news.

However, the ACT Math tends to have fewer steps per question. I sometimes call them "bite-sized" - they are smaller, faster, and shorter questions on average. On the other hand, a typical SAT Math question has a few more steps in the solution process and a bit more complexity than an equivalent ACT Math question.

For example, it's more common for an SAT Math question to mix several unrelated math topics into a single question. On the ACT Math, it's more common for each question to focus more narrowly on a single math concept per question.

The most common student complaint I get about SAT Math questions is "I don't know how to *start* this." The most common complaint I get about ACT Math questions is "I don't *remember* how to do this."

I think these two complaints sum up the essential difference between the two math tests: the SAT Math questions are more complex (so students feel like they don't know where to *begin*), but the ACT Math covers a broader variety of topics (so students feel like they've seen this kind of question before, but can't *remember* what they've learned about it).

Let's also consider the "difficulty curves" of the two tests. The difficulty scaling for the SAT Math test is a gradual but constant increase from easier to harder questions throughout the section. Most students tell me they can feel the SAT Math questions gradually getting more difficult throughout the section. On the other hand, many ACT Math students report feeling "like it was really easy until it suddenly got really hard." In other words, the jump from "easy" to "difficult" seems more abrupt on the ACT Math test.

Personally, I feel that the ACT Math is actually a more challenging experience than the SAT Math. For a long time I believed the opposite. But for several years now, I've felt that the SAT's predictable focus on Algebra 1 and Algebra 2, plus the slower pace and extra time, makes it easier to prepare for. I would rather prepare for slightly more challenging questions in a narrower range of math topics than slightly easier questions with a wider range of topics. There's just less "stuff" to review and remember. But that might just be me.

It is also worthwhile to point out that the ACT Math is only *one* of four test sections, while the SAT Math counts for *two* of four sections. This makes an argument for math skills being more significant to your overall SAT score, since your SAT Math score is 50% of your total test score, instead of just 25% of your total score on the ACT.

At the end of the day, I feel like these comparisons between the SAT and ACT Math tests are somewhat "splitting hairs." Both are demanding and comprehensive challenges for high school students. There is no "free ride" to be had by choosing one test or the other. Overall, the tests are more similar than different - both are comprehensive tests of high school math from arithmetic all the way through Algebra 2 and basic trigonometry. In my experience, it's best to pick a single test and focus on it until you've mastered it.

Prelesson 4: SAT Math Mindset

Over the years, I have noticed the startling difference that the correct mindset makes to SAT Math scores. Keeping a positive mindset with a good big-picture perspective, staying persistent with tough questions, ensuring that you're totally focused on your task, and finding a way to enjoy math are some of the best ways to improve your SAT Math score.

How you handle the frustration of a tough question or the anxiety of a timed math test make a world of difference to your overall performance.

First of all - and I can't emphasize this enough - try to stay positive, *especially* when you are most tempted to be "down on yourself" or frustrated with the test itself.

Maybe I'm lucky, or maybe I'm crazy, but I've always loved math. Perhaps I didn't like a certain *topic*, or a certain *teacher*, but I've always enjoyed the challenge of solving an interesting math problem.

Try to discover what *you* like about math. For example, many math-lovers say that they like that "there's only one right answer" to each question. They like the certainty of getting one final answer and being finished with the question, in contrast to the ambiguities of History and English - subjects where it can feel like "there is no one right answer."

Other students enjoy recognizing math topics they've seen before, applying rules they understand, and following a specific process that gets them the answer. There is something comforting and rewarding about working through the steps of a good math problem.

It also helps to have a sense of curiosity towards each question - to view them as a puzzle or a mystery to solve. Each math question is a riddle - some simple, some complex. They all have their own mysteries to unlock and understand. Try thinking of it as a game!

Next, never tie your self-worth up with your SAT Math score. Take the big-picture perspective. You're only taking the SAT test to serve your long-term goals. It's just a tool to help open up more options for your college applications.

I promise, your SAT score won't "make or break" the rest of your life. Most people never think about the SAT again after high school. Keep a long-term perspective. I know this can be tough when you're in high school and you have pressure from all angles - from parents, teachers, friends, homework, sports, and certainly not least - from *yourself*!

Don't be too hard on yourself. Laugh off your mistakes and frustrations, then learn from them and move forward.

Persistence is also key - both in the big picture of your SAT prep and on the smaller level of individual math questions. It's a weird paradox: when you're studying SAT Math, the more time you're *willing* to spend on a question, the quicker you'll usually figure out the answer.

Of course, the official SAT is a timed test. But your homework and study time is *not* under the same time limit. If you're willing to spend ten times longer on a question before giving up - when you reach that level of *commitment* - you will be surprised at how quickly you start solving even the hardest of questions.

Focus plays a huge role in your results. Even my top math students can have a bad day when they're tired, distracted, or just "not in the mood" to work. On these days, I see their results drop immediately.

Often, these students even recognize what's happening and explain (somewhat apologetically) that "I wasn't very focused when I did this section of homework." Sometimes they'll tell me that they were doing their work in a room where another family member was watching TV. Usually, they don't even need to tell me - it's obvious from the change in their results.

On the other hand, I've seen so-called "weaker" math students do exceptionally well on their homework. When I ask why they got so many questions right this time, they say "this week I just sat down in a quiet room and stayed *really* focused on my work." I have these sorts of conversations about *focus* with students every day.

In addition, *emotional management* plays a major role in your performance. When everything is going well and the questions feel easy, this isn't a problem. But when the questions get tougher (and they will), then your emotional management comes into play.

When I get stuck on a question, I take comfort that I can never have a *breakthrough* unless I'm stuck *first*. I know it feels good to have a breakthrough and I remind myself that feeling stuck is just the first necessary step in reaching that breakthrough.

I also know that once I figure the question out, I'll usually realize it wasn't even hard in the first place. I've also seen this among my students again and again - even the ones who tell me they're "bad" at math. When they get completely stuck on a question - and then I explain the solution - one of the most common reactions is "Oh, that's all?" They realize the question wasn't nearly as hard as they assumed it was.

We're often only one tiny insight away from figuring out the whole math problem. Every day, I've seen that one small hint leads my students to wave me away from the tutoring desk, saying "OK, give me a second - I think I can figure this out now."

When I feel frustrated with a hard question, I remind myself that I'm probably just one small step from figuring out what to do next - and once I'm finished, I'll look back and laugh at myself for ever thinking it was hard.

Overall, the SAT Math questions are *much* easier than most students think. But they defeat themselves before giving themselves a chance to succeed - throwing down their pencil and giving up, or shrugging and pleading that they "just have no idea how to start."

Mindset is the key to success in every aspect of life, including the SAT Math test. Yes, it's essential to understand the individual math topics - that's the main purpose of this book - but the *foundation* of your SAT Math score is your mindset, focus, and persistence.

Prelesson 5: How to Work SAT Math Problems

When you work SAT Math questions, precision and accuracy are extremely important.

Because **Careless Mistakes** lose more points on SAT Math than any other topic, we should dedicate ourselves to anticipating, preventing, and correcting the issues that cause common Careless Mistakes (more on these issues soon).

It's not worth saving a few seconds of time just to fall prey to a simple mistake in your calculations.

Therefore, avoid Mental Math. Instead, write your steps cleanly and neatly on the paper.

As a tutor, it terrifies me when my students start working math problems in their head. Unfortunately, it's often the kids with the strongest math skills who tend to do this.

Personally, I know how tempting it is to do all my work in my head. It feels *nice* to do mental math. Sometimes it seems like I'm saving time - as if I'm wasting time by writing down each of my steps. Sometimes, I just want to show off my skills.

But the truth about most mental math is that it's a sign of overconfidence, often connected with rushing your work, and the SAT Math test *will* punish you for it.

One of the biggest problems with Mental Math is that you can't check and correct any work that's not written down. If you make a mistake in your head, all of the following steps will be ruined, and you won't be able to diagnose the error without working the problem again from scratch.

So, do everything you can to avoid Mental Math. Make a vow *not* to trust the math you do in your head. Get a written setup for the question. Label your diagrams. Show your steps - cleanly and neatly.

When using a calculator, don't do more than one step on the calculator before writing down your results. Calculators are for individual *calculations* (like multiplying or dividing large numbers), not for problem-solving. Like Mental Math, it's difficult to check the work you've done on a calculator, and hard to correct any Careless Mistakes you may have made.

Because Misreading mistakes are a common problem on the SAT, be sure to reread the question carefully before giving your final answer. Many times, right after we finish a long set of calculations, we *assume* that the result of our final step is also the final answer to the SAT question. Don't assume. Double-check the exact wording of the question before you move on.

Prelesson 6: The 4 Careless Math Mistakes

There are four Careless Mistakes that are responsible for a *huge* number of lost points on the SAT Math test. From the weakest to the strongest Math students - wherever you fall on that spectrum - these Careless Mistakes will eat your score alive. They do not discriminate.

What's worse, the SAT *knows* these common mistakes and deliberately writes many of their questions and answer choices to exploit and catch students with these mistakes.

After 11 years of tutoring, I am 110% certain that these are *the* four most common and dangerous Careless Mistakes that *everyone* makes on the SAT Math test. Disregard the following info at your own peril.

1. Negative Signs and Subtraction

The single most common math mistake on the SAT, by far, is Negative Signs and Subtraction.

Sometimes I joke with my students that, as a math tutor, I could replace myself with an audio recording that just says "Did you check your Negative Signs?" on repeat, then lean back and press a button to play it on every question whenever someone is getting the wrong answer.

As a general habit, you should always double- or even triple-check your Negative Signs.

What's more, some questions seem like they're just *made* by the SAT test to draw out Negative Sign mistakes on purpose. They're not hard to notice if you're looking for them - these are questions that have negative numbers or subtraction signs in key parts of the question.

Also look out for Answer Choices are are obviously playing with negative signs - for example:

(A) -5

(B) -3

(C) 3

(D) 5

Answer Choices like these are a dead giveaway that many students *will* make a negative sign mistake on this question, so be sure you're not among them.

The secret to handling this mistake is to *expect* yourself to make it. Trust me, we all fall for it from time to time. Then you've got a little doubt buzzing at the back of your head - "could I have made a Negative Sign mistake?" - that reminds you to double-check.

Any time you feel like something is "off" in your work, or if you just want to know what to prioritize when checking your work at the end of a section, it's Negative Signs. It always is.

2. Distributing and Parentheses

The next most common Careless Mistake is anything that involves Distributing or Parentheses. These are often mistakes from **PEMDAS** or "Order of Operations," which we'll cover in more detail just a few pages from now.

A simple example of distributing is below:

$$3(2x+4)$$

You'd be surprised how many students - when they're stressed, tired, or in a hurry - will accidentally Distribute this as:

$$3(2x+4)=6x+4$$

Of course, the correct version would be:

$$3(2x+4)=6x+12$$

Similar Distributing mistakes often happen during Algebra solutions, when students forget to distribute a step to *everything* on both sides.

Many students also make Parentheses mistakes when squaring a parentheses, as in the example below.

$$(x+3)^2$$

Here's the *mistake* version:

$$(x+3)^2=x^2+9$$

The version above is *wrong* and comes from a misunderstanding of how the ^{2} will apply to the parentheses. This expression must be FOIL'ed (from **Basic Algebra 2**). The correct version is below:

$$(x+3)^2=(x+3)(x+3)$$
$$=x^2+6x+9$$

Just last night, I saw a student with a 750 on the math section make this exact mistake. She has a *750*, close to a perfect score. She knows what she's doing and math is a strong point of hers, and even *she* forgot about it.

Guess what? The SAT test had given a "fake" answer choice to catch her making this exact mistake. Never forget - the test writers *know* the most common mistakes and exploit them.

Distributing and Parentheses can also combine with Negative Signs and Subtraction to create the kind of situations that give me nightmares as a tutor. For example, the situation below makes me bite my fingernails when students are working on it:

$$-(4-x)^2$$

This must first be FOIL'ed out, because **PEMDAS** says **E**xponents happen before **M**ultiplication:

$$-(4-x)^2$$
$$=-(16-8x+x^2)$$

Then the Negative Sign at the front must be distributed to all three terms:

$$\begin{aligned}&-(4-x)^2\\&=-(16-8x+x^2)\\&=-16+8x-x^2\end{aligned}$$

You can imagine how many ways there are to make a careless mistake in situations like the one above.

It's not that this is *hard* math - it's just math that leaves you a lot of room to make your own mistakes, then the SAT follows it up with "fake" answer choices that will catch you with the common errors.

3. Misreading or Incomplete Reading

Misreading the question is another extremely common and deadly - but predictable - Careless Mistake. Naturally, it happens the most on longer, more detailed, and wordier questions. It's also a common error on questions with charts or tables, since these contain a lot of detailed information in a small space.

I've noticed that two types of students tend to make this mistake in two different ways.

One one hand, "Misreading" is usually a trait of *overconfident* students who rush the question or make assumptions. They often see what they *expect* to see, rather than what's actually written in the question.

On the other hand, "Incomplete Reading" is a trait of *under-confident* students who "get stuck" somewhere in the problem. They forget to go back and check the question for other key information that they haven't used yet.

The longer and wordier the question, the greater the chance that you will make a Misreading mistake. When you see long, wordy problems - or questions involving charts and tables - just *slow down* and *read thoroughly.*

It is also helpful to underline or circle key information. Not exactly rocket science, I know, but there's something about marking up the question that helps prevent Misreading mistakes.

If you feel like you're completely stuck on a question - especially a longer, more detailed word problem - make sure you've used everything in the problem before giving up.

4. The Switcheroo

Although technically a type of Misreading mistake, "The Switcheroo" is another distinctive and predictable Careless Mistake that the SAT *purposefully* tries to catch you with.

This Careless Mistake can happen even when you actually solve the math correctly. For example, you finish an Algebra question and find that $x = 5$. Feeling good about your work, you check the answer choices, and right there at the top, it gives you Choice A) 5. So you pick Choice A and move on, feeling confident.

Then when the test comes back, you find out you missed this question. Shocked, you look for a mistake in your work, but there isn't one. You're completely certain that $x = 5$.

And then you reread the question and realize, horrified, that the question asked for the value of $3x$, not x. And there in the answer choices, sitting at the very bottom, is D) 15. It's your original $x = 5$ solution, but multiplied by 3 to give the correct answer that they actually *asked* for.

Congrats, you fell for a Switcheroo Mistake! It's happened to all of us at one point or another.

You solve for Radius, but the question asks for Diameter. You solve for Length, but the question asks for Width. You solve for n, but the question asks for $4n$. You solve for x, but the question asks for y. You solve for Area, but the question asks for Perimeter.

These are all common, everyday setups for Switcheroo mistakes. Remember, they're *deliberately* included by the authors of the SAT. You'll see them everywhere once you start looking for them.

The really devious thing about this mistake is that the SAT *plans* for it to happen to you. That's why Answer Choice A almost always has the "trap" answer up at the top of the list - the first answer you notice - so your eyes jump to it and you assume you're finished.

Then further down near the bottom of the answer choices is the correct answer. But it's too late - you already *assumed* you had your final answer.

Perhaps the saddest thing about this mistake is that you spend all the time and energy to do the problem correctly. It's literally the *very last step* of the question where you make this mistake and throw everything away at the last second.

Like most other Careless Mistakes, the key is to *expect* this sort of trickery on test day. Look for it in your practice questions. It's also similar to Misreading mistakes, so always remember to look for both of them before you commit to a final answer.

Prelesson 7: SAT Math Timing & Pacing

When it comes to SAT and ACT Prep, I always recommend students focus on mastering the test questions one-by-one before they worry about the timer.

My strong recommendation is - for most students with adequate time to prepare (which is ideally a month or more) - *not* to use a timer until you're able to get 90% of SAT Math questions right *without* a timer.

Yes, eventually you must train against the timer. This is a timed test, after all - and time management is a crucial skill. But until you're getting answers *correct*, there's no benefit in rushing to beat the time.

Think of an archer training to compete in the Olympics. If the archer can't hit the bullseye with most of their shots, why would they slap a timer on themselves? They'll just miss more shots - but even faster. That's a lot of time spent walking through the woods looking for missing and broken arrows!

Instead, the archer should focus on mastering their craft - hitting the dead center of the target. Once they can confidently, patiently, and accurately hit bullseye shots - over and over again - then, and *only* then, should they start worrying about how *fast* they can shoot.

So, your first priority - assuming you've got more than a month before your final test date - is to hit the target - to master the content of the Math test on a topic-by-topic basis. That's what this book is for.

At the end of the book, there's a section titled **What to Do Next**. But essentially, once you've used the Posttest in this book to check your mastery of the math content, you'll move onto Volume 2 of *SAT Math Mastery* (available on Amazon at https://amzn.to/2z6hMge), and then to the official SAT practice tests released by the College Board (authors of the SAT test).

Once you move to the official practice tests, you'll continue to focus on working without a timer, proving that you can confidently and accurately answer the questions.

Then you'll start timing yourself, one section at a time. Once you've proven your skills on individual sections, you can start stringing together both a No-Calculator and Calculator section to work on your endurance.

There are exceptions to this rule - for example, if you're a Senior with only a week or two to prepare for your final chance at the SAT test - in which case you have to make compromises, and should start practicing with a timer sooner rather than later.

But the focus of this book is *content mastery*. Put simply, that means knowing what's on the test, knowing how to do it, and feeling rock-solid confident about it.

My top priority in this book is to help you review and master every topic on the SAT Math test. Then, when you see similar questions on test day, you'll be ready to leap into action, solve the question quickly and accurately, and not waste your energy worrying about the timer.

When you're fully prepared for the content of the test, you'll even find yourself finishing the Math sections with time to spare. Use that extra time to check your work, starting with any questions you're particularly uncertain about. Then return to the first questions in the section and look for any Careless Mistakes. Since all the questions are worth the same amount of points, it's usually better to quickly check a bigger number of easy questions than a smaller number of harder questions.

But remember: as I've mentioned, there is no penalty for guessing on the SAT Math test. Wrong answers will not be counted against you.

Ideally, you'll complete this book and enough practice SAT tests afterwards that you never have to worry again about running out of time and being forced to guess. But, it can still happen to the best of us.

If you still have unanswered questions and time is about to end, you should always bubble an answer for every question before time is called. At worst, it won't hurt your score, and at best, you might get lucky and guess a few answers correctly.

A good time to do this is when there are 5 minutes left in the section. It's best to leave a bit of a pad so you're not still bubbling answers when the proctor calls "pencils down." After ensuring that you've bubbled an answer for every question, then with any remaining time, you can keep working on one or two last questions before time is called.

Furthermore, for many of my students, time pressure is a major source of anxiety.

This is another big reason I recommend *forgetting* about the timer and focusing on *content mastery*. It's impossible to feel both confident and anxious at the same time. If you study hard, you'll have so much confidence that it pushes out all the room in your head for anxiety - whether it comes from the timer or from anything else.

Anxiety is a major issue for many SAT test-takers - and one of my favorite topics, since I've had a lot of personal experience with it. There's a lot I could say about it. But perhaps now is not the time.

For now, understand that your top priority should be *content mastery*, followed by timed practice once you've proven that you can solve the math questions accurately.

Again, if you're a Senior with only a couple of weeks to prepare before your final test date, then you're going to have to make some compromises, and start working practice sections against a timer as soon as you can. Everyone else - please put that timer in the closet and forget about it for a while.

Prelesson 8: Backup Strategies

Full disclaimer: after 10+ years of teaching and multiple perfect scores, I strongly believe that *content mastery* is the key to success. In other words, if you know the math topic properly, there's very little need for any "alternative strategies" to "hack the test."

However, there are two important strategies that may be useful from time to time, especially as backup options if you've run out of ideas to solve the question.

I suggest you only start using these strategies on the SAT if you've already gone through the rest of the Math section and solved all the questions that you can. These strategies tend to be relatively slow, and are almost never the optimal use of your time.

Testing the Answer Choices

Since there are only four answer choices for any SAT Math Multiple-Choice question, it's sometimes possible to plug in the answer choices rather than solving the math question.

But before you get too excited and think you can hack through the SAT Math test with this strategy, let's expose the downsides. The following info is not abstract theory - it's the reality of watching hundreds of students try this strategy on literally thousands of SAT questions.

Despite how tempting this strategy may seem, understand that the SAT *knows* that students will try it. Most of the time, they've written "blockers" into the question that prevent this strategy from being effective.

Trust me, I know. When I wrote the hundreds and hundreds of practice questions for this book, I made sure to follow the style of the SAT test and deliberately make this strategy as ineffective as possible. Call it unfair, but that's just the way the test is written.

Even when it works, this strategy is also usually very inefficient - sometimes taking four times longer than the optimal solution, or even more.

It's also especially prone to **Careless Errors**, since you'll typically be in a hurry and you often have to make four times as many calculations to test all four answer choices. That means four times the Negative Signs, four times the Distributing - and you can imagine how quickly the Careless Mistakes start to stack up.

Of course, this strategy also can't be applied to the Free Response questions.

Plugging In Numbers

There are also some questions that provide an opportunity to plug in our own numbers to replace certain variables or unknowns. This strategy can also be called "Making Up Your Own Numbers."

There are questions that include wording, for example, like "for all real values of x" or "for *any* real number b".
In such a case, it *may* be an opportunity to use a number of your choosing to replace the unknown variable.

The best numbers to plug in are usually 0, 2, or 3. These keep the values low and make for relatively simple calculations. It can also be worthwhile to test a negative number like -2.

However, the choice depends on context. Sometimes plugging in 0 is the absolute worst decision you could make. Other times, it's perfect. It takes experience and awareness to know which numbers are better or worse to plug in for any given problem.

Usually, I avoid using "1" as my plug-in value. I've just learned from experience that it doesn't work very well. Perhaps the SAT authors anticipated that this would be the most common choice and set traps for it. Or it might be that plugging in 1 just often doesn't tell you anything interesting about the problem. Regardless of the underlying reason, I just usually choose to avoid using 1 as my plug-in number.

"Strategies" are a Poor Substitute for Content Mastery

Throughout the lessons of this book, I've tried to highlight and explain the rare question types where these strategies may be particularly effective.

Other than those rare and specific cases, you should know that both of these strategies are usually a sign that something is missing in your understanding of the math concepts from the question.

For example, if you frequently find yourself plugging in numbers, it often means that you don't feel confident in your Algebra skills. Instead of solving the question as intended, you're trying to avoid the central point of the question.

Working against the grain of the test typically gets worse results than just "giving the test writers what they want." If the SAT is testing you on Algebra, *use* Algebra in response. The question will probably be easiest if you just solve it the way it was intended, instead of trying to desperately work around the challenges with a backup strategy.

Prelesson 9: Order of Operations (PEMDAS)

Math expressions cannot be worked (or "evaluated") in any order that we want. There is a set of rules called the "Order of Operations" that governs the order of steps we do to work through an expression.

PEMDAS is an acronym that stands for

- **P**arentheses
- **E**xponents
- **M**ultiplication
- **D**ivision
- **A**ddition
- **S**ubtraction

The PEMDAS acronym reminds us what order to evaluate our math in. If you violate this order you will end up with an incorrect answer.

Parentheses always come first. If there are multiple sets of "nested" parentheses, work from the inside of parentheses to the outside. Any work within the parentheses must *also* follow the PEMDAS Order of Operations.

With "big fractions", like $\frac{6+3}{4-1}$, this is the same as writing $(6+3)\div(4-1)$. In other words, we treat the top and bottom as if they are inside Parentheses.

After you work the inside of any Parentheses in the expression, then apply any Exponents found outside the Parentheses.

Multiplication and Division come next - complete any Multiplication or Division steps. Technically, it doesn't matter which you do first. Multiplication and Division both have equal priority with each other.

Then any Addition or Subtraction steps finish out the problem. It also doesn't matter which you do first - Addition and Subtraction have equal priority.

Try the following three practice examples to make sure you're clear on PEMDAS. **Don't use your calculator!**

Evaluate the expression $\frac{-4(7-2)^2}{2}-3$.

First we work the Parentheses:

$$\frac{-4(5)^2}{2}-3$$

Then the Exponent:

$$\frac{-4(25)}{2}-3$$

Now work the top of the "big fraction", which is Multiplication in this case:

$$\frac{-100}{2}-3$$

Now the Division:

$$(-50)-3$$

And finish off with Subtraction to get -53. Now try another one:

$$\text{Evaluate the expression } 2\left(\frac{4-(2\times 3+5)^2}{3}\right)+1.$$

Work from the inside out of the Parentheses. The interior of the Parentheses will also follow PEMDAS. First, multiplication in the innermost Parentheses:

$$2\left(\frac{4-(6+5)^2}{3}\right)+1$$

Then Addition in the innermost Parentheses:

$$2\left(\frac{4-(11)^2}{3}\right)+1$$

We're still working inside the Parentheses of the top of the fraction. Now apply the Exponent:

$$2\left(\frac{4-121}{3}\right)+1$$

Now do the Subtraction on the top of the fraction, since the top and bottom of fractions are treated as if they are each in Parentheses:

$$2\left(\frac{-117}{3}\right)+1$$

Now the Division inside the Parentheses:

$$2(-39)+1$$

And Multiplication:

$$-78+1$$

And finish the addition to get -77.

Let's do one more PEMDAS practice example:

$$\text{Evaluate the expression } 2\left(\frac{4}{(2-4)^2}+\frac{7+(2-3)^3}{5-2^2}\right)-4.$$

First, deal with the innermost Parentheses - there are two sets:

$$2\left(\frac{4}{(-2)^2}+\frac{7+(-1)^3}{5-2^2}\right)-4$$

Then apply the Exponents. Note that $(-1)^3$ is equal to -1:

$$2\left(\frac{4}{4}+\frac{7+(-1)}{5-4}\right)-4$$

Then clean up the top and bottom of the fraction on the right:

$$2\left(\frac{4}{4}+\frac{6}{1}\right)-4$$

Now complete the Divisions inside the Parentheses:

$$2(1+6)-4$$

Now finish the inside of the Parentheses:

$$2(7)-4$$

Now comes Multiplication:

$$14-4$$

And finish the Subtraction to get a final answer of 10.

A final note on PEMDAS: your calculator is extremely literal when it comes to PEMDAS. It is simply unable to forget to follow the rules of PEMDAS.

I often see students enter steps into their calculator that have the right *numbers*, but miscommunicate their intentions to the machine - because the student is ignoring the rules of PEMDAS. This can often end badly when the student means to do one thing, but the calculator does something different.

When you enter work into your calculator, be very attentive to Order of Operations. The calculator will *always* follow PEMDAS rules, even if you intended something different when you entered it in.

Prelesson 10: Fractions

Fractions are a common part of many SAT Math questions and are worked into the fabric of a variety of topics in this textbook.

There are some common Fraction situations we encounter in lessons where I notice my students either hesitating or getting completely mixed-up about how to proceed. These situations include "Adding & Subtracting" with fractions and "Multiplying & Dividing" with fractions. We may also see fractions with **Exponents & Roots**.

Adding & Subtracting Fractions

To add or subtract fractions, they must have a "common denominator", which means the bottom half of both fractions must be the same.

For example, we can't add the two fractions below in their current form:

$$\frac{1}{2}+\frac{1}{5}$$

We have to find a "common denominator", or get the same number on the bottom of both fractions. To do this, we have to choose a number that both 2 and 5 can multiply into. "10" would be a good choice since $2 \times 5 = 10$.

So, let's get the common denominator of 10 on the bottom of both fractions. When we do this, we cannot change the value of our expression. We have to do it by multiplying each fraction by a second fraction that can reduce to $\frac{1}{1}$. Check out how we do this below:

$$(\frac{5}{5})\frac{1}{2}+\frac{1}{5}(\frac{2}{2})$$
$$=\frac{5}{10}+\frac{2}{10}$$

We multiplied our original fractions by $\frac{5}{5}$ and $\frac{2}{2}$, both of which reduce to $\frac{1}{1}$. The effect of this is to get the same denominator (bottom number) for both fractions.

Now we can add the tops of the two fractions. With Fraction Addition and Subtraction, we only add or subtract the *tops* of the Fractions. The bottom does not change:

$$\frac{5}{10}+\frac{2}{10}=\frac{7}{10}$$

Fraction Subtraction works the same way as Fraction Addition. We need the same denominator (bottom number) for both fractions. Then we subtract the tops of the fractions only. Try the practice example below:

What is the value of $\frac{6}{7}-\frac{2}{3}$?

First, we need a common denominator:

$$(\frac{3}{3})\frac{6}{7}-\frac{2}{3}(\frac{7}{7})$$
$$=\frac{18}{21}-\frac{14}{21}$$

Now that we have a common denominator of 21 on bottom, we can subtract the tops of the fractions:

$$\frac{18}{21}-\frac{14}{21}=\frac{4}{21}$$

Multiplying & Dividing Fractions

When it comes to Multiplication, it's actually much easier to multiply fractions than to add or subtract them. To multiply fractions, we just do "Top times top, bottom times bottom."

$$\frac{2}{3}\times\frac{4}{5}=\frac{(2)(4)}{(3)(5)}=\frac{8}{15}$$

Multiplying fractions is as easy as can be. Just do "Top times top and bottom times bottom"!

Division with fractions throws many students off, but it's almost as easy. To divide by a fraction, simply flip it upside down and multiply. We call this "multiplying by the reciprocal."

$$\frac{4}{3}\div\frac{2}{5}=\frac{4}{3}\times\frac{5}{2}$$

Notice I've flipped the dividing fraction upside-down. Now just multiply top times top and bottom times bottom.

$$\frac{4}{3}\div\frac{2}{5}=\frac{4}{3}\times\frac{5}{2}=\frac{20}{6}$$

On a similar note, we will occasionally find fractions with other fractions within them. For example:

$$\frac{(\frac{3}{2})}{(\frac{4}{5})}$$

This is easy if you just view it as a Division problem - which, as we've just learned, is the same as flipping the dividing fraction upside-down, then "multiplying by the reciprocal":

$$(\frac{3}{2})\div(\frac{4}{5})$$
$$=\frac{3}{2}\times\frac{5}{4}$$

Again, we've taken the divisor and just flipped it upside-down.

Now multiply "Top times top and bottom times bottom" for the final answer:

$$(\frac{3}{2}) \div (\frac{4}{5}) = \frac{3}{2} \times \frac{5}{4} = \frac{15}{8}$$

Exponents & Roots with Fractions

We may also have to raise fractions to **Exponents**. In this case, *both* the top and bottom of the fraction must be raised to that power.

$$(\frac{4}{3})^2 = \frac{4^2}{3^2} = \frac{16}{9}$$

Notice that *both* the top and bottom of the fraction must be squared.

As we will study in a later lesson, Roots follow similar rules to Exponents. And likewise, if a root is applied to a fraction, *both* the top and bottom of the fraction must be rooted:

$$\sqrt{\frac{16}{9}} = \frac{\sqrt{16}}{\sqrt{9}} = \frac{4}{3}$$

Again, notice that *both* the top and bottom of the fraction must be square-rooted.

Reducing Fractions

We can also "Reduce" certain fractions. It's good to reduce whenever possible, because it's easier and cleaner to work with smaller numbers instead of bigger numbers. We can reduce fractions whenever a common multiple can be divided out of both top and bottom.

For example, the fraction $\frac{15}{6}$ reduces to $\frac{5}{2}$ by dividing both the top and bottom by 3.

Calculators with Fractions

A note on Calculator usage with Fractions: in many cases, if you have a calculator available, it may be easier to work with Decimals rather than Fractions.

The Texas Instruments ("TI") calculators have a command called "Frac" in the "Math" button menu. This command can convert decimals to fractions in the press of a button. It's very powerful when used at the right moment.

However, you can't rely on your calculator to save you from every question involving Fractions - and furthermore, of course you don't have a calculator available in the No-Calculator section of the SAT test.

Ideally, you should master Fractions without *needing* a calculator - but still know how to save yourself a bit of time when your calculator is handy and available.

Volume 1: Pretest 1

19 Questions

Answers follow the Pretest

Explanations within Lesson 1-19

Volume 1: Pretest 1

DO NOT USE A CALCULATOR ON ANY OF THE FOLLOWING QUESTIONS UNLESS INDICATED.

$$\frac{5-17x+15+12x}{10}=12-3x-9+2x$$

1. What value of x satisfies the equation above?

(A) -6

(B) -2.5

(C) 2

(D) 2.5

2. The noise level n in decibels of a certain amplifier system in a certain rectangular listening room can be calculated by the following equation, where l is the length of the room in feet, w is the width of the room in feet, A is the power of the speaker system in watts, and b is the acoustic coefficient.

$$n=\frac{2A\sqrt{lw-2A^2}}{blw}$$

Which of the following equations can be used to calculate the acoustic coefficient?

(A) $b=\dfrac{2A\sqrt{lw-2A^2}}{nlw}$

(B) $b=\dfrac{2nA\sqrt{lw-2A^2}}{lw}$

(C) $b=\dfrac{lw}{2nA\sqrt{lw-2A^2}}$

(D) $b=\dfrac{nlw}{2A\sqrt{lw-2A^2}}$

3. FREE RESPONSE: If n and t are the two solutions to the equation $2|x-8|+4=12$, what is the value of $n+t$?

4. Geraldine randomly selects marbles from a bag with four colors of marbles in it: blue, orange, pink, and bright-red. Of the marbles Geraldine selected, $\frac{1}{2}$ were bright red and $\frac{1}{5}$ were orange. Of the other marbles Geraldine selected, $\frac{1}{3}$ were pink. If Geraldine selected six pink marbles, how many orange marbles did Geraldine select?

(A) 3

(B) 5

(C) 12

(D) 30

5. (CALCULATOR) FREE RESPONSE: Heather took a four-day hiking trip over a long weekend. On Friday, Heather hiked 18 miles in 6 hours. On Saturday, she hiked 16 miles in 7 hours. On Sunday, she hiked 20 miles in 8 hours. On Monday, she hiked 4 miles in 2 hours. What was Heather's average speed, rounded to the nearest tenth of a mile per hour, for the time she spent hiking?

6. (CALCULATOR) FREE RESPONSE: Twelve divers are competing in a diving competition. The divers are scored on a 1-10 scale. The average score of the twelve divers is 7.6. If the lowest scoring diver is disqualified, the average score of the remaining eleven divers increases to 7.9. What was the lowest-scoring diver's score?

7. (CALCULATOR) Tree A is currently 18 feet tall, and Tree B is currently 24 feet tall. The ratio of heights of Tree A to Tree B is equal to the ratio of the heights of Tree C to Tree D. If Tree C is 28 feet tall, what is the height of Tree D in feet?

(A) $13\frac{1}{2}$

(B) 21

(C) $33\frac{1}{3}$

(D) $37\frac{1}{3}$

8. Jeffery is practicing for an archery competition. While practicing, he shoots an average of n arrows every 12 seconds. Which of the following can be used to solve for h, the average number of arrows he shoots per hour?

(A) $h = 12n$

(B) $h = 300n$

(C) $h = \frac{300}{n}$

(D) $h = \frac{n}{12}$

9. Line A passes through the origin and the point $(-2, 2)$. Line B has a slope of 2 and passes through the point $(5, 4)$. If Lines A and B intersect at the point (t, n), what is the value of $2t - n$?

(A) -6

(B) -2

(C) 2

(D) 6

10. (CALCULATOR) Henry is training for a reality show involving multiple physical competitions. In one of these competitions, Henry will be required to swing across a minimum of 300 monkey bars without stopping. Henry's training plan calls for him to increase the number of monkey bars he can swing across by a constant amount each week. If Henry predicts he will be able to swing across 168 monkey bars by the end of week 3 and be able to complete the competition at the end of week 7, what was the original number of monkey bars that Henry could swing over without stopping?

(A) 68

(B) 69

(C) 70

(D) 71

Peaches	5
Cherries	2
Bananas	6
Apples	9
Grapes	8
Total	30

11. (CALCULATOR) The table above shows the favorite fruits for each student in a class of 30. If a student is picked at random from all students whose favorite fruit is not peaches or apples, what is the probability that their favorite fruit is either cherries or bananas?

(A) $\frac{4}{15}$

(B) $\frac{1}{3}$

(C) $\frac{1}{2}$

(D) $\frac{4}{7}$

		Coloration Style					
		Solid	Mirrored	Swirl	Matte	Two-Tone	Total
Primary Color	Red	11	1	2	3	4	21
	Green	4	5	3	2	6	20
	Purple	2	4	1	4	5	16
	Orange	6	8	5	2	2	23
	Yellow	3	2	6	1	1	13
	Blue	7	5	9	3	3	27
	Total	33	25	26	15	21	120

12. (CALCULATOR) A boy has a collection of marbles in a variety of primary colors and coloration schemes. His collection of marbles is summarized in the table above. What fraction of his green marbles have a swirled coloration pattern?

(A) $\frac{1}{40}$

(B) $\frac{3}{26}$

(C) $\frac{2}{15}$

(D) $\frac{3}{20}$

13. (CALCULATOR) A company lowered the price of a product, Item A, from $12 to $9. To make up for this, the company raised the price of Item B by a percent equal to twice the percent decrease on Item A. There is also a 10% shipping charge on all products the company sells. The original price of Item B was $18. What is the new price of Item B *after* shipping?

(A) $14.85

(B) $22.50

(C) $27.00

(D) $29.70

14. The expression $\dfrac{\sqrt{x^{\frac{7}{3}}y^{-\frac{4}{3}}}}{\sqrt[3]{x^{-\frac{3}{2}}y^{\frac{5}{2}}}}$, where $x > 1$ and $y > 1$, is equivalent to which of the following?

(A) $\dfrac{1}{\sqrt[36]{(xy)^{11}}}$

(B) $\dfrac{\sqrt[3]{x^5}}{\sqrt{y^3}}$

(C) $\dfrac{\sqrt{y^3}}{\sqrt[3]{x^5}}$

(D) $\sqrt[36]{(xy)^{11}}$

15. A certain town's population is currently 30,000 and is forecast to decrease by 10% every six years for the next 60 years. Which of the following expressions represents the population of this town after t years for the next 60 years?

(A) $30{,}000(.1)^t$

(B) $30{,}000(.9)^{6t}$

(C) $30{,}000(.9)^{\frac{t}{6}}$

(D) $30{,}000(.1)^{\frac{t}{6}}$

$$x^2 - 15 = 2(4x + 9)$$

16. FREE RESPONSE: If b is a solution to the equation above and $b < 0$, what is the value of $-b$?

17. What are the solutions to $4x^2+16x-8=0$?

(A) $x=-2\pm\sqrt{6}$

(B) $x=-2\pm\sqrt{3}$

(C) $x=2\pm\sqrt{6}$

(D) $x=8\pm4\sqrt{6}$

18. FREE RESPONSE: If $x>0$, what is one possible solution to the equation $x^5-5x^3+4x=0$?

19. (CALCULATOR) The parabola represented by the equation $y=x^2+x+b$ is graphed in the xy-plane, where b is a constant. The parabola passes through the point $(0,-20)$ and the vertex of this parabola is at the coordinate point (m,n). What is the value of $m+n$?

(A) -20

(B) -20.25

(C) -20.75

(D) -21.25

Volume 1: Pretest 1 Answers

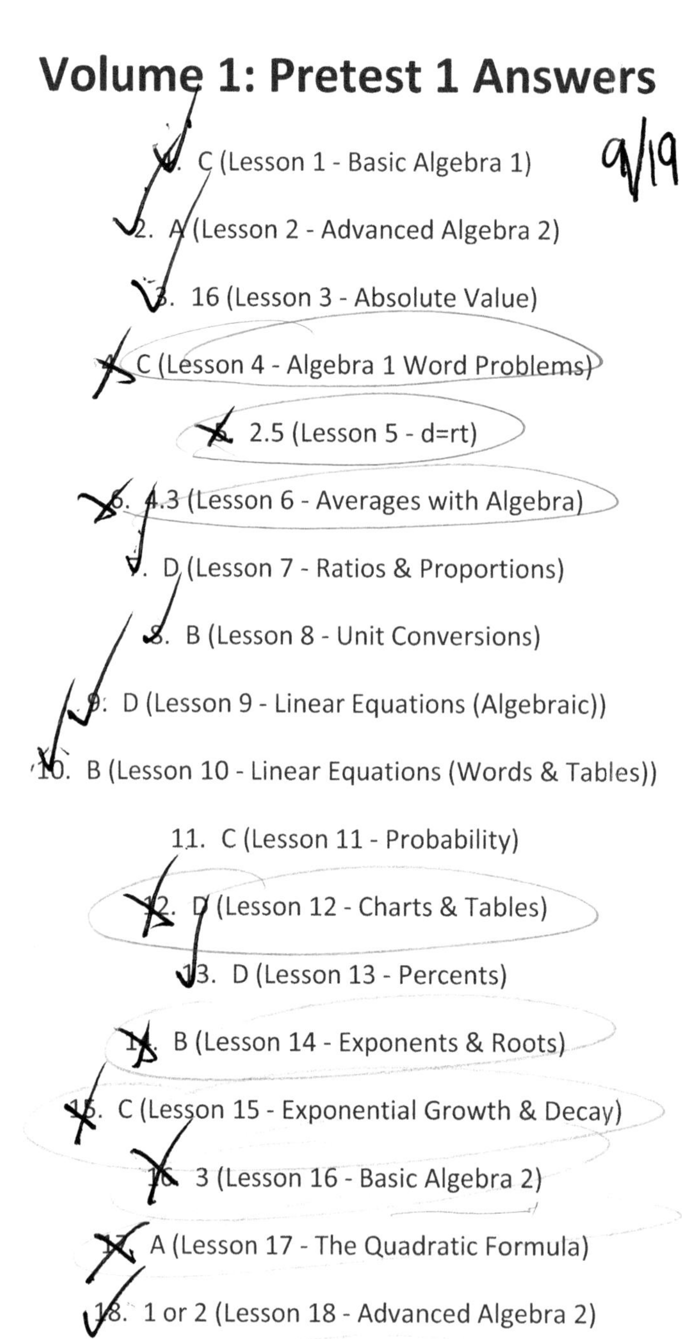

1. C (Lesson 1 - Basic Algebra 1)
2. A (Lesson 2 - Advanced Algebra 2)
3. 16 (Lesson 3 - Absolute Value)
4. C (Lesson 4 - Algebra 1 Word Problems)
5. 2.5 (Lesson 5 - d=rt)
6. 4.3 (Lesson 6 - Averages with Algebra)
7. D (Lesson 7 - Ratios & Proportions)
8. B (Lesson 8 - Unit Conversions)
9. D (Lesson 9 - Linear Equations (Algebraic))
10. B (Lesson 10 - Linear Equations (Words & Tables))
11. C (Lesson 11 - Probability)
12. D (Lesson 12 - Charts & Tables)
13. D (Lesson 13 - Percents)
14. B (Lesson 14 - Exponents & Roots)
15. C (Lesson 15 - Exponential Growth & Decay)
16. 3 (Lesson 16 - Basic Algebra 2)
17. A (Lesson 17 - The Quadratic Formula)
18. 1 or 2 (Lesson 18 - Advanced Algebra 2)
19. C (Lesson 19 - Algebra 2 Parabolas)

Volume 1: Pretest 2

19 Questions

Answers follow the Pretest

Explanations within Lesson 1-19

Volume 1: Pretest 2

DO NOT USE A CALCULATOR ON ANY OF THE FOLLOWING QUESTIONS UNLESS INDICATED.

1. Which value of x is a solution to the equation below?

$$-\frac{1}{2}(44x+24) > 20-30x$$

(A) 5

(B) 4

(C) 1

(D) -5

$$\sqrt{3n^2+x}+x=0$$

2. If $n=2$ in the equation above, what is the value of x?

(A) $\{-3,4\}$

(B) $\{-3\}$

(C) $\{4\}$

(D) There are no solutions to the given equation.

3. Which of the following equations could define the graph shown below?

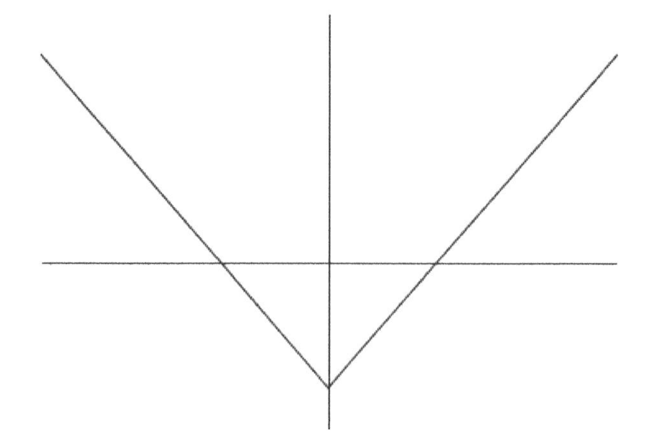

(A) $y=2x-2$

(B) $y=-2x-2$

(C) $y=|2x-2|$

(D) $y=|2x|-2$

4. Helen drives an average of 200 miles each week. Her car can travel an average of 20 miles per gallon of gasoline. Helen would like to reduce her weekly expenditures on gasoline by \$10. Assuming gasoline costs \$3 per gallon, which equation can Helen use to determine how many fewer average miles, d, she should drive each week?

(A) $\frac{20}{3}d = 10$

(B) $\frac{3}{20}d = 10$

(C) $\frac{10}{3}d = 20$

(D) $\frac{3}{20}d = 20$

5. (CALCULATOR) The distance traveled by Mercury in one orbit around the sun is about 29,000,000 miles. Mercury makes one complete orbit around the Sun every 88 Earth days. Of the following, which is closest to the average speed of Mercury, in miles per minute, as it orbits the Sun?

(A) 229

(B) 13,731

(C) 329,545

(D) 823,864

6. If a is the average of 12 and b, b is the average of c and 6, and c is the average of 8 and b, what is the average of a, b, and c in terms of b?

(A) $\frac{5}{12}b + 5$

(B) $\frac{5}{6}b + 10$

(C) $\frac{12}{5b + 60}$

(D) $\frac{5b + 12}{60}$

7. (CALCULATOR) FREE RESPONSE: A construction worker is mixing sand, gravel, cement, and earth in a ratio of $2:3:5:9$ by volume, respectively. If the total amount of mixture is 95 liters in volume, what was the total volume of the sand and earth combined, in liters?

8. FREE RESPONSE: A sarpler was a historical unit used to measure the weight of wool. A sarpler is equivalent to 80 tods, and three tods are equivalent to 84 pounds. A weight of 160 stone is equivalent to 2,240 pounds. How many stone are equivalent to a 40-sarpler weight of wool?

9. Line A passes through the point $(4,1)$ and is perpendicular to the line containing points $(-2,3)$ and $(6,2)$. Inequality B has the equation $y-2x \le 6$. Which quadrants do not contain any solutions for Line A and Inequality B?

(A) I only

(B) II only

(C) II and III only

(D) I, III, and IV only

t	4	7	10
v	95	131	167

10. (CALCULATOR) A sports car is in a straight-line race that begins with the car already moving at a certain speed. Once the race begins, the car accelerates at a constant rate per second. The table above shows a set of values for time t since the race began, in seconds, and the car's resulting velocity v in miles per hour. Which of the following functions best models the relationship of t and v?

(A) $v = 95 + 12t$

(B) $v = 67 + 7t$

(C) $v = 57 + 11t$

(D) $v = 47 + 12t$

Dream Topic	Number of Students
Going to School	10
Surfing	3
Robots	2
Animals	x
Seeing Friends	8
Couldn't Recall	12

11. (CALCULATOR) FREE RESPONSE: A class of students took a survey on the topic of their dreams from the previous night. The table above gives the topics of their dreams with the number of students who dreamed about each topic. Among the students who could remember their dreams, the probability of selecting a student at random who dreamt about either animals or surfing was $\frac{1}{5}$. How many students dreamt about animals?

Time (minutes)	Concentration (milligrams per liter)
0	12.0
3	17.4
6	25.2
9	36.6
12	53.0
15	76.9

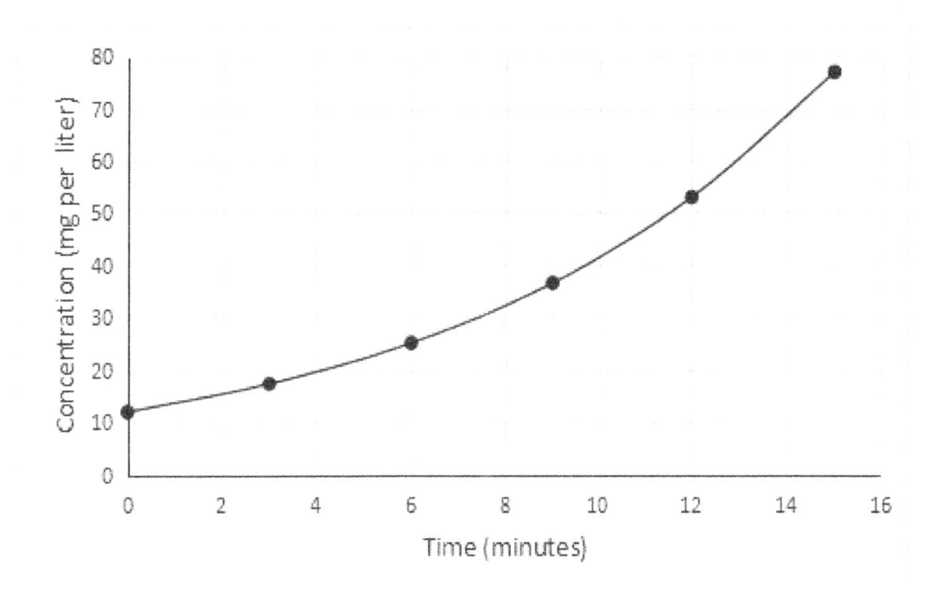

12. (CALCULATOR) FREE RESPONSE: The table and graph above show the concentration over time of a certain painkiller in the bloodstream of a hospital patient who is in surgery. According to the data, how many more milligrams of painkiller are present in 2 liters of the patient's blood after 12 minutes than in 1.5 liters of the patient's blood after 9 minutes? (Round your answer to the nearest tenth)

13. (CALCULATOR) FREE RESPONSE: An aeronautical engineering team has been developing new test aircraft. From Version 1 to Version 2 of the aircraft, the maximum speed increases by 60%. From Version 2 to Version 3, the maximum speed increases by 50%. From Version 3 to Version 4, the maximum speed decreases by 25%. From Version 4 to Version 5, the maximum speed increases by 5%. If Version 5 has a maximum speed of 378 miles per hour, what was the maximum speed of Version 1?

14. If $x-4y=7$, what is the value of $\frac{2^x}{16^y}$?

(A) $(\frac{1}{8})^3$

(B) $8^{\frac{7}{4}}$

(C) 2^7

(D) The value cannot be determined from the information given.

15. (CALCULATOR) FREE RESPONSE: Alex made an initial investment at the beginning of the year in 2008. The value of this investment increased by 40% per year for exactly three years, at which point the investment was worth $6,860. What was the value of Alex's initial investment?

16. FREE RESPONSE: If $x+y=-2$ and $x-y=3$, what is the value of $(x^2-y^2)(x+y)$?

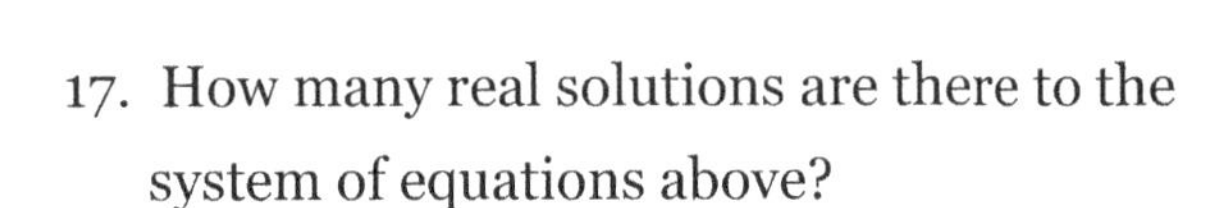

$$y=3x^2+2x+7$$
$$y=x^2-5x-6$$

17. How many real solutions are there to the system of equations above?

(A) There are exactly 4 real solutions.

(B) There are exactly 2 real solutions.

(C) There is exactly 1 real solution.

(D) There are no real solutions.

$$(ax+1)(2x^2+bx-3)=4x^3+4x^2-5x-3$$

18. FREE RESPONSE: The equation above is true for all x, where a and b are constants. What is the value of $a+b$?

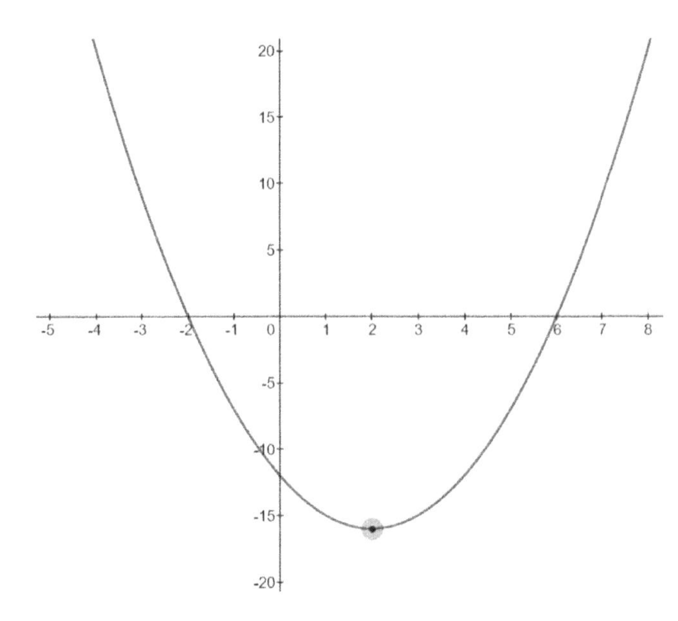

19. The equation of the parabola in the xy-plane above is $y=x^2-4x-12$. Which of the following is an equivalent form of the equation from which the coordinates of the vertex can be identified as constants or coefficients in the equation?

(A) $y=(x-6)(x+2)$

(B) $y=(x+6)(x-2)$

(C) $y=x(x-4)-12$

(D) $y=(x-2)^2-16$

Volume 1: Pretest 2 Answers

1. A (Lesson 1 - Basic Algebra 1)
2. B (Lesson 2 - Advanced Algebra 2)
3. D (Lesson 3 - Absolute Value)
4. B (Lesson 4 - Algebra 1 Word Problems)
5. A (Lesson 5 - d=rt)
6. A (Lesson 6 - Averages with Algebra)
7. 55 (Lesson 7 - Ratios & Proportions)
8. 6400 (Lesson 8 - Unit Conversions)
9. B (Lesson 9 - Linear Equations (Algebraic))
10. D (Lesson 10 - Linear Equations (Words & Tables))
11. 2 (Lesson 11 - Probability)
12. 51.1 (Lesson 12 - Charts & Tables)
13. 200 (Lesson 13 - Percents)
14. C (Lesson 14 - Exponents & Roots)
15. 2500 (Lesson 15 - Exponential Growth & Decay)
16. 12 (Lesson 16 - Basic Algebra 2)
17. D (Lesson 17 - The Quadratic Formula)
18. 3 (Lesson 18 - Advanced Algebra 2)
19. D (Lesson 19 - Algebra 2 Parabolas)

Lesson 1: Basic Algebra 1 & Inequalities

Percentages:

- 6.9% of Test
- 9.5% of No-Calculator Section
- 5.5% of Calculator Section

Prerequisites:

- Careless Mistakes
- Order of Operations

Let me start this lesson with a confession: I think Algebra is cool. It's an extremely *practical* form of math. You can use Algebra in real life to build bridges, start a business, or manage your personal budget. I also think it's fun!

Now, maybe I'm crazy, but I'm also lucky, because the SAT Math test is approximately 85% Algebra-based - and that's the actual data, not just one of those fake statistics people make up on the internet.

Algebra is the foundation of the entire SAT Math test. Any lack of confidence you have in Basic Algebra will be ruthlessly exposed, so let's lock it down right away in this lesson.

Basic Algebra 1 & Inequalities Quick Reference

- Basic Algebra is about finding the value of a single unknown variable (or constant) by doing arithmetic steps in reverse to a balanced equation.
- Only do one step at a time. And don't do any steps in your head - write them all down.
- Show your work neatly and thoroughly: write every step (balanced on both sides) and then write a fresh line to show the result. Use clear handwriting to avoid Careless Mistakes.
- Clean things up (Distribute or Combine Like Terms) before you make any new moves.
- Inequality Algebra has one major difference: we flip the direction of the inequality whenever we multiply or divide both sides by a negative number.
- Avoid common careless mistakes, especially Negative Signs, Distributing, and the Switcheroo (review the Prelesson on **Careless Mistakes**).

What is Basic Algebra?

Basic Algebra is the art of finding the value of an unknown number (such as "x") from a given equation. We call this unknown number either a "variable" or a "constant," depending on the situation. (We will define both of those terms in a moment.)

Algebra is all about untying a mathematical knot by doing the correct arithmetic steps *in reverse*. It's more about *undoing*, not "doing."

In Basic Algebra, the goal is to "isolate" a single variable or constant on one side of an equation. We accomplish this with small, simple steps - making one small move at a time, until the mathematical knot is untangled and we reveal the value of our unknown number.

Constants and Variables

Let's clear up a common question I get: what is the difference between a "variable" and a "constant"?

First, they both have something in common: *variables* and *constants* are both examples of using *letters* to represent unknown *numbers*. That's why we call both of these "unknowns."

A *variable* is an unknown that allow plugging in a *variety* of different numbers. It is literally "variable" - it can "vary" or change between different numbers. I think of variables as a "sliding scale" of numbers.

However, in many math situations, only *one* specific solution value is possible for a variable under the circumstances of the question. We'll be practicing this in the lesson. But broadly speaking, a variable represents a "sliding scale" of numbers, with infinite options for what number you plug in.

On the other hand, a *constant* is an unknown number that has a *specific* and *set* value - we just don't know what that value is yet. Constants do *not* live on a sliding scale: every constant is a *specific number* that just happens to be unknown to us at the moment.

Constants are *specific* numbers and their values cannot change. Variables represent a *range* of possible values. But, both Constants and Variables are "unknowns", or examples of using letters to represent unknown numbers.

Algebra Guidelines and Warnings

Perhaps the single most important Commandment of Algebra is to **SHOW YOUR WORK**. Always use orderly, neat, legible handwriting.

Start by copying the original equation in your own handwriting - this will help you absorb the key elements of the equation into your brain and give a starting place for your work.

Only make ONE move at a time. Write that move on one side of the equation, then copy it to the other side to keep your equation balanced; then write the result of that step on a fresh, clean line underneath. Keep similar terms (like x or y) aligned to make your work look clearer.

Clean written work in Algebra is similar to cleanliness and hygiene in a hospital: it prevents disease and infection from creeping into your math.

When it comes to my students, I've noticed that Careless Mistakes (often resulting from mental math, poor handwriting, or rushed work) cause more lost SAT points than any lack of skill or question difficulty.

Particularly watch your Negative Signs and your Distributing (review the Prelesson on **Careless Mistakes**).

When you give your final answer, always double-check the question to avoid Switcheroo mistakes. You may have solved for the value of "x", but the SAT is always happy to pull a last-second trick and ask for the value of "$3x$" in hopes of sneaking a last-second trap.

Let's try a simple practice question. I recommend trying to work each question on your own first, before looking at my explanation beneath the question.

Find the value of x in the balanced equation below:

$$5x+7=2x-5$$

Here's how we work through this little guy:

$$\begin{aligned} 5x+7 &= 2x-5 \\ -7 \quad & \quad\;\; -7 \\ 5x &= 2x-12 \\ -2x & \;\; -2x \\ \frac{3x}{3} &= \frac{-12}{3} \\ x &= -4 \end{aligned}$$

Notice how I keep my terms vertically aligned to keep things neat and easy to read. This not only helps me solve the equation in the first place, it also helps me check my work or find any errors.

Combining Like Terms

Whenever the chance arises, you should always "clean things up" before you move on. The most common form of "cleaning up" your algebra is to "*Combine Like Terms.*"

A "term" is a single number (such as "5") or variable (such as "x"), or a variable with a *coefficient* (such as "$6x$"). A *coefficient* is simply a fancy word for a "multiplier." In the example of "$6x$", the coefficient is "6."

Examples of Common Terms:

$$\begin{aligned} 3 &\ \text{(a number)} \\ t &\ \text{(a variable)} \\ 2x &\ \text{(variable \& coefficient)} \end{aligned}$$

$$\begin{aligned} -4d &\ \text{(variable \& coefficient)} \\ -12 &\ \text{(a number)} \\ 5y &\ \text{(variable \& coefficient)} \end{aligned}$$

When we Combine Like Terms, we combine numbers with other numbers, and combine variables with similar variables.

For example, "$7+4$" can be combined to "11." Or, "$3x+2x$" should be combined to "$5x$."

Combining Like Terms is a valuable algebraic move in its own right. It will make your work simpler and clearer, and helps reduce common mistakes.

Let's try another practice question that involves combining like terms. As before, try to work the question on your own before looking at my explanation below.

Find the value of n in the balanced equation below:

$$2n+8n-3n+11=15-4n+5n-16$$

Here's a good way to deal with this question, starting by combining like terms:

$$\begin{gathered} 2n+8n-3n+11=15-4n+5n-16 \\ 7n+11=n-1 \\ +1 \qquad +1 \\ 7n+12=n \\ -7n \qquad -7n \\ \frac{12}{-6}=\frac{-6n}{-6} \\ -2=n \end{gathered}$$

Distributing & Basic Fractions

Many Basic Algebra problems will involve the use of "Distributing."

Distributing is when a single multiplication or division step applies to multiple terms. The majority of common distributions involve either parentheses or fractions.

For example, "$2(x+3)$" means "two times $(x+3)$". This will distribute to equal $2x+6$. Notice how both the "x" and the "3" have been multiplied by the "2" outside the parentheses. This is Distribution in action.

Try it in the following practice problem. Be sure to try your best before checking my explanation underneath the problem.

$$4(5-t)+2=-8(t+2)-2$$

What value of t satisfies the equation above?

Here's how we use Distributing as the first step of solving the problem. We then Combine Like Terms before moving on to any actual Algebraic steps:

$$4(5-t)+2=-8(t+2)-2$$
$$20-4t+2=-8t-16-2$$
$$22-4t=-8t-18$$
$$+18 \qquad\qquad +18$$
$$40-4t=-8t$$
$$+4t \quad +4t$$
$$\frac{40}{-4}=\frac{-4t}{-4}$$
$$-10=t$$

Distribution can also happen in fractions.

Fractions are ugly. It's generally a good policy to get rid of fractions in your Algebra problems.

When your algebra problem contains a fraction, it's usually best to multiply both sides in a way that gets rid of the fractions. If the fraction is dividing by 3, then multiply both sides by 3. If the fraction is divided by "$x+5$", then multiply both sides by "$x+5$".

Dealing with fractions in this way will often require a small amount of simple distribution on the non-fraction side of the equation.

For example, to solve the practice problem below, we will need to multiply both sides by something that destroys the fraction on the left side, and as a result we will have to distribute multiplication on the right side.

Try solving it on your own before you look at my explanation.

$$\frac{y-7}{3+y}=2$$

What value of y satisfies the equation above?

Here's my explanation. I'll start by multiplying both sides to destroy the fraction.

$$(3+y)\frac{y-7}{3+y}=2(3+y)$$

$$y-7=6+2y$$

$$+7 \quad +7$$

$$y=13+2y$$

$$-2y \qquad -2y$$

$$-y=13$$

$$(-1)-y=13(-1)$$

$$y=-13$$

Pretest Example #1

Let's try an example from the Pretest.

$$\frac{5-17x+15+12x}{10}=12-3x-9+2x$$

What value of x satisfies the equation above?

(A) -6

(B) -2.5

(C) 2

(D) 2.5

Here's how we deal with this. Notice that the first step is to clean up by Combining Like Terms. Then we multiply both sides by 10 to eliminate the fraction, which requires some Distribution on the right side of the equation.

$$\frac{5-17x+15+12x}{10}=12-3x-9+2x$$

$$\frac{20-5x}{10}=3-x$$

$$(10)\frac{20-5x}{10}=(3-x)(10)$$

$$20-5x=30-10x$$

$$-30 \qquad -30$$

$$-10-5x=-10x$$

$$+5x \qquad +5x$$

$$\frac{-10}{-5}=\frac{-5x}{-5}$$

$$2=x$$

The correct answer is Choice C. Ah, the glories of Basic Algebra!

Inequality Algebra

Basic Algebra can also be done with Inequalities.

An "Inequality" (meaning "a relationship that is *not* equal") is commonly encountered as a "greater than" (>) or "less than" (<) sign. These signs give a *range* of possible values, such as "x is any value greater than 6" or "x is any value less than 7", instead of definitive final values (such as "x equals 3").

Notice that these two inequality symbols do *not* include the value itself. In "$x < 7$", only numbers *less than* 7 would be included: "7" itself is *not* an acceptable solution for this inequality.

We will also encounter the "greater than or equal to" (≥) and "less than or equal to" (≤) signs. These signs *do* include the value itself, so in the example of "$x \geq 7$", x *can* equal 7 (or any value greater than 7).

Common Inequality Symbols on the SAT Test:

> (greater than)
< (less than)
≥ (greater than *or* equal to)
≤ (less than *or* equal to)

In Inequality Algebra, everything is done almost exactly the same as before when we were using an equal sign.

The first difference is that instead of an equal sign ("="), your equation will use one of the inequality symbols (such as "<" or "≥"). We'll deal with the other difference in just a moment.

For now, try the Inequality Algebra practice problem below.

$$2x - 5 \leq 9$$

What are all the values of x that satisfy the inequality above?

Here's how to handle this question:

$$2x - 5 \leq 9$$
$$+5 \ +5$$
$$\frac{2x}{2} \leq \frac{14}{2}$$
$$x \leq 7$$

This tells us that x can be 7 or *any* number that is lower than 7.

Flipping the Inequality Sign

The second difference in Inequality Algebra - and this is *very* y important - is that you *must* flip the direction of the inequality sign *any time* you divide or multiply the equation by a negative number.

Try this out on the practice problem below. **REMEMBER:** Switch the direction of the inequality whenever you multiply or divide both sides by a negative.

$$2n-3(4n+8)>4n+4$$

Find the range of values that n can represent in the inequality above.

Here's how to solve it:

$$2n-3(4n+8)>4n+4$$
$$2n-12n-24>4n+4$$
$$-10n-24>4n+4$$
$$+24 \qquad +24$$
$$-10n>4n+28$$
$$-4n \quad -4n$$
$$-14n>28$$
$$\frac{-14n}{-14}>\frac{28}{-14}$$
$$n<-2$$

Notice the direction of the inequality sign changes direction when we divide both sides by -14 in the last step.

The final result tells us that n can equal any value less than -2.

Here's one more practice problem that puts together several of the techniques you've been learning. Try your hand before you look at my explanation.

$$-6-t\geq\frac{-7-3t+4}{-2}$$

Find the range of values that t can represent in the inequality above

Here's how we can solve this problem:

$$-6-t \geq \frac{-7-3t+4}{-2}$$

$$-6-t \geq \frac{-3-3t}{-2}$$

$$-2(-6-t) \geq \frac{-3-3t}{-2}(-2)$$

$$12+2t \geq -3-3t$$

$$+3 \qquad +3$$

$$15+2t \geq -3t$$

$$-2t \quad -2t$$

$$15 \geq -5t$$

$$\frac{15}{-5} \geq \frac{-5t}{-5}$$

$$-3 \leq t$$

Again, we flip the direction of inequality sign in the last step, when we divide both sides by a negative number.

Pretest Example #2

Here's an example of an Inequality we first encountered on the Pretest. Give this question another try if you got it wrong the first time:

Which value of x is a solution to the equation below?

$$-\frac{1}{2}(44x+24) > 20-30x$$

(A) 5

(B) 4

(C) 1

(D) -5

Here's how to work out this Inequality.

$$-\frac{1}{2}(44x+24)>20-30x$$
$$-22x-12>20-30x$$
$$+12 \quad +12$$
$$-22x>32-30x$$
$$+30x \qquad +30x$$
$$8x>32$$
$$\frac{8x}{8}>\frac{32}{8}$$
$$x>4$$

The correct answer is **Choice A**, the only value that is greater than 4. (Choice B doesn't work because the inequality sign says "x is *greater than* 4," not "x is greater than *or equal to* 4."

This is not the only order of steps you can follow to get a solution. If your solution process includes dividing or multiplying by a negative number at any point, be sure to flip the direction of the inequality sign.

Review & Encouragement:

Basic Algebra is an extremely powerful field of math that forms the foundation of more than 80% of the SAT Math test. It is not something to fear, but to embrace and even enjoy.

Basic Algebra occurs in a huge amount of common problem types. In fact, many complex SAT Math problems end up becoming Basic Algebra problems as you approach the final steps of the problem.

The most important things in Algebra are to keep a positive mindset, be crystal-clear in your written work, avoid mental math, and pay special attention to avoiding careless errors on your negative signs and distributing.

Remember to always flip the inequality sign any time you divide or multiply by a negative number!

Basic Algebra 1 & Inequalities Practice Questions

NOTE: Do NOT work these practice problems by plugging in the answers choices. Our purpose is practicing your Algebra. Use Basic Algebra to *solve* each equation step by step. I repeat - do *not* just plug in all the choices to see which ones work - yes, you may get the right answers, but it's a waste of your valuable time and you won't learn anything useful!

DO NOT USE A CALCULATOR ON ANY OF THE FOLLOWING QUESTIONS.

1. Solve for the value of x in the equation $4x-7=5-2x$.

   (A) -2

   (B) -1

   (C) 1

   (D) 2

2. Solve for the value of n in the equation $-4n-18=-7n+9$.

   (A) -9

   (B) -3

   (C) 3

   (D) 9

3. Solve for the value of x in the equation $4x+4+8x+8=8+x+3x-12$.

   (A) -2

   (B) -1

   (C) 1

   (D) 2

4. Solve for t in the equation $4(t+7)=2(3(2+t)$.

   (A) -8

   (B) 2.5

   (C) 8

   (D) 20

5. Solve for the value of x in the equation below:

$$\frac{-5x-5}{10}=5-x$$

   (A) 2

   (B) 11

   (C) 15

   (D) 20

6. Consider the equation below:

$$\frac{21n+18+24n}{4+n}=3(1+5)$$

What is the value of $3n$?

(A) 2

(B) 4

(C) 6

(D) 12

7. What is a possible value of y in the inequality below?

$$y+7<2y+4$$

(A) 4

(B) 2

(C) -2

(D) -4

8. Which of the following values is a solution to the inequality below?

$$7-x<\frac{-4x-2}{2}$$

(A) -9

(B) -8

(C) -7

(D) -6

9. Which of the following values is a possible solution to the inequality below?

$$1-x<5+3x<17-x$$

(A) -3

(B) 1

(C) 3

(D) 5

10. Which of the following values is a solution for the inequality below?

$$2(6n-4)\leq 4(4+2n)\leq 6(2+2n)$$

(A) 10

(B) 5

(C) -1

(D) -6

Basic Algebra 1 & Inequalities Answers

1. D
2. D
3. A
4. C
5. B
6. C
7. A
8. A
9. B
10. B

Basic Algebra 1 & Inequalities Explanations

1. **D.** Let's work through this. It's just the bare basics of Algebra.

$$4x-7=5-2x$$
$$+7+7$$
$$4x=12-2x$$
$$+2x \qquad +2x$$
$$\frac{6x}{6}=\frac{12}{6}$$
$$x=2$$

2. **D**. More Basic Algebra.

$$-4n-18=-7n+9$$
$$+18 \qquad +18$$
$$-4n=-7n+27$$
$$+7n \quad +7n$$
$$\frac{3n}{3}=\frac{27}{3}$$
$$n=9$$

3. **A**. Be sure to Combine Like Terms at the beginning to clean things up before you get started.

$$4x+4+8x+8=8+x+3x-12$$
$$12x+12=4x-4$$
$$+4 \qquad +4$$
$$12x+16=4x$$
$$-12x \qquad -12x$$
$$\frac{16}{-8}=\frac{-8x}{-8}$$
$$-2=x$$

4. **C**. Notice the "double distribution" on the right side. We always work parentheses from inside to outside (review the Prelesson on **Order of Operations**).

So, do the distribution for the inner parentheses first, then do the distribution for the outer parentheses.

$$4(t+7)=2(3(2+t)$$
$$4t+28=2(6+3t)$$
$$4t+28=12+6t$$
$$-4t \qquad -4t$$
$$28=12+2t$$
$$-12-12$$
$$\frac{16}{2}=\frac{2t}{2}$$
$$8=t$$

5. **B**. Here's how to work through the Algebra. Notice that my first step is to multiply both sides by 10 to get rid of the fraction. Be sure to distribute it on the right side:

$$\frac{-5x-5}{10}=5-x$$
$$(10)\frac{-5x-5}{10}=(5-x)(10)$$
$$-5x-5=50-10x$$
$$+10x \qquad +10x$$
$$5x-5=50$$
$$+5 \quad +5$$
$$\frac{5x}{5}=\frac{55}{5}$$
$$x=11$$

6. **C.** Here's how to work this question. We begin by Combining Like Terms before we do any new work.

$$\frac{21n+18+24n}{4+n}=3(1+5)$$
$$\frac{45n+18}{4+n}=3(6)$$
$$\frac{45n+18}{4+n}=18$$
$$45n+18=18(4+n)$$
$$45n+18=72+18n$$
$$-18 \quad -18$$
$$45n=54+18n$$
$$-18n \qquad -18n$$
$$27n=54$$
$$\frac{27n}{27}=\frac{54}{27}$$
$$n=2$$
$$3n=6$$

Notice the Switcheroo in the question asking for "$3n$", not "n". The fake answer choice A was a total setup: the first thing you see when you get "$n=2$" is Choice A for "2". Review the Prelesson on **Careless Mistakes** if you fell for this trap!

7. **A.** Here's how to work this inequality:

$$y+7<2y+4$$
$$-y \qquad -y$$
$$7<y+4$$
$$-4 \qquad -4$$
$$3<y$$

Notice that y can be any value greater than 3, which leaves us with Choice A as the only option.

If you get Choice D, it's possible you solved with a different order of solution steps. In that case, you may have forgotten to flip the direction of the inequality if you multiplied or divided both sides by a negative number.

8. **A.** Here's how to work through this inequality.

$$7-x<\frac{-4x-2}{2}$$
$$2(7-x)<-4x-2$$
$$14-2x<-4x-2$$
$$+2 \qquad\qquad +2$$
$$16-2x<-4x$$
$$+2x \quad +2x$$
$$\frac{16}{-2}<\frac{-2x}{-2}$$
$$-8>x$$

Notice the inequality flips direction in the last step because we divide both sides by a negative number. Because x must be *less* than -8, only Choice A works.

9. **B.** Here's something new: inequalities on both sides of our equation. The good news is everything works exactly the same as before, except we have three "sides" to balance instead of two. It's easy, though. Follow my work:.

$$1-x<5+3x<17-x$$
$$+x \qquad +x \qquad +x$$
$$1<5+4x<17$$
$$-5 \; -5 \qquad -5$$
$$\frac{-4}{4}<\frac{4x}{4}<\frac{12}{4}$$
$$-1<x<3$$

This resulting inequality "sandwich" tells us that x can be any number more than -1 and less than 3.

10. **B.** This is another "inequality sandwich" question, and we deal with it the same way as Question 9. Focus on Combining Like Terms and "capturing" the n in the middle of the inequality. Here's one way to solve it:

$$2(6n-4)\le 4(4+2n)\le 6(2+2n)$$
$$12n-8\le 16+8n\le 12+12n$$
$$-12n \qquad -12n \qquad -12n$$
$$-8\le 16-4n\le 12$$
$$-16 \; -16 \qquad -16$$
$$\frac{-24}{-4}\le\frac{-4n}{-4}\le\frac{-4}{-4}$$
$$6\ge n\ge 1$$

Be careful to catch both inequality signs flipping direction at the end, when you divide all three "sides" by -4.

Lesson 2: Advanced Algebra 1

Percentages:

- Whole Test 4.8%
- No Calculator Section 6.5%
- Calculator Section 3.9%

Prerequisites:

- Basic Algebra 1 & Inequalities
- Careless Mistakes
- Order of Operations

In this lesson, we'll look at what happens when you take **Basic Algebra 1** to the next level. Compared to Lesson 1, we'll encounter a wider variety of topics and scenarios that are mixed into our existing Basic Algebra skills.

These types of questions can include new symbols (like square roots), new moves (such as cross-multiplying), new formats (we'll see more word problems), and more traps (Switcheroos and Negative Signs are common **Careless Mistakes** that students fall for).

The questions may be a little more intimidating in appearance, but fear not: they're still based on the same essential Algebra 1 concepts. Note that this lesson does not include any major topics from Algebra 2 - only Algebra 1.

Advanced Algebra 1 Quick Reference

- Intermediate & Advanced Algebra may have more variety than the previous chapter, but it's still firmly based on **Basic Algebra 1**.
- Be sure you understand how and when to use Cross-Multiplying.
- Basic Factoring is the *reverse* of Distributing.
- You will encounter rearrangement of complicated-looking equations with word problems. Refer to the Answer Choices for clues about what variable you're supposed to isolate.
- Square Roots & False Solutions are some of the most dangerous questions in this lesson. However, they're extremely easy if you just plug in the answer choices to test them.
- It's always best to know the proper Algebraic steps, but Testing Answer Choices can be a useful backup strategy - or even a primary solution method, in certain cases.
- Advanced Algebra 1 problems often rely on intimidation, causing students to give up too soon. They are typically easier than they look, but don't rush, and watch out for the common **Careless Mistakes**!

Cross-Multiplying & Fractions

What is "Cross-Multiplying," and when should we use this valuable technique?

The perfect time to cross-multiply is when you're solving an Algebraic equation that has two fractions on opposite sides of an equal sign. Here's an example of that situation:

$$\frac{3-x}{2}=\frac{-x+4}{6}$$

Notice that there are two fractions, separated by an equal sign. This is the perfect time to use Cross Multiplication. There are other ways to solve this equation, but Cross Multiplication is the most efficient.

How do we do it? It's simple, although it's easier to *see* it worked out than to explain in words. Still, I will explain it first, then show you a visual example in just a moment.

With Cross Multiplication, we multiply in an X-pattern across the equal sign: bottom-left times top-right, and top-left times bottom-right. This creates two multiplication products that will be separated by the equal sign.

In our example above, the "2" on the bottom of the left side will multiply the "$-x+4$" on the top of the right side, and the "$3-x$" on the top of the left side will multiply the "6" on the bottom of the right side.

Let me show you what I mean. The first step is Cross-Multiplication:

$$\frac{3-x}{2}=\frac{-x+4}{6}$$
$$2(-x+4)=6(3-x)$$

Remember, we always Cross-Multiply in an *x*-pattern *across an equal sign*. Never get this mixed up with Fraction Multiplication, which is when we multiply "top times top" and "bottom times bottom." (review the Prelesson on **Fractions** if this is confusing to you).

Now we can finish solving the equation:

$$2(-x+4)=6(3-x)$$
$$-2x+8=18-6x$$
$$-8 \quad -8$$
$$-2x=10-6x$$
$$+6x \qquad +6x$$
$$4x=10$$
$$\frac{4x}{4}=\frac{10}{4}$$
$$x=2.5$$

Basic Factoring

Now we need to introduce a topic that will be explored in greater depth in a later lesson, when we take a closer look at **Basic Algebra 2** & Factoring/FOILing techniques.

This topic is called "Factoring," and it is the *reverse* of Distributing. Instead of Distributing one multiplier *to* several terms, we will factor out a common multiple *from* several terms.

Here are some examples of Basic Factoring, going from left to right:

$$5x+10 \text{ -> } 5(x+2)$$

$$x-2nx \text{ -> } x(1-2n)$$

$$5t-10tw \text{ -> } 5t(1-2w)$$

$$24d-12dx \text{ -> } 3d(8-4x)$$

$$3n+6n^2+9xn+12bn^3 \text{ -> } 3n(1+2n+3x+4bn^2)$$

Notice something interesting in each of the examples above: if you start on the right side and turn the arrow the other direction, you would be Distributing, and you would end up returning to the same equation you started with on the left side. Go ahead - try it right now for yourself, Distributing from right to left!

You see that both the left and the right sides are identical. From left to right, we are Factoring, but from right to left, we are Distributing.

That's why I say "Factoring is the *reverse* of Distributing."

In some situations, the *only* way to solve an Algebra 1 question is to use Factoring.

When working with my tutoring students, I notice that they often forget about the option of Factoring. We should always *remember* that Factoring is an available tool, and *look* for opportunities to use it. When you start actively *looking* for opportunities to Factor, you'll start seeing it more often.

Word Problems with Rearrangement

Now let's look at a common type of Advanced Algebra 1 question that tries to overwhelm you with information, in the form of a very wordy Word Problem or a complicated-looking equation (often both).

A certain cruise ship calculates its remaining fuel according to the formula below, where a represents the average height of the waves in feet, d is the number of nautical miles the ship has sailed since leaving port, n is the number of passengers aboard, and x is the remaining fuel in liters.

$$x = \frac{a^2 - 4d - (n - a)}{2d + 5}$$

Which of the following equations can be used to calculate the number of nautical miles the ship has sailed since leaving port?

(A) $d = \frac{a^2 - a - n - 5x}{4x + 4}$

(B) $d = \frac{a^2 + a - n - 5x}{4x + 4}$

(C) $d = \frac{a^2 - n + a - 5x}{2x + 4}$

(D) $d = \frac{a^2 - n - a - 5x}{2x + 4}$

The secret to this type of question is to use the answer choices to quickly identify exactly what you are solving for. In many cases, you can completely ignore the word problem!

For questions like this one, look at the four answer choices. They should show you what variable or constant to isolate on one side.

In this question, the answer choices reveal that all we have to do is isolate d. The word question itself doesn't tell us anything useful - it's just an intimidating distraction.

Note that we don't even need to solve for an actual numerical value - we just rearrange stuff until we get d all by itself. Try it yourself, then check my explanation on the next page.

Remember, we're just trying to get d by itself on one side. The first half of my solution focuses on moving all the d terms to the left side. The second half focuses on isolating d by itself on the left side. Keep an eye out for any opportunities to factor during the problem.

There are a lot of steps here, so focus on understanding each step one at a time, then zoom out and look at the bigger picture of how I move all the d terms to the left side, and then isolate a single d with Factoring.

$$x = \frac{a^2 - 4d - (n-a)}{2d+5}$$

$$(2d+5)x = \frac{a^2 - 4d - n + a}{2d+5}(2d+5)$$

$$(2d+5)x = a^2 - 4d - n + a$$

$$2dx + 5x = a^2 - 4d - n + a$$

$$+4d \qquad +4d$$

$$2dx + 4d + 5x = a^2 - n + a$$

$$-5x \qquad -5x$$

$$2dx + 4d = a^2 + a - n - 5x$$

$$d(2x+4) = a^2 + a - n - 5x$$

$$\frac{d(2x+4)}{(2x+4)} = \frac{a^2 + a - n - 5x}{(2x+4)}$$

$$d = \frac{a^2 + a - n - 5x}{(2x+4)}$$

Notice that I factored out a d on the left side in the third-to-last step. This was an essential step to finish the problem! Without this Factoring move, there's no way to get d by itself.

The correct answer is **Choice C**. The terms of our solution are in a slightly different order, but nevertheless, they're all the same. Double-check your Algebra, negative signs and your Distributing if you picked any other choice.

Two-Variable Solutions

The SAT Math section likes to test Algebra equations that have two different variables (such as "x" and "y") within a single equation.

Despite having two variables, these are *not* **Systems of Equations**, which we will explore in a later lesson. The main difference is that a System of Equations would also have multiple *equations*, not just multiple *variables*.

The solution approach is also slightly different from other problems we've done so far. Instead of solving for individual variables, as we do in most Basic Algebra 1 questions, here we solve for a *ratio* ("ratio" essentially means "fraction" in this context) or some other *relationship* between the two variables.

Here's an example of such a question. Try it yourself first and see where you get.

If $\frac{2x+3y}{4} = \frac{3x+2y}{2}$, what is the value of $\frac{x}{y}$?

In this practice example, we'll be solving for a *ratio* of variables, instead of an *individual* variable. In a way, this question has something in common with the idea of Algebraic Rearrangement that we explored earlier in this lesson.

My goal is to isolate "$\frac{x}{y}$" on one side. Instead of isolating one variable, then the other, I'll need to rearrange everything until I have $\frac{x}{y}$ on one side, and my resulting value on the other side.

Here's how I work it out. My first step will be Cross-Multiplication. Try to follow each step I'm making. I've put a few more notes below the algebra.

$$\frac{2x+3y}{4}=\frac{3x+2y}{2}$$
$$2(2x+3y)=4(3x+2y)$$

My next moves are designed to Combine Like Terms, gathering together all the x terms with the other x terms, and all the y terms with other y terms.

$$\begin{aligned} 2(2x+3y)&=4(3x+2y)\\ 4x+6y&=12x+2y\\ -4x\quad&\quad-4x\\ 6y&=8x+2y\\ -2y\quad&\quad-2y\\ 4y&=8x \end{aligned}$$

Finally, I focus on isolating $\frac{x}{y}$ on one side. In the third step below you can see me multiply both sides by $\frac{1}{y}$ to cancel the y on the left side and move it under the x on the right side to produce my desired $\frac{x}{y}$ result:

$$\begin{aligned} 4y&=8x\\ \frac{4y}{8}&=x\\ (\frac{1}{y})\frac{4y}{8}&=x(\frac{1}{y})\\ \frac{4}{8}&=\frac{x}{y}\\ \frac{1}{4}&=\frac{x}{y} \end{aligned}$$

So, the value of the two-variable solution $\frac{x}{y}$ is $\frac{1}{4}$.

Now let's try one of the Pretest Questions. Try it yourself before you look at my explanation below.

Pretest Question #1

The noise level n in decibels of a certain amplifier system in a certain rectangular listening room can be calculated by the following equation, where l is the length of the room in feet, w is the width of the room in feet, A is the power of the speaker system in watts, and b is the acoustic coefficient.

$$n = \frac{2A\sqrt{lw - 2A^2}}{blw}$$

Which of the following equations can be used to calculate the acoustic coefficient?

(A) $b = \frac{2A\sqrt{lw - 2A^2}}{nlw}$

(B) $b = \frac{2nA\sqrt{lw - 2A^2}}{lw}$

(C) $b = \frac{lw}{2nA\sqrt{lw - 2A^2}}$

(D) $b = \frac{nlw}{2A\sqrt{lw - 2A^2}}$

This is the question type I described as a "Word Problem with Rearrangement." With this type of question, recognize how little understanding of the word question is actually required!

We don't need to understand anything about the room, the speaker system, or the "acoustic coefficient" - we just have to glance at the answer choices to see that our only job is to isolate the letter b. The word problem itself is a total smoke screen for this question. It's only there to make us feel intimidated and to waste our time.

Here are the steps I will take to solve it. Notice how quickly it can be accomplished with the power of Algebra!

$$(b)n = \frac{2A\sqrt{lw - 2A^2}}{blw}(b)$$

$$bn = \frac{2A\sqrt{lw - 2A^2}}{lw}$$

$$(\frac{1}{n})bn = \frac{2A\sqrt{lw - 2A^2}}{lw}(\frac{1}{n})$$

$$b = \frac{2A\sqrt{lw - 2A^2}}{nlw}$$

The answer is **Choice A.**

Strategy: Testing Answer Choices

With Basic Algebra it's usually smarter to actually *do* the Algebra, but in some problems it can be wise to test the given answer choices.

The main benefits of this strategy are that it can unlock questions when you're totally stuck, it's a good backup plan when you don't know what else to do, and it *occasionally* saves time when used on the right questions.

The main downsides are that this strategy usually takes longer that just "doing the Algebra" the way your math teacher would, it is very prone to Careless Errors, and it's not available for Free Response questions because you won't have any multiple-choice answers to test.

A Testing Answer Choices strategy can work on complex Algebra problems when you're solving for one specific variable, and when the multiple-choice answers give you four possible values for that variable.

The way you accomplish this strategy is exactly the way it sounds: just try plugging in the four answer choices into the given equation. Replace the target variable with the answer choices, work the numbers, and see if the resulting equation is true and balanced.

For example, if you test an answer choice of $x = 5$ and get a result of "$2 = 6$", that would *not* be a balanced result, and proves that *x* does *not* equal 5. But if you test $x = 3$ and get a final result like "$4 = 4$", that *would* indicate that you've selected the correct answer choice.

We'll use a version of this strategy in just a moment as we tackle the next type of Advanced Algebra 1.

Square Roots

Square Roots are the reverse of "squaring" a number. If $4^2 = 16$, then $\sqrt{16} = 4$. If $7^2 = 49$, then $\sqrt{49} = 7$.

Square Roots can make some Basic Algebra problems look more difficult than they really are. Take a look at the practice problem below and try it yourself first:

If $\sqrt{25} - \sqrt{x} = \sqrt{9}$, what is the value of x?

This problem is very easy if we just clean things up first by calculating the values of the square roots:

$$\sqrt{25} - \sqrt{x} = \sqrt{9}$$
$$5 - \sqrt{x} = 3$$
$$-5 \qquad -5$$
$$-\sqrt{x} = -2$$
$$(-1)(-\sqrt{x}) = -2(-1)$$
$$\sqrt{x} = 2$$
$$(\sqrt{x})^2 = (2)^2$$
$$x = 4$$

Notice how we cleaned up, then isolated $\sqrt{x}$ on one side, then squared both sides. Squaring a square root will cancel out the root and reveal the x underneath it.

False Solutions and Square Roots:

Unfortunately, square roots can also cause a specific problem during Algebra known as "False Solutions." On the SAT test, you don't have to understand *why* False Solutions exist; you just have to know *when* to watch for them.

The concept is that a square root will never produce a negative number as its result. All square roots can only result in *positive* numbers. That means that any square root that returns a negative value is a False Solution.

On the SAT Math, the signs of a possible False Solution can be easily noticed and handled with a minimum of fuss. I'll show you how in just a moment.

First, take a look at the following practice question. Try it yourself, if you're feeling bold!

What is the set of all solutions to the equation $\sqrt{2x+8} = x$?

(A) $\{-2,4\}$

(B) $\{-2\}$

(C) $\{4\}$

(D) There are no solutions to the given equation.

The giveaway signs you should notice are that the question contains a square root in the Algebra equation, and most importantly, the *Answer Choices* include a set of options like "both -2 and 4," or "only -2", or "only 4".

This is exactly what a False Solutions question will look like on the SAT. They all look *exactly* like this, with only the smallest of variations between them. There's a square root in your algebra equation, and the answer choices provide a set of either "two solutions" or just one of those solutions, then the other solution.

Testing Answer Choices is actually the fastest and best approach to these questions. See, if we actually work out all the Algebra the "correct" way, we will end up with two possible solutions (this is a topic we cover in more depth in the lesson on **Basic Algebra 2**).

The irony is that, even after doing all steps of the the "correct" work, we'll still have one of the False Solutions - and we won't even know which one is false or true until we plug the solutions back in and test them.

In other words, we could use Algebra methods to find the two possible solutions to the equation, but we'll just end up back where we started - with the answer choices they give us in the multiple choice.

So, why don't we just skip all the Algebra and just go ahead and test the solutions they've given us? It's much easier and more efficient.

Here's how we use that strategy on the previous example question:

What is the set of all solutions to the equation $\sqrt{2x+8}=x$?

Notice that our answer choices give us the options of -2 and 4. We'll just test each of those two values and see what happens. First, let's try plugging in -2 for x:

$$\sqrt{2x+8}=x$$
$$\sqrt{2(-2)+8}=-2$$
$$\sqrt{-4+8}=-2$$
$$\sqrt{4}=-2$$
$$2\neq-2$$

Notice that this equation is *not* true: "2" does NOT equal "-2"! This reveals that $x=-2$ is a False Solution.

Now let's try plugging in the other option, 4, for x.

$$\sqrt{2x+8}=x$$
$$\sqrt{2(4)+8}=4$$
$$\sqrt{8+8}=4$$
$$\sqrt{16}=4$$
$$4=4$$

This produces a valid result. 4 *does* equal 4. This is how we know that $x=4$ is a valid solution. Our only possible multiple choice answer is **Choice C**. See how much this strategy can help?

Now let's look at a similar question that we encountered on the Pretest. Try it yourself first!

Pretest Question #2

$$\sqrt{3n^2+x}+x=0$$

If $n=2$ in the equation above, what is the value of x?

(A) $\{-3,4\}$

(B) $\{-3\}$

(C) $\{4\}$

(D) There are no solutions to the given equation.

The easiest way to work this question is simply by testing answer choices. First plug in $n = 2$, as given in the words of the question. Then try plugging in -3 for x and solving. Then try plugging in 4 for x and solving. You will find that only the answer choice -3, or **Choice B**, provides a true equation.

If you solve this the "long way" through the use of **Basic Algebra 2** techniques, you will be led to the wrong conclusion that both -3 and 4 can satisfy the equation. But simply testing these two possibilities will reveal that the false solution of 4 will create a negative value from a square root, which is impossible.

Here's what I mean about the "long way" giving a false solution to this equation (again, these techniques come from a future lesson on **Basic Algebra 2** and they aren't essential to understand right now):

$$\sqrt{3n^2 + x} + x = 0$$
$$\sqrt{3(2)^2 + x} + x = 0$$
$$\sqrt{3(4) + x} + x = 0$$
$$\sqrt{12 + x} + x = 0$$
$$-x \quad -x$$
$$\sqrt{12 + x} = -x$$
$$(\sqrt{12 + x})^2 = (-x)^2$$
$$12 + x = x^2$$
$$-12 - x \quad -12 - x$$
$$0 = x^2 - x - 12$$
$$0 = (x - 4)(x + 3)$$
$$x = 4, -3$$

However, one of these answers is wrong if we actually test it. Plugging back in the "solution" of 4 yields the equality "$4 = -4$", which is clearly false:

$$\sqrt{12 + 4} = -4$$
$$\sqrt{16} = -4$$
$$4 \neq -4$$

That's why, on these type of questions, we skip all the Algebra 2 and just plug in the answer choices to test.

Review & Encouragement

Despite the variety of interesting topics in this lesson, hopefully you can see how these question types aren't much harder than **Basic Algebra 1**.

We've reviewed how to Cross-Multiply, the use of Basic Factoring, how to rearrange complicated-looking equations and word problems to isolate single variables, and how to approach single equations with multiple variables in them. We've also covered False Solutions to Square Roots and when to watch out for them.

These questions often rely on intimidation to make you feel like you have no idea what to do, but you're usually only one move away from unlocking the entire question. Remember that, and stay confident and patient while you work on them.

Advanced Algebra 1 Practice Questions

DO NOT USE A CALCULATOR ON ANY OF THE FOLLOWING QUESTIONS.

1. What is the value of x in the equation below?

$$\frac{2.5x+3.5}{4}=\frac{1.25+2.25x}{2}$$

(A) -1

(B) $-.5$

(C) $.5$

(D) 2.5

$$10a-25b>25$$

2. Which of the following inequalities is equivalent to the inequality above?

(A) $5a-5b>5$

(B) $2a-5b>5$

(C) $2a-5b>25$

(D) $5a-5b>25$

3. What is the set of all solutions to the equation $\sqrt{3x+4}=-x$?

(A) $\{-1,4\}$

(B) $\{-1\}$

(C) $\{4\}$

(D) There are no solutions to the given equation.

4. If a is a constant and $y=ax$ when $y=4$ and $x=2$, what is the value of $y+2$ when $x=4$?

(A) 2

(B) 4

(C) 8

(D) 10

$$\frac{t-z}{z}=q$$

5. In the equation above, if t is positive and z is negative, which of the following must be true?

(A) $q<0$

(B) $q=-1$

(C) $q=1$

(D) $q>0$

6. What is the set of all solutions to the equation $\sqrt{10-3x}=x$?

(A) $\{-5,2\}$

(B) $\{-5\}$

(C) $\{2\}$

(D) There are no solutions to the given equation.

7. The estimated mass m in tons of a certain extinct dinosaur can be calculated by the equation below, where h is the estimated height in feet of the dinosaur at its shoulders, l is the length in feet of its tail, and b is the constant of dinosaur mass.

$$m = \frac{h(h^3 + l)}{b\sqrt{2h + 2l}}$$

Which of the following equations can be used to calculate the value of the constant of dinosaur mass?

(A) $b = \dfrac{h(h^3 + l)}{m\sqrt{2h + 2l}}$

(B) $b = \dfrac{mh(h^3 + l)}{\sqrt{2h + 2l}}$

(C) $b = \dfrac{m\sqrt{2h + 2l}}{h(h^3 + l)}$

(D) $b = \dfrac{\sqrt{2h + 2l}}{mh(h^3 + l)}$

8. If $\dfrac{y-2}{2} = \dfrac{x-8}{8}$ what is the value of $\dfrac{y}{x} + 1$?

(A) .25

(B) 1.25

(C) 4

(D) 5

9. The velocity v of a newly-discovered particle in a testing chamber can by calculated by the formula below, where g is a constant, n is the number of similar particles in the testing chamber, and x is the coefficient of atmospheric friction at a temperature of 270 degrees Celsius.

$$v = \frac{\sqrt{g - n^2} + 17n}{3x - 12n}$$

Which of the following equations can be used to calculate the coefficient of atmospheric friction?

(A) $x = \dfrac{\sqrt{g - n^2} + 17n - 12nv}{v}$

(B) $x = \dfrac{\sqrt{g + n^2} - 17n + 12nv}{v}$

(C) $x = \dfrac{\sqrt{g - n^2} + 17n + 12nv}{3v}$

(D) $x = \dfrac{\sqrt{g + n^2} - 17n - 12nv}{3v}$

10. The future value v in dollars of a certain speculative investment in electronic currency can be predicted by the equation below, where i is the initial investment in dollars, p is the constant of market volatility, n is the number of competing electronic currencies, and D is the current value of the Dow-Jones Industrial Average.

$$v = \frac{D\sqrt{pn^2 + 100i}}{i\left(\frac{1-\sqrt{i-n^3}}{p-20i}\right)}$$

Which of the following equations can be used to calculate the value of the Dow-Jones Industrial Average?

(A) $D = \frac{\sqrt{pn^2 + 100i}}{v\left(\frac{1-\sqrt{i-n^3}}{p-20i}\right)}$

(B) $D = \frac{vi\sqrt{pn^2 + 100i}}{\left(\frac{1-\sqrt{i-n^3}}{p-20i}\right)}$

(C) $D = \frac{i\left(\frac{1-\sqrt{i-n^3}}{p-20i}\right)}{v\sqrt{pn^2 + 100i}}$

(D) $D = \frac{vi\left(\frac{1-\sqrt{i-n^3}}{p-20i}\right)}{\sqrt{pn^2 + 100i}}$

Advanced Algebra 1 Answers

1. C
2. B
3. B
4. D
5. A
6. C
7. A
8. B
9. C
10. D

Advanced Algebra 1 Explanations

1. **C.** Start this question with cross-multiplying. Then distribute and combine like terms.

$$\frac{2.5x+3.5}{4}=\frac{1.25-2.25x}{2}$$
$$2(2.5x+3.5)=4(1.25-2.25x)$$
$$5x+7=5+9x$$
$$-5x \qquad -5x$$
$$7=5+4x$$
$$-2-2$$
$$\frac{2}{4}=\frac{4x}{4}$$
$$\frac{1}{2}=x$$

And of course, $\frac{1}{2}=.5$.

2. **B.** Barely any work to do here - just divide both sides by 5!

$$\frac{10a-25b}{5}>\frac{25}{5}$$
$$2a-5b>5$$

3. **B.** This has all the signs of a "Square Roots & False Solutions" question. So, just test the answer choices:

$x=-1$ is a successful answer choice:

$$\sqrt{3(-1)+4}=-(-1)$$
$$\sqrt{-3+4}=1$$
$$\sqrt{1}=1$$
$$1=1$$

$x=4$ is NOT a successful answer choice:

$$\sqrt{3(4)+4}=-(4)$$
$$\sqrt{12+4}=-4$$
$$\sqrt{16}=-4$$
$$4\neq-4$$

4. **D.** The first thing to do is plug in the given starting values of $y=4$ and $x=2$, then solve for a, like this:

$$y=ax$$
$$4=2a$$
$$\frac{4}{2}=\frac{2a}{2}$$
$$2=a$$

The next thing to do is set up the equation again, this time knowing that the constant a is always 2.

$$y=ax$$
$$y=2x$$

Then plug in the new value, $x=4$:

$$y=2x$$
$$y=2(4)$$
$$y=8$$

Finally, avoid falling for a Switcheroo. They're asking for the value of $y+2$, NOT the value of y. Since $y=8$ the value of $y+2=10$.

5. **A.** Try making up your own values. The question says "which of the following *must* be true," so any values should work, as long as they fit the descriptions of positive and negative numbers in the question.

Let's use a positive value for t, like 6, and a negative value for z, like -2.

$$\frac{t-z}{z}=q$$
$$\frac{6-(-2)}{-2}=q$$
$$\frac{6+2}{-2}=q$$
$$\frac{8}{-2}=q$$
$$-4=q$$

Notice that the value of q is a negative, so we can safely assume that $q<0$. Don't believe me? Try using any positive value for t and any negative value for z. You'll always get a negative value for q.

6. **C.** This has all the signs of a "Square Roots & False Solutions" question. Simply test the answer choices, just like we did in Question 3.

First test $x=-5$:

$$\sqrt{10-3(-5)}=-5$$
$$\sqrt{10+15}=-5$$
$$\sqrt{25}=-5$$
$$5\neq-5$$

Note that "5" does NOT equal "-5", so this is a false solution.

Then test $x=2$:

$$\sqrt{10-3(2)}=2$$
$$\sqrt{10-6}=2$$
$$\sqrt{4}=2$$
$$2=2$$

This creates a true equality, so $x=2$ is the only valid solution for this equation.

7. **A.** This is a "Word Problem with Rearrangement" question, just like we explored earlier in the chapter. In these cases, remember that understanding the word problem itself is not necessary to finish the problem.

All of the answer choices show us that our only job is to isolate b on the left side of the equation. That's easier than it looks - follow my steps, starting with the original equation:

$$m=\frac{h(h^3+l)}{b\sqrt{2h+2l}}$$
$$(b)m=\frac{h(h^3+l)}{b\sqrt{2h+2l}}(b)$$
$$bm=\frac{h(h^3+l)}{\sqrt{2h+2l}}$$
$$(\frac{1}{m})bm=\frac{h(h^3+l)}{\sqrt{2h+2l}}(\frac{1}{m})$$
$$b=\frac{h(h^3+l)}{m\sqrt{2h+2l}}$$

8. **B.** This question tests our Cross-Multiplying skills, and also requires us to find the combined variable fraction (or "ratio") of $\frac{y}{x}$, instead of just finding the value of x or y by themselves.

Here are the steps:

$$\frac{y-2}{2}=\frac{x-8}{8}$$
$$8(y-2)=2(x-8)$$
$$8y-16=2x-16$$
$$+16 \qquad +16$$
$$8y=2x$$
$$\frac{8y}{x}=\frac{2x}{x}$$
$$8(\frac{y}{x})=2$$
$$(\frac{1}{8})8(\frac{y}{x})=2(\frac{1}{8})$$
$$\frac{y}{x}=\frac{2}{8}$$
$$\frac{y}{x}=\frac{1}{4}$$

Don't forget to add "+1" at the end since the question asks for $\frac{y}{x}+1$, to avoid the sneaky Switcheroo mistake they're trying to cause.

$$\frac{y}{x}+1$$
$$=\frac{1}{4}+1$$
$$=1.25$$

9. **C.** This is another "Word Problem with Rearrangement" question, which means the answer choices reveal what constant or variable to isolate, then we ignore the rest of the question since it's only meant to distract us.

In this case, the answer choices show us that we're meant to isolate x. Let's do it! Starting with the original equation, focus on getting x by itself:

$$v = \frac{\sqrt{g-n^2}+17n}{3x-12n}$$

$$(3x-12n)v = \frac{\sqrt{g-n^2}+17n}{3x-12n}(3x-12n)$$

$$3xv-12nv = \sqrt{g-n^2}+17n$$

$$+12nv \qquad\qquad +12nv$$

$$3xv = \sqrt{g-n^2}+17n+12nv$$

$$\frac{3xv}{3v} = \frac{\sqrt{g-n^2}+17n+12nv}{3v}$$

$$x = \frac{\sqrt{g-n^2}+17n+12nv}{3v}$$

In addition to matching the correct answer to your final equation, I also recommend eliminating some of the answer choices that don't work. For example, several of the answer choices have negative signs in the wrong places This reduces how much mental processing you need to do and reduces careless errors.

10. **D.** Here's one last "Word Problem with Rearrangement" question. We've seen a lot of these by now. This one looks particularly ugly, so keep your eyes on the prize: we're supposed to isolate D on one side.

Also, view this ugly fraction as a "big fraction" with a "smaller fraction" on the bottom of it. It may help to review the Prelesson on **Fractions**.

Notice in the last step of my solution, we flip both the left and right sides upside down at the same time.

Here are the moves, starting with the original equation.

$$v = \frac{D\sqrt{pn^2+100i}}{i(\frac{1-\sqrt{i-n^3}}{p-20i})}$$

$$(\frac{1}{D})v = \frac{D\sqrt{pn^2+100i}}{i(\frac{1-\sqrt{i-n^3}}{p-20i})}(\frac{1}{D})$$

$$\frac{v}{D} = \frac{\sqrt{pn^2+100i}}{i(\frac{1-\sqrt{i-n^3}}{p-20i})}$$

$$(\frac{1}{v})\frac{v}{D} = \frac{\sqrt{pn^2+100i}}{i(\frac{1-\sqrt{i-n^3}}{p-20i})}(\frac{1}{v})$$

$$\frac{1}{D} = \frac{\sqrt{pn^2+100i}}{vi(\frac{1-\sqrt{i-n^3}}{p-20i})}$$

$$D = \frac{vi(\frac{1-\sqrt{i-n^3}}{p-20i})}{\sqrt{pn^2+100i}}$$

Lesson 3: Absolute Value

Percentages:

- Whole Test 0.9%
- No-Calculator Section 1%
- Calculator Section 0.8%

Prerequisites:

- Basic Algebra
- Careless Mistakes
- (Recommended) Linear Equations (Algebraic)

In this lesson, we'll explore the concept of Absolute Value - on its own, and in combination with Basic Algebra problems and some simple graphing.

Absolute Value Quick Reference

- Absolute Value means "distance from zero." Both positive and negative numbers can have the same distance from zero.
- All Absolute Value operations will return positive numbers.
- Solving for an x value in an Absolute Value equation requires a special setup and will typically end with two solutions, which I refer to as "positive side" and a "negative side" solutions.
- Know what graphs of Absolute Value functions look like and how to recognize them.
- Be extremely careful of your negative signs when working Absolute Value questions.

What is Absolute Value?

A pair of vertical bars are the symbol for Absolute Value.

$$|x| \text{ means "the absolute value of } x\text{".}$$

They are similar to parentheses because the expression contained within the bars must be evaluated first. In other words, Absolute Value bars act the same way as parentheses in regards to PEMDAS (review the Prelesson on **Order of Operations**).

Absolute Value can be imagined in several different ways. One of my favorite ways is the concept of "distance from zero." A distance is always positive. You can't take a "negative 10 mile drive", can you? No - whether you go west to east, or east to west, you're still traveling a distance of 10 miles - always a positive number.

For example, consider the numbers "5" and "-5." Both of these numbers are the same distance (5) from zero. Therefore the absolute values of both "5" and "-5" are the same: 5.

$$|5| = 5$$
$$|-5| = 5$$

Another way to consider Absolute Value is that anything inside the Absolute Value bars will emerge as a positive number. Think of Absolute Value bars as a sort of "jail" - any number can be positive or negative while *inside* the jail, but when they leave the jail, they always emerge as positive numbers.

Let's see an example of this in action. Try it yourself before you look at my explanation.

$$\text{What is the value of } 7 + |n - 5| \text{ if } n = 3?$$

Here's how we can work this basic problem. First we plug in the given value of $n = 3$, then go from there:

$$\begin{aligned} &7 + |3 - 5| \\ &= 7 + |-2| \\ &= 7 + 2 \\ &= 9 \end{aligned}$$

Simple! In this case there wasn't much to do - we just evaluated a simple expression with an Absolute Value in it. Notice how the Absolute Value of "-2" changes to a value of $+2$ when it leaves the Absolute Value bars.

Let's try one more example:

$$\text{What is the value of } (5-3)(|1 - 7(t - |-2|)|) \text{ when } t = 3?$$

This one looks a little more complicated - but only because we've used some Absolute Values inside other Absolute Values.

Remember, the bars act like parentheses - you need to do what's inside them first, from the inside to the outside (review the Prelesson on **Order of Operations**).

Here's how we work it out. First we plug in the given value of $t=3$. Then work from the inside out. Whenever you apply the Absolute Value bars, make sure to turn any negative numbers into their positive version.

$$
\begin{aligned}
&(5-3)(|1-7(3-|-2|)|) \\
&=(2)(|1-7(3-2)|) \\
&=2(|1-7(1)|) \\
&=2(|1-7|) \\
&=2(|-6|) \\
&=2(6) \\
&=12
\end{aligned}
$$

Notice how we obeyed the Order of Operations with our Absolute Value bars and parentheses.

Algebra with Absolute Values

Now let's look at the use of Absolute Value within a **Basic Algebra 1** problem. If you're feeling confident, give the following question a try before you read my explanation below.

If a and b are the two solutions to the equation $|x-4|=5$, what is the value of $a+b$?

At first glance, this may seem a little strange. How can one algebra equation, with a single variable x, have *two* solutions?

The key to this riddle lies in the way we solve Absolute Value equations. We will break it into two branching paths - which I call a "positive side" solution and a "negative side" solution.

Remember that Absolute Values are a special operation: whatever is inside the Absolute Value bars comes out as a positive number, *regardless of whether it was a positive or negative number while it was inside the bars.*

If you consider this, you may realize that *we don't know* whether the value inside the bars was a positive or negative number - we only know that it was positive *after* it left the Absolute Value.

Therefore, we must consider both possibilities in our Algebra solution process. In the practice question above, the value of "$x+4$" could be *either* $+5$ or -5. We just don't know for sure either way.

That's why we have to branch our solution for $|x-4|=5$ into two possibilities: one for a result of $+5$, and another for -5, like so:

$$x-4=5 \qquad\qquad x-4=-5$$

This split into a "positive side" and "negative side" is the all-important maneuver that allows us to solve for *both* possibilities. Let's continue with **Basic Algebra 1** to get our two solutions:

$$
\begin{aligned}
x-4&=5 \\
+4&+4 \\
x&=9
\end{aligned}
\qquad\qquad
\begin{aligned}
x-4&=-5 \\
+4&\;\;+4 \\
x&=-1
\end{aligned}
$$

So, we can see that x can equal *both* $+9$ and -1. Surprised? Test them both out and see what results:

Testing $x=9$

$$|x-4|=5$$
$$|9-4|=5$$
$$|5|=5$$
$$5=5$$

Testing $x=-1$

$$|x-4|=5$$
$$|-1-4|=5$$
$$|-5|=5$$
$$5=5$$

See? Both possibilities are equally true: x can indeed equal either $+9$ or -1.

Let's not forget to finish the rest of the question:

If a and b are the two solutions to the equation, what is the value of $a+b$?

We know that our two solutions are $+9$ and -1. All that's left is to add the two values: $9+(-1)=8$. Our final answer, therefore, is "8".

Pretest Question #1

Here's an Absolute Value Algebra question from the Pretest. Try it again now if you got it wrong the first time.

FREE RESPONSE: If n and t are the two solutions to the equation $2|x-8|+4=12$, what is the value of $n+t$?

The first thing I'm going to do is use **Basic Algebra** to clean up the equation and isolate the Absolute Value before I split it into the "positive side" and "negative side" solutions.

$$2|x-8|+4=12$$
$$-4 \quad -4$$
$$2|x-8|=8$$
$$\frac{2|x-8|}{2}=\frac{8}{2}$$
$$|x-8|=4$$

Now the equation is in good shape to split into two "positive side" and "negative side" branches.

$$x-8=4$$
$$+8 \quad +8$$
$$x=12$$

$$x-8=-4$$
$$+8 \quad +8$$
$$x=4$$

Now that we've found our two solutions, $x = 12$ and $x = 4$, we can finish up the problem that was asked.

If n and t are the two solutions to the equation, what is the value of $n + t$?

If we add our two solutions, we get $12 + 4 = 16$, giving our final answer of **16**.

Is Absolute Value combined with Basic Algebra 1 starting to make sense to you now?

It's all about getting the Absolute Value by itself, then splitting into the two branches of "positive side" and "negative side" solution, and finally solving for both possibilities.

Graphing Absolute Value

Using Absolute Values in graphs can cause some interesting effects. Since Absolute Values cannot produce negative values, neither can their graphs.

This creates a distinctive visual shape. Here's the basic shape of an Absolute Value graph that you're most likely to see on the SAT math test:

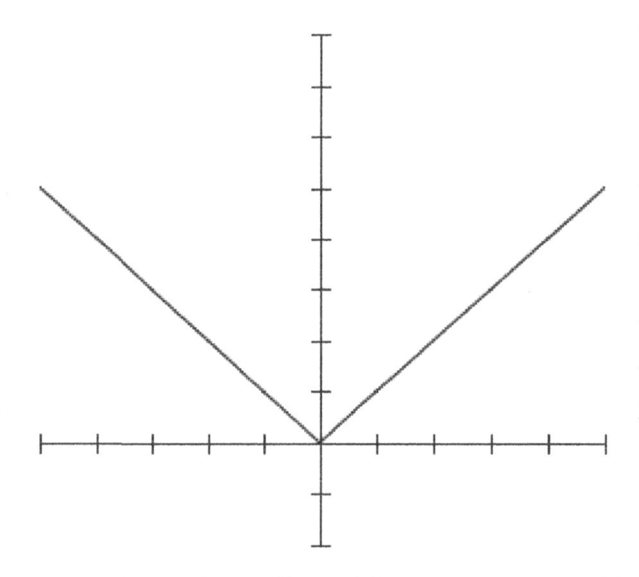

This is an basic graph of the function $y = |x|$. It's nearly identical to the basic **Linear Equation** $y = x$ (study the upcoming lessons on **Linear Equations** for more information). But, there's one major difference: anywhere the line would normally produce a negative *y*-value, it will now produce a *positive* *y*-value.

To me this graph shape has always given the impression of a laser beam bouncing off a mirror. The line "wants" to go in a straight line down below the *y*-axis - down where the *y* values are less than zero - but it's not allowed to because of the Absolute Value, which only permits positive results. Instead it "reflects" off the *y*-axis and produces the equivalent positive values instead - like a beam reflecting off a mirror.

However, it is still possible to see Absolute Value graphs that drop below 0.

For example, let's consider the graph of the equation $y = |x| - 3$.

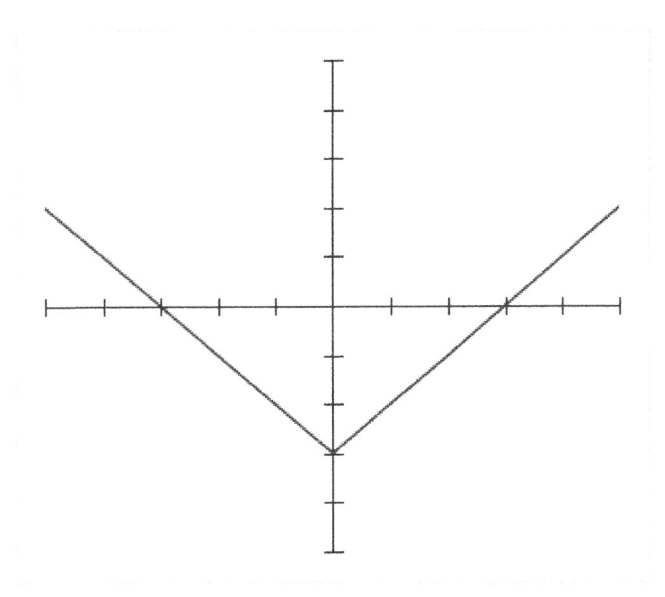

Notice that this line, while still based on an Absolute Value equation, *can* drop below zero. That's because we've taken the original line and shifted it *down* by 3: observe the "-3" outside the Absolute Value bars in our equation $y = |x| - 3$.

True, the Absolute Value itself must return a positive value. But *after* it produces that positive value, the "-3" will pull that value downwards. So, for example, if $x = -1$, then the Absolute Value would return "$+1$". But then the "-3" on the outside will pull that down result to -2. You can find this point on the graph at (-1,-2).

So you see, it's still possible for Absolute Value graphs to go below zero in some cases.

Pretest Question #2

Let's look at another question from the Pretest. If you missed it the first time, try again now.

Which of the following equations could define the graph shown below?

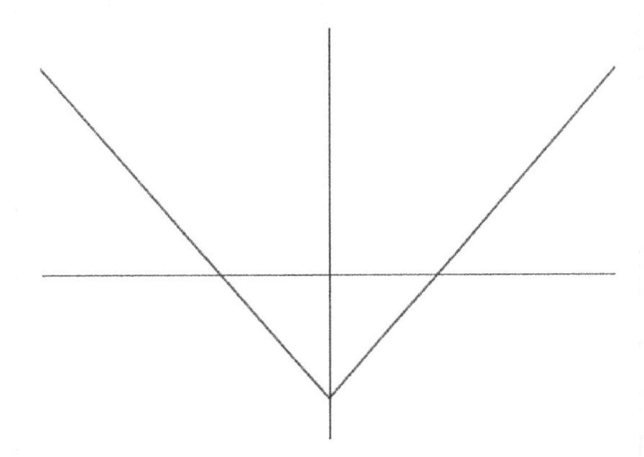

(A) $y = 2x - 2$

(B) $y = -2x - 2$

(C) $y = |2x - 2|$ ✗

(D) $y = |2x| - 2$

At first glance, this question may seem impossible: how are you supposed to choose the correct equation for a graph when you're not even given any values or tick marks to judge the scale by?

There are two keys to this: first of all, recognizing the signature V-shape of an Absolute Value graph. Secondly, using a process of Elimination on the Answer Choices. Let's try this out now.

First of all, we *know* by the shape that this graph must be based on an Absolute Value equation. There's simply no other type of equation that produces that sharp corner, like a laser beam reflecting off a mirror. That's how we can quickly get rid of Choices A and B, which don't include any Absolute Values.

Unfortunately, both Choices C and D make use of Absolute Value. How to decide? Well, Choice C is an equation that's unable to dip below $y = 0$. Notice that the entire equation is surrounded by Absolute Value bars. That means it can *never* go below 0 at any point, no matter what.

On the other hand, the correct **Choice D** has a "-2" *outside* of the Absolute Value bars. Although the Absolute Value is guaranteed to return a positive value, the "-2" on the outside can still drag the entire equation down below $y = 0$. And that is, in fact, what is happening.

Note that this question also overlaps with the upcoming lessons on **Linear Equations**, so you'll probably also want to check those lessons out as well.

Review & Encouragement

Absolute Value is not the *most* common topic on the SAT Math test, but it definitely shows up from time to time - about once every test or so.

The most common mistakes I see my students make with Absolute Value questions are either A) **Careless Mistakes** related to negative signs or B) failing to properly set up for the *two* solutions to Absolute Value Algebra.

When it comes to graphs involving Absolute Value, look out for the signature "V-Shape" where the direction of the graph abruptly shifts from sloping down to sloping up. As I mentioned, it reminds me of a laser beam reflecting off a mirror.

I've never seen a particularly difficult Absolute Value question on the SAT Math. These questions tend to be relatively easy - and when you understand the rules laid out in this lesson, you'll find them quite simple.

Absolute Value Practice Questions

DO NOT USE A CALCULATOR ON ANY OF THE FOLLOWING QUESTIONS.

1. If $x < 0$, what is the value of x in the equation below?

$$|x-6|=10$$

(A) $x=16$

(B) $x=-4$

(C) $x=-6$

(D) $x=-16$

$$3+|n|=12$$

2. What is the set of solutions for n in the equation above?

(A) $\{9,-15\}$

(B) $\{-9\}$

(C) $\{-15\}$

(D) $\{9,-9\}$

$$|-4-2t|=14$$

3. FREE RESPONSE: What is the value of t in the equation above, if $t>0$?

4. What is the solution set for the equation below?

$$|-2x+3|=|-9|$$

(A) $\{3\}$

(B) $\{-6,3\}$

(C) $\{-3,6\}$

(D) There are no solutions to the given equation.

$$|x-4.5|=3.5$$

5. If a and b are solutions to the equation above, what is the value of $a+b$?

(A) 1

(B) 7

(C) 8

(D) 9

$$|4x+12|=4$$

6. FREE RESPONSE: If n and t are solutions to the equation above, what is the value of $|n-t|$?

7. Which of the following equations could produce the graph below?

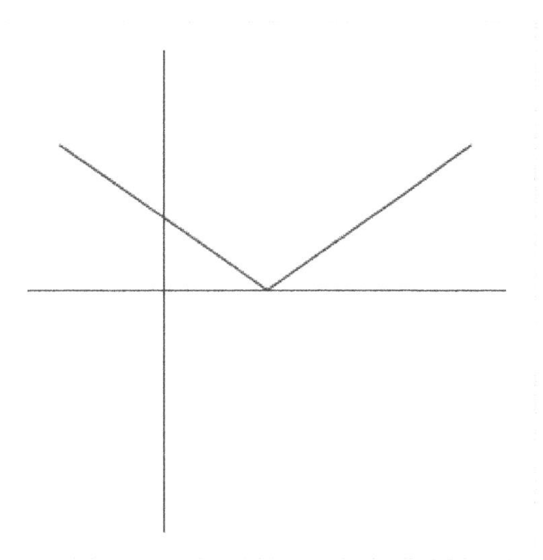

(A) $y = x + 4$

(B) $y = |x| - 4$

(C) $y = |x - 4|$

(D) $y = x - 4$

8. A number line contains two different points A & B. These points are both 7 units from the point with coordinate -2. The solutions to which of the following equations gives the coordinates of both points?

(A) $|x - 2| = 7$

(B) $|x + 2| = 7$

(C) $|x - 7| = 2$

(D) $|x - 7| = -2$

9. Which of the following equations could produce the graph below?

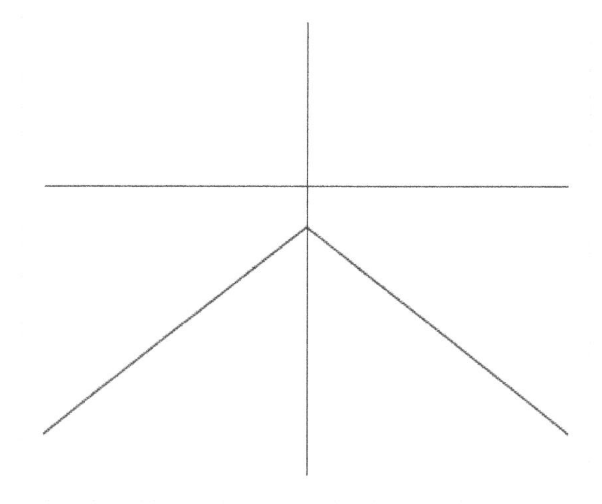

(A) $y = -|x| - 1$

(B) $y = |x| - 1$

(C) $y = |-x| - 1$

(D) $y = |-x - 1|$

$$|s - t|$$

10. FREE RESPONSE: If G and H are points on a number line that are 12.5 units from the point with coordinate -6, and the coordinate of point G is s and the coordinate of point H is t, then what is the value of the expression above?

Absolute Value Answers

1. B
2. D
3. 5
4. C
5. D
6. 2
7. C
8. B
9. A
10. 25

Absolute Value Explanations

1. **B.** This problem gives us a Basic Algebra problem combined with Absolute Value. Remember to set up both the "positive side" and "negative side" solutions, as we learned in this lesson. That means considering the possibility that the equation should equal 10 *or* -10:

$$\begin{aligned} x-6&=10 \\ +6\ \ &+6 \\ x&=16 \end{aligned} \qquad\qquad \begin{aligned} x-6&=-10 \\ +6\ \ &\ \ +6 \\ x&=-4 \end{aligned}$$

There are two solutions for *x*, but the question asked us for a value of *x* that is less than zero. Our only option is $x=-4$.

2. **B.** Another Basic Algebra problem combined with Absolute Value. As we learned in the lesson on **Basic Algebra 1**, it's best to Combine Like Terms before doing any more complicated steps. We'll do that first:

$$\begin{aligned} 3+|n|&=12 \\ -3\ \ \ \ &\ \ -3 \\ |n|&=9 \end{aligned}$$

Now that the basic equation is cleaned up, consider both the "positive side" and "negative side" solutions. There's not much to do, though. If $|n|=9$ then $n=9$ or $n=-9$.

3. **5.** More Basic Algebra combined with Absolute Values. Remember to set up and solve for *both* the "positive side" and "negative side" solutions (both $+14$ and -14 on the right side).

Also, watch your negative signs. Just because $-4-2t$ is inside Absolute Value bars does *not* mean that all those negative signs and subtraction magically disappear. Absolute Value bars act like parentheses - whatever takes place within them still needs to happen first, regardless of whether the terms are positive or negative.

$$\begin{aligned} -4-2t&=14 \\ +4\ \ \ \ \ \ &\ \ +4 \\ \frac{-2t}{-2}&=\frac{18}{-2} \\ t&=-9 \end{aligned} \qquad\qquad \begin{aligned} -4-2t&=-14 \\ +4\ \ \ \ \ \ &\ \ \ \ +4 \\ \frac{-2t}{-2}&=\frac{-10}{-2} \\ t&=5 \end{aligned}$$

This Free Response question asks us for a value of t that is greater than zero. The only possibility is the value $t=5$.

4. **C.** More **Basic Algebra 1** combined with Absolute Values. The first thing to do is clean it up: the absolute value of "-9" is always 9, so we can simplify the right side of the equation as $|-2x+3|=9$.

Then solve for both the "positive side" and the "negative side" of the equation, just as we've done in the previous three problems.

$$\begin{aligned} -2x+3&=9 \\ -3\ &-3 \\ \frac{-2x}{-2}&=\frac{6}{-2} \\ x&=-3 \end{aligned} \qquad\qquad \begin{aligned} -2x+3&=-9 \\ -3\ &\ -3 \\ \frac{-2x}{-2}&=\frac{-12}{-2} \\ x&=6 \end{aligned}$$

Our solution set is both values: $x=\{-3,6\}$

5. **D.** You guessed it, more **Basic Algebra 1** with Absolute Value. Start off by setting up the "positive side" and "negative side" of the equation.

$$x - 4.5 = 3.5$$
$$+4.5 \quad +4.5$$
$$x = 8$$

$$x - 4.5 = -3.5$$
$$+4.5 \quad +4.5$$
$$x = 1$$

Remember that the question asks $a + b$, where a and b are the solutions of the equation. Finish the question with "$8 + 1$", giving a final value of 9.

6. **2.** More **Basic Algebra 1** with Absolute Value. First let's run through the essentials: set up both the "positive side" and "negative side" of the equation:

$$4x + 12 = 4$$
$$-12 \quad -12$$
$$\frac{4x}{4} = \frac{-8}{4}$$
$$x = -2$$

$$4x + 12 = -4$$
$$-12 \quad -12$$
$$\frac{4x}{4} = \frac{-16}{4}$$
$$x = -4$$

But, we're not done yet. The question asked for the value of $|n - t|$, where n and t are solutions to the equation (we know the solutions are "-2" and "-4"). Finish it off:

$$|-2 - (-4)|$$
$$|-2 + 4|$$
$$|2| = 2$$

Notice that it doesn't matter which solution is which. Whether you assign -2 or -4 to n or t, you'll get the same final result. Don't believe me? Try both ways yourself!

7. **C.** This is obviously an Absolute Value graph because of it's V-shape (looks like "a laser reflecting off a mirror"). Therefore we can eliminate Choices A and D, neither of which involve an Absolute Value.

But, we're left with Choices B and C. How to tell them apart? Well, Choice B has a "-4" *outside* the Absolute Value bars, meaning that the graph of that equation should still dip below $y = 0$. However, our graph never passes below the *x*-axis.

That leaves us only with the correct answer, Choice C, which has the entire equation contained inside Absolute Value bars - thereby preventing the graph from ever dropping into negative *y*-value territory.

8. **B.** This is an interesting question! We're asked to consider a number line. I suggest your first move is to sketch one. Just make a medium-length straight horizontal line with arrows at both ends, then mark a point near the center with the coordinate "$x = -2$". Here's an example (forgive my lack of artistry):

<---------- (-2) ------------->

OK, now count 7 units up from -2 (otherwise known as $-2+7$). This gives you a point at $x = 5$. Then count 7 units *down* from -2 (or just calculate $-2-7$). This gives you another point at $x = -9$. Here's our number line with the new values filled in:

<- (-9) -------- (-2) -------- (5) ---->

The easiest option at this point is just to test all four multiple-choice answers. You should be able to plug in *either* "-9" or "5" and get a true and balanced equation both times. Only Choice B will succeed for both values.

Alternately - if you're feeling more confident - you can work this a quicker way. The concept of "distance" is also a concept of "difference," and "difference" is just another word for "subtraction." Take any two coordinate points, subtract one from the other, and take the Absolute Value of the result - you'll have the distance between the two points.

We know the point "-2" is our first coordinate, and we can use the variable x to represent our other unknown coordinates.

Follow the instructions above and you can set up this equation: $|x-(-2)|=7$. In other words, the distance of point x from -2 is 7 units. This equation can be simplified to the correct answer: $|x+2|=7$.

9. **A.** As with the other Graphing questions in this lesson, you first should recognize the characteristic shape of an Absolute Value graph. However, something's different about this one: it's facing upside-down!

What's the reason for this? Well, this is what happens when you apply a negative sign to the *outside* of the Absolute Value bars that contain the x variable. It inverts ("turns upside-down" the shape of the graph).

Think of it like this: the Absolute Value "forces" the value of x to come out as a positive, but then the negative sign on the outside "forces" those same positive values to become negative. That negative sign on the outside of the Absolute Value bars is how you end up with an Absolute-Value-shaped graph that heads *downwards* instead of upwards. (This also relates to upcoming lessons on **Linear Equations**).

None of other choices (B, C, and D) have a negative sign on the outside of the Absolute Value, so their "V-shapes" would all face upwards, and cannot be the right answers.

10. **25.** Another Absolute Value Number Line question. Call me crazy, but I really enjoy these problems.

Best thing to do is probably to sketch out a number line again, as in our solution to Question 8, for the helpful visual cues.

Then you'll need to both add 12.5 and subtract -12.5 from the starting coordinate of -6. That gives $-6+12.5=6.5$ and $-6-12.5=-18.5$. Now we know our values of s and t: 6.5 and 18.5.

Finish the question by completing the final expression $|s-t|$:

$$|6.5-(-18.5)|$$
$$|6.5+18.5|$$
$$=|25|$$
$$=25$$

Again, it doesn't matter which value is assigned to which value - you'll get the same answer either way. As always, if you don't believe me, just try it for yourself (it's a good learning exercise!)

Lesson 4: Algebra 1 Word Problems

Percentages:

- 10.7% of Whole Test
- 7.5% of No-Calculator Section
- 12.4% of Calculator Section

Prerequisites:

- Basic Algebra 1
- Advanced Algebra 1
- Careless Mistakes

At an incredibly-high 11% frequency of appearance, Algebra 1 Word Problems are the second-most-common question type you'll see on the SAT Math test - more than one per ten questions! Only **Linear Equations** are more common.

There are a wide variety of different Algebra 1 Word Problems that you will encounter. As a result, our goal is *not* to prepare you for every single possibility (which would be almost impossible) but rather to give you a large variety of experience across a wide range of common Algebra 1 Word Problems.

The majority of challenges in Algebra 1 Word Problems come down to one of two things: setting up the words into an Algebra equation, and/or plugging the correct values into the proper variables, all without misreading anything important.

Algebra 1 Word Problems Quick Reference

- There are a wide variety of Algebra 1 Word Problems. Sometimes you'll set up equations, other times you'll plug in given values, and frequently you'll do both. There are other types of questions as well.
- It may seem obvious, but *reading carefully* is the most important element of Algebra 1 Word Problems.
- To reduce overwhelm, practice identifying key values, variables, and equations. You can circle them or write them down in your own handwriting. This will help you organize your thoughts and trim the excess fat of the word problem.
- If you're desperate, you may be able to try alternate strategies of Making Up Values or Testing Answer Choices. However, this usually takes more time than the "ideal" solution process.
- Ensure that you always finalize the question by providing exactly which value they've asked you for.

Plugging In Values

"Plugging In Values" word problems are usually one of the easier types of Algebra 1 Word Problems you'll encounter.

These word questions provide you with an equation, and your main job is to pull out the correct numbers from within the words of the problem, plug them into the right places in the equation, and work from there.

Here's an example of this type of question. Try it yourself before looking at my explanation.

$$Q = 80h + 95$$

The equation above gives the amount Q, in dollars, a certain plumber charges for a job that takes h hours. James and Jonathan each hire this plumber. The plumber worked for 5 hours longer on Jonathan's job than on James's job. How much less did the plumber charge James than Jonathan?

In this Algebra 1 Word Problem, you've been provided with an equation - and you can be sure that you'll need it to solve the problem. The only number given within the word problem is "5 hours longer" on Jonathan's job than on James's job.

There are two ways to solve this. First, you could simply choose any number of hours (like "10 hours") for James's job and calculate James's cost. Then, use a value five hours higher for Jonathan's job (say "15 hours") and calculate *his* cost. The difference between these resulting costs would give you the answer.

It doesn't matter how long the jobs take; as long as Jonathan's job takes 5 hours longer, the numbers will work out. Try it yourself!

Another way to solve this question is to think more abstractly about the algebra formula. Both men will have to pay the constant "95 dollars" regardless of how long the job takes. However, Jonathan will have to pay for 5 more hours of labor. Each hour of labor costs an additional $80, as we can see from the equation. Therefore, simply multiply $80/hr by 5 hours and you'll find the difference in their two costs (**$400**). In fact, this question would also fit perfectly into the lesson on **Linear Equations (Word Problems)**.

One takeaway about Algebra 1 Word Problems is that there is typically *more than one way* to find a valid answer. I wish more of my tutoring students understood this better, because they would be more confident trying whichever ideas come to their mind - rather than panicking, getting frustrated, or giving up too early.

Turning Words Into Equations

Another common task in Algebra 1 word problems is to translate word problems into Algebra setups, and then solve the resulting Algebra equation.

Turning words into Algebra equations is both an art and a science. I view this as similar to translating from one language into another. We are interpreters, translating the language of English into the language of Math.

There are many words in English that have direct equivalents in Math symbols. Here are some common ones, though not an exhaustive list:

$=$ means "is, equals, results in..."

$+$ means "sum, total, in all, increase, more than..."

$-$ means "less than, take away, fewer than, difference..." (Note: "distance" is the **Absolute Value** of a subtraction result.)

× means "of, product"

÷ means "quotient, how many can fit into..."

"per" can mean "multiply" or "divide", depending on the context.

x means "what value, an unknown value, how many..."

"Minimum / Maximum possible..." means when asked for the *minimum* value of one variable, plug in the *maximum* possible value of another variable, and vice versa.

Here's an example of this type of "Turning Words Into Equations" question. Give it a try!

When 7 times a number is added to 14, the result is 28. What number results when five times this number is added to 7?

First, we need to translate the first sentence of the question into a **Basic Algebra** setup and solve it for x:

$$7x+14=28$$
$$-14\ -14$$
$$7x=14$$
$$x=2$$

Now translate the second sentence of the question and plug in your newly-found value of x.

$$5x+7$$
$$=5(2)+7$$
$$=10+7$$
$$=17$$

Our final answer value is **17**.

Complicated Algebra 1 Word Problems

On more complicated Word Problems, when my student is getting overwhelmed by a complicated word problem, I will repeat this to them calmly and clearly, like a meditation mantra:

> "Translate *their* words into *your* words; translate *your* words into math symbols."

I use this technique to first "trim the fat" from the word problem by selecting the essential information and translating the test's word problem into a simpler form in my own words.

From there, it's typically easier to take another small step and translate my own trimmed-down version into math symbols.

This is in contrast to what many students try to do, which is to attempt translating the words of the test's problem *directly* into math symbols and numbers.

The biggest problem with that approach is the feeling of "information overwhelm," which can rapidly become confusion, frustration, and worst of all, the risk of giving up on the word problem entirely.

In most cases, there is a lot of useless filler material within the word problem itself. Don't get distracted by back-stories about factories making computers, boats sailing in various directions, or the specific chemical solutions that a student is mixing. The numbers, equations, and variables are all that really matter.

In other words, if you set up the right equation, and you put the right numbers in the right places, then understanding the backstory of the word problem usually doesn't provide any additional benefits.

You only need the words to identify key information, set up the problem and transition into doing "math work" instead of "reading work."

That's why "trimming the fat" and summarizing the key information, equations, and values in your own handwriting is so effective.

Secret Tricks & Common Mistakes on Algebra Word Problems

The number-one "trick" for Algebra Word Problems is to make sure you use all the given numbers and equations in the word problem. It's very rare that they give you any numbers or equations that you don't use.

If you're feeling totally stuck, look back over the word problem and make sure you've used all the given numbers and equations for something.

On that note, some of my best tutoring students make a habit of circling all useful numbers, variables, and equations within the word problem. This is a good strategy.

An even better strategy, as I mentioned above, is to copy down all the critical information from the word problem in your own handwriting. This forces you to decide what's important, and how to organize, label, and arrange it. Rewriting the basic information will help you unlock the next steps of the word problem.

As always, avoid the common careless mistakes. If I had a dollar for every time my students misread a word problem, I could probably buy myself a private island! The most common mistake on these questions is (of course) Misreading, but Negative Signs, Distributing, and the Switcheroo also show up frequently - so review the Prelesson on **Careless Mistakes**.

Regarding the Switcheroo mistake, don't make the classic error of solving your math work correctly and then giving the wrong final answer to the word question. Never assume you remember what the question is asking for; always double-check the word problem for your final value before moving on.

Unfortunately, there's no "one secret trick" to make these word problems easy every time. I wish there was, so I could tell you about it. Instead, I recommend staying focused, patient, and reading carefully.

It's their endless variety that makes Algebra 1 Word Problems dangerous. From the easiest questions to the most complicated problems, you can never fall asleep on a word problem.

Pretest Question #1

Here's an ugly-looking Algebra 1 Word Problem from the Pretest. If you missed it the first time, try it again now. My explanation is below.

> Geraldine randomly selects marbles from a bag with four colors of marbles in it: blue, orange, pink, and bright-red. Of the marbles Geraldine selected, $\frac{1}{2}$ were bright red and $\frac{1}{5}$ were orange. Of the other marbles Geraldine selected, $\frac{1}{3}$ were pink. If Geraldine selected six pink marbles, how many orange marbles did Geraldine select?
>
> (A) 3
> (B) 5
> (C) 12
> (D) 30

This problem presents an "information overwhelm" situation filled with various fractions and colors. The question is, where can we *start* this problem and feel like we're making progress, instead of freezing up and getting stuck?

The key is to find and begin with the only actual *value* we've been given: "six pink marbles." Since it's the only specific number we're provided, it is very valuable.

We know the pink marbles are $\frac{1}{3}$ of "the other marbles" (notice this is not $\frac{1}{3}$ of the whole bag). Let's set this situation up using a mix of words and math symbols:

$$\frac{1}{3}(\text{the other marbles}) = \text{pink marbles}$$

Now plug in the value "6" for pink marbles and solve:

$$\frac{1}{3}(\text{the other marbles}) = 6$$

$$(3)\frac{1}{3}(\text{the other marbles}) = 6(3)$$

$$\text{the other marbles} = 18$$

But now there's a new issue: these "18 other marbles" don't represent the whole bag. Instead, these "other marbles" are what are left over *after* Geraldine already selected $\frac{1}{2}$ bright red marbles and $\frac{1}{5}$ orange marbles.

What we need to do next is figure out what fraction of the bag the "18 other marbles" were. We'll do this by figuring out what fraction of the bag were bright red and orange combined. This also requires some use of fractions and finding the same denominators so we can add them (review the Prelesson on **Fractions**).

$$\frac{1}{2}\text{ are bright red} + \frac{1}{5}\text{ are orange}$$

$$= \frac{5}{10} + \frac{2}{10}$$

$$= \frac{7}{10}\text{ of Geraldine's marbles are bright red or orange}$$

This is interesting. If $\frac{7}{10}$ of the marbles or bright red or orange, then we know that the "18 other marbles" must be the rest ($\frac{3}{10}$) of her entire selection (because all the fractions of marbles taken together must add up to 1). Now we can set up a new situation and solve it:

$$\frac{3}{10}(\text{all marbles}) = 18\ (\text{other marbles})$$

$$\left(\frac{10}{3}\right)\frac{3}{10}(\text{all marbles}) = 18\left(\frac{10}{3}\right)$$

$$\text{all marbles} = \frac{180}{3}$$

$$\text{all marbles} = 60$$

Great! We've figured out something very valuable: Geraldine selected a total of 60 marbles.

We're almost done. Now finish the question by giving them what they asked for in the Word Problem - the number of orange marbles. The question told us that orange marbles were $\frac{1}{5}$ of all the marbles Geraldine selected. Here's the final step:

$$\frac{1}{5}(60\text{ marbles})$$

$$= 12\text{ orange marbles}$$

Phew! We're finished - I think. Double-check what the question asked for. Yup, they wanted the number of orange marbles that Geraldine selected, and now we have it.

The final answer is 12 orange marbles or **Choice C**.

Strategy: Making Up Values

A certain type of Algebra 1 Word Problem will ask you to set up an Algebra equation from a complex word problem. Surprisingly, you may not have to *solve* the equation for a final answer - just set it up. However, these questions can still be on the difficult side.

Luckily, if you're not able to confidently set up an equation from the word problem, you may be able to try a backup strategy: Making Up Values.

Here are the steps to this strategy:

1. Make up your own values to plug in for variables in the word problem.
2. Figure out the final value of the word problem *based on the values you chose*.
3. Test those same made-up values by inputting them into the answer choices.
4. Eliminate any answer choices that don't return the same value you got in Step 2.

If more than one answer choice survives your first elimination round, you go back to Step 1: choose a new set of values, and repeat the process - but now you only have to check any answer choices that still remain after your first pass.

In the end you will be left with only one possible answer choice.

Note that this strategy *cannot prove the right answer, it can only eliminate wrong answers*. It's a time-consuming strategy, because you must test all 4 answer choices with your made-up numbers (and sometimes you must test several choices more than once).

This backup strategy only makes sense if the following criteria are met:

- The question is a word problem with unknown variables embedded in the words.
- The answer choices are equations based on the same variables as in the question.
- You can't think of a better way to complete the problem.
- You're patient enough to check all four answer choices, possibly more than once.
- There aren't any other, easier questions you can work on first.

Pretest Question #2

Here's another Algebra 1 Word Problem from the Pretest. This one's pretty tough for most students. Be sure to try it again before looking at my explanation if you missed it the first time.

Helen drives an average of 200 miles each week. Her car can travel an average of 20 miles per gallon of gasoline. Helen would like to reduce her weekly expenditures on gasoline by $10. Assuming gasoline costs $3 per gallon, which equation can Helen use to determine how many fewer average miles, d, she should drive each week?

(A) $\frac{20}{3}d = 10$

(B) $\frac{3}{20}d = 10$

(C) $\frac{10}{3}d = 20$

(D) $\frac{3}{20}d = 20$

We're going to solve this question two different ways: first, by testing the answer choices, and then the more "math teacher" way.

Testing the Answer Choices:

Here's how we can test the answer choices. First, we'll have to solve the word problem on our own:

If Helen needs to save $10, and gas costs $3 per gallon, we can easily figure out how many fewer gallons Helen needs to buy each week. That would be $\frac{10}{3}$ gallons, or $3\frac{1}{3}$ gallons. If she cut her gas usage by this much at a cost of $3 per gallon, she'd save the $10 per week.

The next step: how many *miles* do those $3\frac{1}{3}$ gallons represent? That's easy, too. Just multiply the $3\frac{1}{3}$ gallons by her 20 miles per gallon:

$$\frac{10}{3}\text{gallons(20 miles per gallon)}$$
$$=\frac{200}{3}\text{miles}$$

So, now we know that Helen must drive $\frac{200}{3}$ miles fewer per week. If only this answer was one of our answer choices! Instead, we're forced to test this value of $\frac{200}{3}$ in each of the four answer choices. If we're lucky, it will only work with one of the choices.

Now I'll try plugging in this value ($\frac{200}{3}$) for *d* in each of the four answer choices:

(A)
$$\frac{20}{3}d = 10$$
$$\frac{20}{3}(\frac{200}{3}) = 10$$
$$\frac{4000}{9} \neq 10$$

(C)
$$\frac{10}{3}d = 20$$
$$\frac{10}{3}(\frac{200}{3}) = 20$$
$$\frac{2000}{9} \neq 20$$

(B)
$$\frac{3}{20}d = 10$$
$$\frac{3}{20}(\frac{200}{3}) = 10$$
$$\frac{600}{60} = 10$$
$$10 = 10$$

(D)
$$\frac{3}{20}d = 20$$
$$\frac{3}{20}(\frac{200}{3}) = 20$$
$$\frac{600}{60} = 20$$
$$10 \neq 20$$

We can see that only **Choice B** returns a true equality. This is one way of determining that Choice B is the right answer.

Solving With Algebra:

Now let's start the question over and solve it the "math teacher" way, using Algebra. First, I'll set up a pair of equations. These equations use the weekly miles driven, divided by miles per gallon, times the cost per gallon, to arrive at weekly cost_1 (before savings) and weekly cost_2 (after savings):

$$\frac{\text{weekly miles}_1}{\text{mpg}}(\text{\$ per gallon}) = \text{weekly cost}_1$$

$$\frac{\text{weekly miles}_2}{\text{mpg}}(\text{\$ per gallon}) = \text{weekly cost}_2$$

The difference of cost_1 and cost_2 must be an amount equal to \$10. Since difference is subtraction, we use the difference between the two weekly costs to set up the following equation:

$$\text{weekly cost}_1 - \text{weekly cost}_2 = \text{savings}$$
$$[\frac{\text{miles}_1}{\text{mpg}}(\text{\$ per gallon})] - [\frac{\text{miles}_2}{\text{mpg}}(\text{\$ per gallon})] = \$10$$

And, we have several values taken from the word problem that we can plug in:

$$[\frac{\text{miles}_1}{20}(3)] - [\frac{\text{miles}_2}{20}(3)] = \$10$$

Clean it up:

$$\frac{(3)\text{miles}_1}{20} - \frac{(3)\text{miles}_2}{20} = \$10$$

Now combine the **Fractions**, since they already have the same denominator:

$$\frac{(3)\text{miles}_1 - (3)\text{miles}_2}{20} = \$10$$

Now factor out the 3 on top (we covered basic factoring in **Advanced Algebra 1**):

$$\frac{3(\text{miles}_1 - \text{miles}_2)}{20} = \$10$$

Consider that the *difference* in miles_1 and miles_2 must be d, the variable for "decrease in miles" that was given in the question. We can rewrite as $\text{miles}_1 - \text{miles}_2 = d$. Make this substitution:

$$\frac{3(d)}{20} = \$10$$

Now rewrite the fraction to match the format offered in the answer choices:

$$\frac{3}{20}d = 10$$

And again, we arrive at the correct Choice B, using a method based more on Algebra setups than testing the answer choices. Remember, there is often more than one way to finish an Algebra 1 Word Problem!

Review & Encouragement

From watching so many students, I *know* how intimidating these word problems can be - especially because of their never-ending variety. I promise, if you give them a chance, you can get the hang of it.

You should *expect* a wide variety of word problems on test day. From that point of view, it's a little easier to stay confident when the word question looks "weird" at first. Of course it looks weird! That's normal, because each word problem is a little different from the last one.

What always stays the same is the need to break the problem down and convert the words into meaningful math symbols. Identify the equations, variables, and values provided in the word problem. "Trim the fat" from the words and focus on translating the essential math into written Algebra form. If you're stuck, make sure you've used all of the given equations, values, and variables.

Remember your backup strategies of "Testing the Answer Choices" and "Making Up Values", which can be lifesavers in certain situations. Sometimes they're even faster than doing it "the math teacher way."

Avoid careless mistakes, and double-check that your final answer matches to the question they've actually asked.

Most importantly, practice with *all* the word problems in the Practice Questions section of this lesson!

Algebra 1 Word Problems Practice Questions

DO NOT USE A CALCULATOR ON ANY OF THE FOLLOWING QUESTIONS UNLESS INDICATED.

1. If $4q-15$ is 19 more than 26, what is the value of $3q$?

   (A) 7.5

   (B) 15

   (C) 22.5

   (D) 45

2. If $21+7x$ is 12 less than 68, how many times does $4x$ divide into 60?

   (A) 7

   (B) 5

   (C) 3

   (D) 2

3. One pint of raspberries costs $3. At this rate, how many dollars will r pints of raspberries cost?

   (A) $\frac{r}{3}$

   (B) $3r$

   (C) $\frac{3}{r}$

   (D) $12r$

$$T = 100c + 50s$$

4. (CALCULATOR) FREE RESPONSE: A factory manufactures completed guitar amplifiers from a speaker cabinet and a speaker cone. The equation above shows the total cost T in dollars of making the amplifiers from c units of speaker cabinets and s units of speaker cones. If the total cost of manufacturing was $2000 and the factory produced 10 speaker cabinets, how many speaker cones did the factory produce?

5. (CALCULATOR) FREE RESPONSE: Membership at a swimming center costs a one-time sign-up fee of $50 plus m dollars for each month. If a swimmer paid $230 for the first 12 months, including the sign-up fee, what is the value of m?

6. Karl's house sits exactly at Mile Marker 6 on a road that runs directly North to South. He walked due south from his house to Mile Marker 5. Then he walked due north to Mile Marker 9. Then he walked south again to Mile Marker 3. From there he walked directly home. How many miles did Karl walk?

   (A) 7

   (B) 14

   (C) 19

   (D) 23

7. (CALCULATOR) If 40 Antarctic Ants were stacked directly on top of each other, the column would be approximately $4\frac{2}{5}$ inches tall. Which of the following is closest to the number of Antarctic Ants it would take to make a 10-inch column?

(A) 9

(B) 89

(C) 90

(D) 91

8. The equation below shows the number of minutes r that Christian spends jogging each week and the number of hours w that he spends walking each week. In the equation, what does the number 300 represent?

$$300 = r + 60w$$

(A) The number of minutes spent jogging each week.

(B) The total number of hours spent jogging and walking each week.

(C) The total number of minutes spent jogging and walking each week.

(D) The difference in minutes spent jogging and walking each week.

$$.15x + .25y = .2(x + y)$$

9. (CALCULATOR) Christian will mix x liters of a 15% by volume peroxide solution with y liters of a 25% by volume peroxide solution in order to create a 20% by volume peroxide solution. The equation above represents this situation. If Christian uses 10 liters of the 25% by volume peroxide solution, how many liters of the 15% by volume peroxide solution must he use?

(A) 10

(B) 15

(C) 100

(D) 150

10. Ian has $225 to spend on video games and controllers. Video games cost $50 each, and controllers cost $30 each. If there is no tax on this purchase and he buys 3 video games, what is the maximum number of controllers he can buy?

(A) One

(B) Two

(C) Three

(D) Four

$$2520 = 8r + 14t$$

11. (CALCULATOR) A warehouse stores two types of tires: race car tires and tractor tires. Because of space limitations, the number of tires the warehouse is capable of storing can be calculated by the equation above, where r is race car tires and t is tractor tires. If the warehouse currently holds 110 race car tires, what is the maximum number of tractor tires that the warehouse can currently store?

(A) 117

(B) 118

(C) 122

(D) 123

12. A manufacturer is considering an upgrade to a new type of machinery that will allow their factory to produce shaving cream canisters at a lower cost per unit, saving $1.50 per canister. The cost of installing this new type of machinery is $20,000. If the factory produces c shaving canisters per month, which of the following inequalities can be solved to find y, the number of months after which the total savings on unit production will exceed the installation cost?

(A) $20{,}000 > \frac{1.5}{cy}$

(B) $20{,}000 < 1.5cy$

(C) $20{,}000 < \frac{1.5c}{y}$

(D) $20{,}000 > \frac{1.5y}{c}$

13. FREE RESPONSE: The density of an object is equal to its mass divided by its volume. What is the volume, in cubic centimeters, of a meteorite with a density of 9 grams per cubic centimeter and a mass of of 54 grams?

14. A student estimates that an end-of-semester project for history class will take h hours to complete, where $h < 30$. The goal is for the estimate to be within 5 hours of the time it will actually take to complete the project. If the student meets the goal and it takes x hours to complete the project, which of the following inequalities represents the relationship between the estimated time and the actual completion time?

(A) $x + h < 10$

(B) $x > 30 - h$

(C) $h < 30 + x$

(D) $-5 < x - h < 5$

15. At a certain food truck, s servings of omelets are made by adding g pinches of shredded cheese to a frying pan full of scrambled eggs. If $s = g + 3$, how many additional pinches of shredded cheese are needed to make each additional serving of omelette?

(A) One

(B) Two

(C) Three

(D) Four

QUESTIONS 16, 17, AND 18 REFER TO THE FOLLOWING INFORMATION:

When designing a skyscraper, a construction firm can use the skyscraper dimensions formula $4h + p = 170$, where h is the height in feet of a single story of the skyscraper, and p is the perimeter in feet of the skyscraper's base. For any given skyscraper, the height of each story and the perimeter of the base are always the same.

16. Which of the following expresses the height of a single story in terms of perimeter of the base?

(A) $h = 4(170 + p)$

(B) $h = \dfrac{170 + p}{4}$

(C) $h = \dfrac{4}{170 - p}$

(D) $h = \dfrac{170 - p}{4}$

17. (CALCULATOR) The building code of a certain city requires that, for all skyscrapers, the height of a single story must be no less than 10 feet and the perimeter of the skyscraper's base must be at least 80 feet. According to the skyscraper dimensions formula, which of the following inequalities represents the set of all possible values for the height in feet of a single story that meets this code requirement?

(A) $10 < h$

(B) $10 < h < 17.5$

(C) $10 < h < 22.5$

(D) $10 < h < 62.5$

18. (CALCULATOR) A construction firm wants to use the skyscraper dimensions formula to design a skyscraper with a total height of 780 feet, a base perimeter of at least 86 feet but no more than 90 feet, and an odd number of floors. Within these constraints, which of the following must be the perimeter, in feet, of the skyscraper?

(A) 86

(B) 87

(C) 88

(D) 90

19. Aaron has a bag of Halloween candy containing n pieces of candy to share with his friends at a party. If he gives each of his friends 4 pieces of candy, he will have 7 pieces of candy left over. In order to give each friend 5 pieces of candy, he will need an additional 6 pieces. How many friends does Aaron have at the party?

(A) 8

(B) 13

(C) 23

(D) 30

20. (CALCULATOR) FREE RESPONSE: The force of gravitational attraction between two bodies can be determined by the equation

$F = G\frac{m_1 m_2}{d^2}$, where F is the force in Newtons, G is a constant, m_1 and m_2 are the masses of the two bodies in kilograms, and d is the distance between the centers of the two objects in meters. An astronomer finds three celestial bodies with exactly equal masses but different distances from each other. If the force of gravitational attraction between the two closer bodies is exactly 2.25 times greater than the force of gravitational attraction between the two farther bodies, what is the ratio of the farther distance to the closer distance?

Algebra 1 Word Problems Answers

1. D
2. C
3. B
4. 20
5. 15
6. B
7. D
8. C
9. A
10. B
11. A
12. B
13. 6
14. D
15. A
16. D
17. C
18. D
19. B
20. 1.5

Algebra 1 Word Problems Explanations

1. **D.** This is just a basic "Translating Words into Algebra" question. Set it up and solve as follows:

$$4q - 15 = 26 + 19$$
$$4q - 15 = 45$$
$$+15 \quad +15$$
$$\frac{4q}{4} = \frac{60}{4}$$
$$q = 15$$

Now we know $q = 15$, but don't forget to finish the question, which asked for the value of $3q$.

$$3(15) = 45$$

2. **C.** Another "Translating Words into Algebra" question like the previous. Set it up and solve as follows:

$$21 + 7x = 68 - 12$$
$$21 + 7x = 56$$
$$-21 \qquad -21$$
$$\frac{7x}{7} = \frac{35}{7}$$
$$x = 5$$

But we're not done yet. Although $x = 5$, we've also been asked "how many times does $4x$ divide into 60?"

Since $4x = 4(5) = 20$, we can ask "how many times does 20 divide into 60?" The answer is $\frac{60}{20} = 3$.

3. **B.** It's a pretty simple word problem - don't overthink it. Each pint will cost \$3, and there will be r pints. It costs three more dollars for each additional pint. That means our total price equation will just be \$3 per r, or $3r$.

4. **20.** This word problem is about using the provided equation and plugging the correct numbers into the proper variables, then solving. The cost of manufacturing, T, is 2000. The value of c is given as 10 speaker cabinets.

Here's our setup and solution:

$$T = 100c + 50s$$
$$2000 = 100(10) + 50s$$

Now solve for s:

$$2000 = 100(10) + 50s$$
$$2000 = 1000 + 50s$$
$$-1000 \quad -1000$$
$$\frac{1000}{50} = \frac{50s}{50}$$
$$20 = s$$

So the value of s, the number of speaker cones, is 20.

5. **15.** The first thing we need to do with this Word Problem is turn it into an Algebra equation. Let's make up our own variable T to represent the total cost of membership and use x to represent the number of months we sign up for. We'll have to pay our \$50 sign-up fee and m dollars per month. Here's what that equation looks like:

$$T = 50 + mx$$

Now let's plug in the values we are given in the word problem: \$230 is our total T value, and 12 is our x value. We're solving for m.

$$230 = 50 + 12m$$
$$-50 \quad -50$$
$$180 = 12m$$
$$\frac{180}{12} = \frac{12m}{12}$$
$$15 = m$$

6. **B.** This Word Problem is about distance. As I mentioned in the lesson, distance is "the absolute value of subtraction." Another way of saying this is "distance is the difference between two points, and it is always a positive value."

We could use Absolute Value to solve this problem in one giant setup, but I think it's overkill. Instead, just go a step at a time and write down each stage of his journey. From home at Mile Marker 6 to Mile Marker 5 is $|6-5|=$ *1 mile*. Then from Mile Marker 5 to Mile Marker 9 is $|5-9|=$ *4 miles*. From Mile Marker 9 to Mile Marker 3 is $|9-3|=$ *6 miles*. From Mile Marker 3 to home at Mile Marker 6 is $|3-6|=$ *3 miles*.

Total up all the miles he's walked and you'll have $1+4+6+3=14$ miles.

7. **D.** Use the word problem to create a setup. I'm going to use decimals instead of fractions to make my life easier - it's just a personal preference. The decimal value of $4\frac{2}{5}$ inches is 4.4 inches. I've also chosen a new variable, A, to represent the height of a single individual ant.

$$40A=4.4$$
$$\frac{40A}{40}=\frac{4.4}{40}$$
$$A=.11$$

Now I know an individual ant has a height of .11 inches, and I can set up another equation for the 10-inch column. This time I'll use the variable x to represent the number of ants:

$$10=.11x$$
$$\frac{10}{.11}=\frac{.11x}{.11}$$
$$90.9...=x$$

Although I get an ugly repeating decimal for the value of x, I've truncated at "90.9" because we're asked to round to the "closest number," which will be 91 ants.

8. **C.** This question asks us to interpret the real-world meaning of the terms in an Algebra equation.

Here's how I see it. On the right side of the equation we have two terms being totaled (added) to each other. It's reasonable to assume that one of those terms is Christian's walking time and the other term is his jogging time. In that case, "300" would be the *total* of all Christian's time spent jogging and walking.

The only other issue is the units, because the word problem uses "minutes" for r and "hours" for w. But notice that w is being multiplied by 60, which is exactly how we would convert w hours into an equivalent time in minutes. Therefore, it's reasonable to assume that "300" is the *total* number of *minutes* spent jogging and walking each week.

9. **A.** This Word Problem provides us an Algebra equation to use. It can be a bit confusing to find the useful data in all those words, but notice the "10 liters of the 25% by volume solution" near the end. 10 is a value we can plug in for the variable y (be sure to plug it in for y on *both* the left and the right side of the equation).

$$.15x+.25y=.2(x+y)$$
$$.15x+.25(10)=.2(x+10)$$

That creates a complete Algebra setup that we can solve, as follows:

$$.15x+.25(10)=.2(x+10)$$
$$.15x+2.5=.2x+2$$
$$-2 \qquad\qquad -2$$
$$.15x+.5=.2x$$
$$-.15x \qquad -.15x$$
$$\frac{.5}{.05}=\frac{.05x}{.05}$$
$$10=x$$

The value of x, the liters of 15% by volume solution, is 10.

10. **B.** This Word Problem gives us the opportunity to set up an Algebra inequality for Ian's purchase. I'll make up the variable g to represent the number of video games, and c to represent the number of controllers:

$$225 > 50g + 30c$$

Notice my use of an inequality to show that Ian's total payment must be *less than* his maximum budget of $225.

Now plug in the given value of "3 video games" for g and solve the inequality:

$$\begin{aligned} 225 &> 50g + 30c \\ 225 &> 50(3) + 30c \\ 225 &> 150 + 30c \\ -150 & \quad -150 \\ 75 &> 30c \end{aligned}$$

Notice that I've stopped right before I finish. Why? Because I'm not allowed to use my calculator, I don't want to actually divide 75 by 30. However, we can easily calculate that 2 controllers would cost $60 and remain within budget, but 3 controllers would cost $90 and go over-budget. Therefore, Ian cannot buy more than 2 controllers with his current budget.

11. **A.** This word problem provides us with an equation - we simply have to plug in the given values and solve. The value of r is 110 race car tires, as given in the question. Plug in and solve for t:

$$\begin{aligned} 2520 &= 8r + 14t \\ 2520 &= 8(110) + 14t \\ 2520 &= 880 + 14t \\ -880 & \quad -880 \\ \frac{1640}{14} &= \frac{14t}{14} \\ 117.14... &= t \end{aligned}$$

Note that we get an ugly decimal value for t. That's OK - the question asked for the *maximum* number of tractor tires t. Of course, there's no such thing as ".14 of a tire", so we must round down to 117.

12. **B.** This Word Problem requires that we translate the words into an Algebra setup with an inequality. We want to find the length of time it will take for the factory to break even on their new $20,000 investment. Therefore, we can start with the following:

$$20000 < \text{amount saved}$$

This inequality says we're looking for the moment when the amount saved is *greater* than the $20,000 investment.

Now we just need to set up the right side of the equation. The factory produces c canisters per month and each canister produced will save $1.50. We'll also need to multiply by the number of months of production. Let's fill this information in:

$$20000 < \$1.5c(\text{months of production})$$

We're told to use the variable y for months of production, so we can now finish our setup, and we're done!

$$20{,}000 < 1.5cy$$

13. **6.** This question isn't so bad. We translate the words into an equation. I'll make up some common-sense variables: m will be "mass," v will be "volume," and d will be "density."

Here's what I come up with from turning the word problem into an equation:

$$d = \frac{m}{v}$$

Now I just plug in the given values from the word problem and solve:

$$\begin{aligned} 9 &= \frac{54}{v} \\ (v)9 &= \frac{54}{v}(v) \\ 9v &= 54 \\ \frac{9v}{9} &= \frac{54}{9} \\ v &= 6 \end{aligned}$$

14. **D.** This Word Problem asks us to set up an inequality representing the timeframe for completing a project.

A key concept is staying "within 5 hours of the actual time." Another way of saying this is: "the *difference* between the *estimated* time and the *actual* project time must be less than 5 hours." Remember that "difference" means subtraction, so $x-h$ could represent the difference between actual time and estimated time.

But, there's an important point to make: the question didn't state whether we needed to be *over* or *under* the estimate. We need to account for both possibilities. That means either "-5" or "$+5$" can be the difference between our two times. This is why the correct answer has an inequality on both sides of the $x-h$, which covers both "5 hours over estimate" and "5 hours under estimate."

Note that we never need to use the $h<30$ information for anything, which is unusual, but sometimes happens.

Also understand that either $x-h$ or $h-x$ will work. Both will calculate the difference (subtraction) between the estimated time and the actual completion time. Either way, the difference must be between -5 and 5 hours. This could also be set up with an **Absolute Value**.

15. **A.** This question seems so easy, but I've noticed a lot of my students have a surprising amount of difficulty with it.

One way to deal with it is to simply Make Up Numbers and plug them in. For example, what if we use just one pinch of cheese - in other words, if $g=1$? Plugging it into the equation would produce $s=4$, or four servings of omelets. In other words, 1 pinch of cheese will make 4 servings.

Now increase g by one, to $g=2$, and plug it in. The result is $s=5$. Notice that by adding 1 pinch of cheese, we got one additional serving of omelets. That's one way to see that for each additional serving, you need one more pinch of cheese.

Another way to view this question is as a Linear Equation of the form $y=mx+b$. From that point of view, you can see that the slope of this line is "1", which means the pinches of cheese and servings of omelette will increase together at a constant 1-to-1 relationship.

16. **D.** This first problem in the set of three is the "gimme" problem - the easiest out of all three "Skyscraper" questions. All of the answer choices show us that we should solve for h. Just take the original equation, $4h+p=170$, and do **Basic Algebra 1** to isolate the variable h. Here are the steps:

$$4h+p=170$$
$$-p \qquad -p$$
$$4h=170-p$$
$$\frac{4h}{4}=\frac{170-p}{4}$$
$$h=\frac{170-p}{4}$$

17. **C.** This question is somewhat more difficult. We must create an inequality for all the possible values of h.

First, the easy part: at a minimum, h must be "no less than 10 feet," or $10<h$. Unfortunately, that doesn't eliminate any answer choices. We still need to find the maximum value of h.

Now, the harder part. We need to use the value "80" that was given in the word problem, and plug it in to the correct variable (p for perimeter of the base). Then solve for h. I will reuse our rearranged equation from Question 16 to save time, but we could also use the original equation from the question:

$$h=\frac{170-p}{4}$$
$$h=\frac{170-(80)}{4}$$
$$h=\frac{90}{4}$$
$$h=22.5$$

This solution, $h=22.5$,is the value of the floor height when the base perimeter is at its *minimum* value of 80 feet. Remember that "if you want to *maximize* one variable, you must *minimize* all the other variables." And that's what we've just done.

Now we know the floor height cannot exceed a maximum of 22.5 feet without violating the minimum base perimeter requirement.

Our final inequality can be set up as $10<h<22.5$.

18. **D.** This is the hardest question (by far) from the set of three "skyscraper" questions. Where to even begin? Well, of all the values in this word problem, the *only* given values that can plug directly into our equation are the base perimeter requirements: "at least 86 feet but no more than 90 feet."

What I recommend doing is calculating the minimum and maximum floor heights by using *both* values of 86 and 90. Yes, you'll have to work through the equation two times. Note, I'll use the rearranged equation from Question 17 just to save a couple of steps, but you could also use the original equation from the word problem:

$$h_1 = \frac{170 - p}{4} \qquad h_2 = \frac{170 - p}{4}$$

$$h_1 = \frac{170 - (86)}{4} \qquad h_2 = \frac{170 - (90)}{4}$$

$$h_1 = \frac{84}{4} \qquad h_2 = \frac{80}{4}$$

$$h_1 = 21 \qquad h_2 = 20$$

Taken together, these two h values tell us the minimum floor height for this building is 20 feet, and the maximum is 21 feet. What we will do next is divide the total height of the skyscraper (given in the question as 780 feet) by *both* possible floor heights. This will tell us the minimum and maximum number of *floors* to build this building.

$$\frac{780}{20} = 39 \qquad \frac{780}{21} = 37.14...$$

Now we know that the skyscraper must have between a minimum of "37.14" floors and a max of 39 floors. The word problem also tells us that the skyscraper must have an odd number of floors. Also, floors must be a whole number (naturally, since you can't have "part of a floor" - you either have one, or you don't). Since 37.14 is a *minimum*, we can't round down to 37. The next option, 38 floors, is not an odd number. Only 39 floors is an odd number of floors that fits within our minimum and maximum number of floors.

Now we can take our 780 foot skyscraper and divide it by the 39 floors to determine how tall each floor must be in feet:

$$\frac{\text{780 feet tall building}}{\text{39 floors}} = \text{20 feet per floor}$$

Almost done. Now plug the "20 feet per floor" back into the original equation for h and solve for p:

$$4h + p = 170$$

$$4(20) + p = 170$$

$$80 + p = 170$$

$$-80 \qquad -80$$

$$p = 90$$

Voila! We have finally determined that the base perimeter of the skyscraper must be 90 feet, which also fits within the original constraints of the project ("no less than 86 feet but no more than 90 feet).

19. **B.** This question might look like an Algebra setup, but I actually find it easier to solve if I just *think* about it. After giving out 4 pieces of candy to each friend the first time, Aaron has 7 pieces of candy left over. Now he wants to give each friend *one more* piece of candy (going from 4 per friend to 5 per friend). To do this, we know he'll need another 6 candies.

Combine the 7 leftover pieces he already has with the additional 6 pieces he needs, and you'll get 13 pieces. Once Aaron has those 13 pieces of candy, he's able to give everyone exactly one more piece of candy. Therefore, he must have 13 friends at the party - one extra piece of candy per friend.

20. **1.5.** The final question of this section - and it's a doozy. This will get pretty "mathy," so hold on to your hats. First of all, all three celestial bodies have "exactly equal masses". That's good news - it means we can replace m_1, m_2 etc. with just one identical variable, m.

Now let's set up two separate equations - one for the "closer" gravitational force, and one for the "farther" force. Notice my use of "sub-labels" to keep track of my variables:

$$F_{\text{closer}} = G\frac{(m)(m)}{(d_{\text{closer}})^2}$$

$$F_{\text{farther}} = G\frac{(m)(m)}{(d_{\text{farther}})^2}$$

We also know that the number "2.25" must be important. It represents how many times greater F_{closer} is than F_{farther}. The "closer" force is stronger, so this 2.25 multiplier goes on the "farther" side to help it balance the scales. We can write this mathematically:

$$F_{\text{closer}} = 2.25(F_{\text{farther}})$$

Now let's put all these equations together. Plug the first two equations into their respective places in the third equation that we just made:

$$F_{\text{closer}} = 2.25(F_{\text{farther}})$$
$$G\frac{(m)(m)}{(d_{\text{closer}})^2} = 2.25[G\frac{(m)(m)}{(d_{\text{farther}})^2}]$$
$$G\frac{(m)(m)}{(d_{\text{closer}})^2} = 2.25G\frac{(m)(m)}{(d_{\text{farther}})^2}$$

This looks ugly, but a lot of terms can cancel. The G constants can be divided out of both sides. So can the $(m)(m)$ terms on top of both sides. Now we're left with this:

$$(1)\frac{(1)(1)}{(d_{\text{closer}})^2} = 2.25(1)\frac{(1)(1)}{(d_{\text{farther}})^2}$$
$$\frac{1}{(d_{\text{closer}})^2} = 2.25\frac{1}{(d_{\text{farther}})^2}$$
$$\frac{1}{(d_{\text{closer}})^2} = \frac{2.25}{(d_{\text{farther}})^2}$$

From here lets' square root both sides to cancel the exponents,

$$\frac{1}{(d_{\text{closer}})^2} = \frac{2.25}{(d_{\text{farther}})^2}$$
$$\sqrt{\frac{1}{(d_{\text{closer}})^2}} = \sqrt{\frac{2.25}{(d_{\text{farther}})^2}}$$
$$\frac{\sqrt{1}}{\sqrt{(d_{\text{closer}})^2}} = \frac{\sqrt{2.25}}{\sqrt{(d_{\text{farther}})^2}}$$
$$\frac{1}{d_{\text{closer}}} = \frac{1.5}{d_{\text{farther}}}$$

I used my calculator to determine the value of $\sqrt{2.25}$, by the way. It looks much cleaner now, right?

Now cross-multiply, then rearrange into a ratio ("fraction") of the farther distance to the closer distance (which can be expressed as $\frac{d_{\text{farther}}}{d_{\text{closer}}}$):

$$\frac{1}{d_{\text{closer}}} = \frac{1.5}{d_{\text{farther}}}$$
$$1.5(d_{\text{closer}}) = 1(d_{\text{farther}})$$
$$\frac{1.5(d_{\text{closer}})}{(d_{\text{closer}})} = \frac{d_{\text{farther}}}{(d_{\text{closer}})}$$
$$1.5 = \frac{d_{\text{farther}}}{d_{\text{closer}}}$$

We have finally finished, demonstrating that the ratio of the farther distance to the closer distance is 1.5.

Lesson 5: d=rt (Distance Equals Rate Times Time)

Percentages

- 1.4% of Whole Test
- 0% of No-Calculator Section
- 2.1% of Calculator Section

Prerequisites

- Basic Algebra
- Algebra 1 Word Problems
- Unit Conversions

This lesson will cover common SAT questions about distance, rate, and time. It's a very easy and straightforward topic that uses the same Basic Algebra equation on every question - a breath of fresh air after the complexity and variety of the questions we encountered in the previous lesson on **Algebra 1 Word Problems!**

d=rt Quick Reference

- $d = rt$ questions are always given as Word Problems. These questions only seem to appear on the Calculator section of the SAT Math test.
- Recognizing these questions quickly is key. Look for keywords about speeds, distances, and times.
- The $d = rt$ equation is not provided - you need to memorize it. Luckily it's very simple and the "DERT" acronym makes it easy to remember.
- These questions will sometimes involve **Unit Conversions** of time or distance - be careful.
- Always double-check your end result against the question - with particular attention to units - before committing to a final answer.

The DERT Equation

This entire lesson can be summed up in a single equation:

$$\text{distance} = (\text{rate})(\text{time})$$

My physics teacher in high school used to call this the "DERT" equation (pronounced "dirt") because "**D** Equals **R T**". I've never forgotten it since.

These questions are easy to recognize and to solve. Since the DERT equation has three main components (distance, rate, and time), you can expect to be given *two* of those values, and asked to *find* the third one.

For example, you may be given a speed and distance, and asked how long it takes to travel the distance at that speed. Or, you could be given a time and a speed, and asked how far you will travel in that time - and so forth.

All you have to do is notice the peed, time, and distance keywords, then write down the $d = rt$ equation, plug in the given values from the question, and solve for whatever is left using **Basic Algebra 1**. Easy!

Here's a simple question to see how this equation works. Try it yourself first - it's easy!

> If a man on a bicycle is traveling 10 miles per hour, how many hours will it take him to travel 25 miles?

All we have to do is set up a $d = rt$ equation with the given values, then solve, like so:

$$d = rt$$
$$25 = (10)t$$
$$\frac{25}{10} = \frac{10t}{10}$$
$$2.5 = t$$

It will take this man **2.5 hours** to travel the distance. There aren't any units to worry about, because we stayed consistent throughout the entire problem (miles, hours, and miles per hour).

Pretest Question #1

Now let's try a question from the Pretest. If you missed it the first time, try it yourself before looking at my explanation below.

> (CALCULATOR) FREE RESPONSE: Heather took a four-day hiking trip over a long weekend. On Friday, Heather hiked 18 miles in 6 hours. On Saturday, she hiked 16 miles in 7 hours. On Sunday, she hiked 20 miles in 8 hours. On Monday, she hiked 4 miles in 2 hours. What was Heather's average speed, rounded to the nearest tenth of a mile per hour, for the time she spent hiking?

This question may seem like a lot of work - won't you have to calculate *four* $d = rt$ equations, and then take the average of them all?

Well, no. It's much easier than that. Instead, we'll add up the *total* distance Heather hiked in four days, and the *total* time she spent hiking, and then we'll set up a single $d = rt$ equation with our total distance and total time.

First, how many total miles did Heather hike in the four days?

$$18 + 16 + 20 + 4 = 58 \text{ miles}$$

Next, how many total hours did Heather hike for?

$$6 + 7 + 8 + 2 = 23 \text{ hours}$$

Now we can set up the $d = rt$ equation with the two values above:

$$\begin{aligned} d &= rt \\ 58 &= 23r \\ \frac{58}{23} &= \frac{23r}{23} \\ 2.52... &= r \end{aligned}$$

The final decimal value is rather ugly, but remember - the free-response question asked us to "round to the nearest tenth of a mile per hour," so our final answer must be entered as **2.5**.

Unit Conversions in d=rt Questions

You may need to watch out for Unit Conversions in these problems (study the upcoming lesson on **Unit Conversions** for more information). Luckily, these conversions tend to be very simple for $d = rt$ questions.

For example, take a look at this practice example. You can use your calculator if you want:

> Thumper the rabbit runs 220 feet in 10 seconds to escape a wolf. What is Thumper's average speed for that distance, rounded to the nearest tenth of a mile per hour?
> (Note: one mile is 5,280 feet)

We'll still use the classic $d = rt$ equation (notice the keywords about distance, time and speed). But this time we'll need to convert our units, because we start with *feet* and *seconds* but end in *miles per hour*.

I find it *much* easier to convert to the final units *before* I work the $d = rt$ equation. It's just easier to keep track of everything when the units are correct before I start my **Basic Algebra**. First, let's convert feet into miles (Again, we also cover this technique in detail in the upcoming lesson on **Unit Conversions**).

$$\begin{aligned} &220 \text{ feet}\left(\frac{1 \text{ mile}}{5280 \text{ feet}}\right) \\ &= \frac{220}{5280} \text{miles} \\ &= \frac{1}{24} \text{miles} \end{aligned}$$

So, Thumper runs $\frac{1}{24}$ of a mile in *distance*. Next, we'll convert seconds into hours.

$$10 \text{ seconds}\left(\frac{1 \text{ minute}}{60 \text{ seconds}}\right)\left(\frac{1 \text{ hour}}{60 \text{ minutes}}\right)$$
$$= \frac{10}{(60)(60)} \text{hours}$$
$$= \frac{10}{3600} \text{hours}$$
$$= \frac{1}{360} \text{hours}$$

Now we know that Thumper's running *time* is $\frac{1}{360}$ of an hour. If any of the previous steps were confusing, check out the lesson on **Unit Conversion**.

Finally we'll set up the $d = rt$ equation with our converted values and solve for the rate, r.

$$d = rt$$
$$\frac{1}{24} = r\left(\frac{1}{360}\right)$$
$$(360)\frac{1}{24} = r\left(\frac{1}{360}\right)(360)$$
$$\frac{360}{24} = r$$
$$15 = r$$

A couple unit conversions, the $d = rt$ equation, and we're done! Thumper is running at **15 miles per hour**.

Pretest Question #2

Here's another question from the Pretest. Try it yourself before you look at my explanation below the question.

(CALCULATOR) The distance traveled by Mercury in one orbit around the sun is about 29,000,000 miles. Mercury makes one complete orbit around the Sun every 88 Earth days. Of the following, which is closest to the average speed of Mercury, in miles per minute, as it orbits the Sun?

(A) 229

(B) 13,731

(C) 329,545

(D) 823,864

As always, we can recognize this as a $d = rt$ question because it's a word problem with keywords about distance, time, and speed. However, this question is going to involve a bit of **Unit Conversion** again.

As in the previous example question, I think it's easier to convert my starting units to the final units *before* I start using the $d = rt$ equation. Our final units are "miles per minute", so we should use a distance in *miles* and a time in *minutes*.

Luckily, the distance is already given in miles. It's only the time conversion that we have to deal with. We'll have to convert 88 days into minutes. (If the following steps are confusing, be sure to study the lesson on **Unit Conversion**):

$$88 \text{ days}(\frac{24 \text{ hours}}{1 \text{ day}})(\frac{60 \text{ minutes}}{1 \text{ hour}})$$
$$= 88(24)(60) \text{ minutes}$$
$$= 126{,}720 \text{ minutes}$$

Now let's set up the $d = rt$ equation:

$$d = rt$$
$$29{,}000{,}000 = r(126{,}720)$$
$$\frac{29{,}000{,}000}{126{,}720} = \frac{r(126{,}720)}{126{,}720}$$
$$228.85... = r$$

Now we know the speed of Mercury is approximately 229 miles per minute as it orbits the Sun. That's crazy-fast! The correct answer is **Choice A**.

Review & Encouragement

These $d = rt$ questions are easily-recognizable and follow a simple formula. Recognize the keywords for distance, time, and speed in the word question. Check the final units the question requires, and make all your unit conversions *before* you set up the $d = rt$ equation. There's nothing to worry about!

Now, complete the practice problem set for this lesson and you'll great about $d = rt$ questions.

d=rt Practice Questions

YOU MAY USE YOUR CALCULATOR ON ALL OF THE FOLLOWING QUESTIONS.

1. FREE RESPONSE: A jogger sets out on her daily run, which is always four miles long. She jogs at a consistent speed of 5 miles per hour. How long does her daily jog take her in hours?

2. It is currently 6:30 pm and Manuel needs to get home by 7:30 pm. If he can run at seven miles per hour without slowing, and his home is eight miles away, will Manuel make it home on time - and if so, how many minutes early will he be? If not, how many minutes late will he be?

   (A) Yes, Manuel will be approximately 9 minutes early.

   (B) Yes, Manuel will arrive exactly on time.

   (C) No, Manuel will be approximately 1 minute late.

   (D) No, Manuel will be approximately 9 minutes late.

3. If a tiger can run at 20 miles per hour for up to thirty minutes at a time, how far can a tiger travel in that time in miles?

   (A) 10

   (B) 40

   (C) 600

   (D) 1200

4. A certain race car can sustain a top speed of 250 miles per hour on a straight road. If this car had an infinitely-long straight road to drive upon, how many miles could this race car travel at top speed in 45 seconds?

   (A) .555

   (B) 3.125

   (C) 187.5

   (D) 11,250

5. Christian took a three-day motorcycle road trip. On Monday, Christian rode his motorcycle 200 miles in 3 hours. On Tuesday, he rode 250 miles in 6 hours. On Wednesday, he rode 100 miles in 2 hours. What was Christian's average speed, to the nearest mile per hour, for the time he spent riding his motorcycle on his road trip?

(A) 45

(B) 50

(C) 53

(D) 55

6. FREE RESPONSE: On September 15th, 2012, Hannah took a glider plane trip. If the glider plane sailed a distance of 15 miles in 25 minutes, what would the average speed of the glider have been, to the nearest tenth of a mile per hour?

7. FREE RESPONSE: Harold walks 600 feet in 2 minutes. At that rate, which of the following is closest to the number of feet he will walk in 132 seconds?

8. A small model car is powered by an electric motor. After starting from rest, the car travels n inches in t seconds, where $n = 8t\sqrt{2t}$. Which of the following gives the average speed r of the car, in inches per second, over the first t seconds after it starts?

(A) $r = 8t\sqrt{2t}$

(B) $r = \frac{t\sqrt{2t}}{8}$

(C) $r = 8\sqrt{2t}$

(D) $r = 8t^2\sqrt{2t}$

QUESTIONS 9 AND 10 RELATE TO THE FOLLOWING TABLE AND WORD PROBLEM:

Sarah's Family Vacation

Segment of trip	Distance (miles)	Average Speed (mph)
From home to Airport A	15	60
From Airport A to Airport B	1375	550
From Airport B to hotel	8	25

Sarah and her family are taking a vacation. On the journey to their destination, they drive from their home directly to Airport A, then board an airplane that flies directly to Airport B, and then they take a cab directly from Airport B to their hotel.

9. What was Sarah's family's average speed in miles per hour for their entire one-way journey for the time during which they were traveling ? Note: round your answer to the nearest mile per hour.

10. Airplanes can often fly faster or slower than normal depending on the strength and direction of the prevailing winds. If their airplane had encountered a headwind and flown at an average speed of 520 miles per hour, how many more minutes would this one-way journey have taken the family? (Round your answer to the nearest minute.)

d=rt Answers

1. .8 or $\frac{4}{5}$
2. D
3. A
4. B
5. B
6. 36
7. 660
8. C
9. 455
10. 9

d=rt Explanations

1. **.8 or $\frac{4}{5}$.** This is a very basic $d = rt$ question. Set up the equation and plug in; the units are already correct. The distance is 4 miles and the rate is 5 miles per hour.

$$d = rt$$
$$4 = 5t$$
$$\frac{4}{5} = \frac{5t}{5}$$
$$.8 = t$$

The jogger's run will take her .8 of an hour, which you could also answer as $\frac{4}{5}$ of an hour.

2. **D.** Let's set up a $d = rt$ equation, based on miles per hour, for the 8-mile distance than Manuel needs to cover at a rate of 7 miles per hour.

$$d = rt$$
$$8 = 7t$$
$$\frac{8}{7} = \frac{7t}{7}$$
$$\frac{8}{7} = t$$

We see that Manuel needs $\frac{8}{7}$ of an hour to get home. Unfortunately, this is more than one hour, and he only has from 6:30 until 7:30 to arrive. Clearly, he will NOT make it home on time; he'll be late by $\frac{1}{7}$ of an hour. That lets us eliminate Choices A and B.

However, we also need to figure out how many minutes late Manuel will be. How many minutes is $\frac{1}{7}$ of an hour? That will require a unit conversion:

$$\frac{1}{7}\text{hours}\left(\frac{60\text{ minutes}}{1\text{ hour}}\right)$$
$$= \frac{60}{7}\text{minutes}$$
$$= 8.57\ldots\text{ minutes}$$

From our answer choices, the closest option is Choice D, "Manuel will be approximately 9 minutes late."

3. **A.** This is another basic $d = rt$ question, with a simple unit conversion. We're asked to work in miles per hour, but the question gives us a time in minutes.

Luckily, converting 30 minutes into hours is extremely simple; as you know, 30 minutes is just half an hour, or ".5 hours." We might just do that in my head on test day, but you could also show it like this:

$$30\text{ minutes}\left(\frac{1\text{ hour}}{60\text{ minutes}}\right)$$
$$= \frac{30}{60}\text{hours}$$
$$= \frac{1}{2}\text{hour}$$

Now we can set up our $d = rt$ equation, plug in $\frac{1}{2}$ for t and 20 for r, and solve for d:

$$d = rt$$
$$d = 20\left(\frac{1}{2}\right)$$
$$d = 10$$

4. **B.** This is a basic $d = rt$ question, with a bit of unit conversion. We're given "miles per hour" and asked for a final distance in miles. Therefore, it will be simplest to do the work if we convert "45 seconds" into hours before we begin our $d = rt$ equation. Let's make that Unit Conversion now (remember to review the lesson on **Unit Conversions** if the following steps are confusing to you):

$$45\text{ seconds}\left(\frac{1\text{ minute}}{60\text{ seconds}}\right)\left(\frac{1\text{ hour}}{60\text{ minutes}}\right)$$
$$= \frac{45}{(60)(60)}\text{hours}$$
$$= \frac{45}{3600}\text{hours}$$
$$= .0125\text{ hours}$$

Now that our units are consistent with the word problem, we can plug into the $d = rt$ equation and solve:

$$d = rt$$
$$d = 250(.0125)$$
$$d = 3.125$$

The car can travel a distance of 3.125 miles in 45 seconds at top speed.

5. **B.** This question is very similar to Pretest Question #1 in this lesson. All the signs of a $d = rt$ question are present: keywords about distances, times, and speeds.

Although it might seem like we need to calculate a separate $d = rt$ equation for each of the three days of the road trip, it's actually much easier and quicker than all that. All we do is *total* the distances and total the times, then enter the totals into the $d = rt$ equation.

The total distance is $200 + 250 + 100 = 550$ miles.

The total time is $3 + 6 + 2 = 11$ hours.

Now let's plug into the $d = rt$ equation and solve for r.

$$\begin{aligned} d &= rt \\ 550 &= 11r \\ \frac{550}{11} &= \frac{11r}{11} \\ 50 &= r \end{aligned}$$

Christian's average speed was 50 miles per hour for the time he spent riding on his trip.

6. **36.** Another basic $d = rt$ equation, with some unit conversions. As always, it's easiest to convert to the correct units *before* plugging values into the equation.

We need to answer in miles per hour, so let's make sure our distance is in miles and our time is in hours. Luckily, our distance is already in miles, but unfortunately, our time is in minutes. Let's make a **Unit Conversion**:

$$25 \text{ minutes}\left(\frac{1 \text{ hour}}{60 \text{ minutes}}\right)$$
$$= \frac{25}{60} = \frac{5}{12} \text{ hour}$$

Now we can enter our values into the $d = rt$ equation and do some **Basic Algebra 1** to solve for r.

$$\begin{aligned} d &= rt \\ 15 &= r\left(\frac{5}{12}\right) \\ \left(\frac{12}{5}\right)15 &= r\left(\frac{5}{12}\right)\left(\frac{12}{5}\right) \\ \frac{180}{5} &= r \\ 36 &= r \end{aligned}$$

Hannah was traveling at a rate of 36 miles per hour during her glider flight.

7. **660.** This question is based around simple $d = rt$ methods, but there's tiny bit of extra work. First, let's consider that our final answer must be based on "132 seconds." Therefore, it will be more convenient to work in units of seconds than in minutes. On that basis, let's convert the "2 minutes" into "120 seconds."

Now set up a $d = rt$ equation to determine Harold's walking rate in feet per second:

$$\begin{aligned} d &= rt \\ 600 &= 120r \\ \frac{600}{120} &= \frac{120r}{120} \\ 5 &= r \end{aligned}$$

Now we know that Harold's walking rate is 5 feet per second. So, let's set up a second $d = rt$ equation to find out how far Harold walks in 132 seconds:

$$\begin{aligned} d &= rt \\ d &= 5(132) \\ d &= 660 \end{aligned}$$

In 132 seconds, Harold will walk 660 feet.

8. **C.** This question looks a bit fancier than it really is. All the keywords ("travels n inches" in "t seconds", "average speed") point directly to a $d = rt$ question.

We know we're solving for a speed, so the r in our $d = rt$ equation will remain as an unknown. The word problem requires us to use t as our time value.

The only confusing part is the distance traveled, but just follow the question's lead: the distance, n, is equivalent to $8t\sqrt{2t}$. A little abstract for a distance, compared to something more standard such as "10 miles," for example - but this little snippet of algebra will still function equally well as a distance to plug into $d = rt$.

Put it all together, do some Basic Algebra, and we get:

$$\begin{aligned} d &= rt \\ n &= rt \\ 8t\sqrt{2t} &= rt \\ \frac{8t\sqrt{2t}}{t} &= \frac{rt}{t} \\ 8\sqrt{2t} &= r \end{aligned}$$

9. **455.** This first question is a bit harder than it looks. It has an element similar to Pretest Question 1 and Practice Problem 5, in that we'll first *total* the entire distance traveled, then total the entire time spent traveling, and then calculate our $d = rt$ equation from those totals.

First, the total distance, which is easy to take directly from the table:

$$15 + 1375 + 8 = 1{,}398 \text{ miles.}$$

However, before we can total their travel time, we have to figure out how much time each leg of the journey took. That means we *will* have to do three separate $d = rt$ equations.

First, from home to Airport A:

$$d_1 = r_1 t_1$$
$$15 = 60t_1$$
$$\frac{15}{60} = \frac{60t_1}{60}$$
$$.25 = t_1$$

The first leg took .25 hours. Now calculate time for the second leg:

$$d_2 = r_2 t_2$$
$$1375 = 550t_2$$
$$\frac{1375}{550} = \frac{550t_2}{550}$$
$$2.5 = t_2$$

The second leg took 2.5 hours. Now calculate time for the third leg:

$$d_3 = r_3 t_3$$
$$8 = 25t_3$$
$$\frac{8}{25} = \frac{25t_3}{25}$$
$$.32 = t_3$$

The third leg of the trip took .32 hours.

Now, the total time for all three legs of the trip:

$$.25 + 2.5 + .32 = 3.07 \text{ hours.}$$

The entire journey was 1,398 miles long and took 3.07 hours of travel. Set this up into a new $d = rt$ equation:

$$d = rt$$
$$1398 = 3.07r$$
$$\frac{1398}{3.07} = \frac{3.07r}{3.07}$$
$$455.37... = r$$

The average speed for their entire journey would round to 455 miles per hour.

10. **9.** This Word Problem only asks us to consider the portion of the journey that took place by airplane. We already know (from our work on the previous Question #9) that the airplane trip took 2.5 hours, or 150 minutes:

$$2.5 \text{ hours}\left(\frac{60 \text{ minutes}}{1 \text{ hour}}\right) = 150 \text{ minutes}$$

But, what if the plane had only been traveling at 520 miles per hour instead of 550? We'll need to recalculate our travel time with another $d = rt$ equation:

$$d = rt$$
$$1375 = 520t$$
$$\frac{1375}{520} = \frac{520t}{520}$$
$$2.6442 = t$$

This equation gives us the new travel time *in hours*. Notice that I've kept the decimal value accurate to 4 places, to avoid any miscalculations when we convert hours to minutes, below:

$$2.6442 \text{ hours}\left(\frac{60 \text{ minutes}}{1 \text{ hour}}\right) = 158.652 \text{ minutes}$$

Now we must calculate the difference between the two flight times (remember from the lesson on **Algebra 1 Word Problems** that "difference" means "subtraction"):

$$158.652 - 150 = 8.652$$

We must round this value up to the nearest minute, as the question requires. Our final answer is a total difference of 9 minutes between the two flight times.

Lesson 6: Averages with Algebra

Percentages

- 0.9% of Whole Test
- 0.3% of No-Calculator Section
- 1.3% of Calculator Section

Prerequisites

- Basic Algebra 1
- Algebra 1 Word Problems
- Fractions

This lesson will cover the concept of Averages and *expand* upon basic Averages by highlighting several common Algebra-based techniques the SAT Math test uses on this topic.

These "Averages with Algebra" questions primarily appear on the Calculator Math section of the SAT - although they occasionally show up on the No-Calculator section as well.

We will encounter Averages again in the lesson on **Basic Statistics**, but here we'll cover a specific and common subset of Algebra-based Average questions.

Averages with Algebra Quick Reference

- You may *think* you've mastered Averages already, but my experiences as an SAT tutor suggests that there's still more for most students to learn.
- Identify these questions by the keywords "average" or "mean" within the word problem. "Mean" and "average" are the exact same thing.
- The Average Equation is expressed as $\text{Avg} = \frac{\text{sum}}{\#}$.
- The key to Averages with Algebra on the SAT is to understand Averages as a balanced *equation* with two sides, rather than merely a *calculation*.
- These problems always begin with setting up one (or more) Average Equations, then filling in as many details as the word problem provides you with.
- Expect these questions to draw on **Basic Algebra 1**.

The Average Equation

Most of my students *think* they already know everything about calculating Averages, but the truth is that they only know half of the story.

When I ask my tutoring students about Averages, I usually get the same response: "add everything up and divide by how many there are." And yes, that's absolutely correct - with one major problem.

The SAT writers *know* that every high school student understands this simple concept, so they frequently put in a certain "twist" to their Average questions: they make you solve the Average question "backwards" by *starting* you with final value of the average, then forcing you to set up and solve a **Basic Algebra 1** equation to determine an unknown value from the original data set.

Not realizing this, tons of students get completely stumped by a question type they *think* they've already mastered - leading to a lot of frustration.

These Average questions don't have to be hard or frustrating. In fact, the basic idea is extremely simple. Probably, the only difference between you and me is that you think of an average as a *calculation*, but I think of it as a balanced *equation*.

If you think of averages as a calculation, you can only go in one direction. But when you understand that averages are *equations*, you can set up to go either forwards or backwards to fit the requirements of the problem.

Here's how I express the Average Equation:

$$\text{Avg} = \frac{\text{sum}}{\#}$$

I would not be surprised if this seems very familiar to you. It simply says that the average ("avg") is calculated by adding all the values in the data set ("sum") and dividing by the number of values in the data set ("#").

This simple equation is all we need to handle each and every Averages with Algebra question the SAT Math section will throw at us.

Pretest Question #1

Let's take a look at our first Pretest question on this topic. Try it yourself if you got it wrong the first time.

> (CALCULATOR) FREE RESPONSE: Twelve divers are competing in a diving competition. The divers are scored on a 1-10 scale. The average score of the twelve divers is 7.6. If the lowest scoring diver is disqualified, the average score of the remaining eleven divers increases to 7.9. What was the lowest-scoring diver's score?

Here's how we'll handle this Average with Algebra question. First, call upon the almighty Average Equation:

$$\text{Avg} = \frac{\text{sum}}{\#}$$

Now, start setting it up based on the details of the word problem. Notice I've sub-labeled my sum as "the sum of scores 1 through 12" to help keep track of things.

$$7.6 = \frac{\text{sum}_{1\text{-}12}}{12}$$

The first problem we encounter is that we don't know the value of the "sum_{1-12}". Luckily, we can easily *find* it by treating it as an unknown value (which it is) and doing **Basic Algebra 1**.

$$(12)7.6 = \frac{\text{sum}_{1\text{-}12}}{12}(12)$$
$$91.2 = \text{sum}_{1\text{-}12}$$

Now we know something useful: the *sum* of the first twelve scores was 91.2 points. But how is that any use? Well, hold your horses while we set up *another* Average Equation - this time, for the eleven scores *after* the lowest-scoring diver is disqualified.

$$7.9 = \frac{\text{sum}_{1\text{-}11}}{11}$$

Notice the key changes. We've entered the new average of 7.9 points. Also, there are only 11 data points this time, so I've changed the bottom of the fraction to represent 11 scores instead of 12, and re-titled the sum "1 through 11" represent the new unknown sum of the 11 scores. Now let's calculate $\text{sum}_{1\text{-}11}$ using more Algebra:

$$(11)7.9 = \frac{\text{sum}_{1\text{-}11}}{11}(11)$$
$$86.9 = \text{sum}_{1\text{-}11}$$

Now we have two very useful bits of information. The sum of the first 12 scores was 91.2, and the sum of the 11 scores (without the lowest score) was 86.9. The difference (subtraction) of these two sums will tell us what the lowest score was:

$$91.2 - 86.9 = 4.3$$

The lowest-scoring diver must have received a score of **4.3 points**.

Let's review how we solved this question. First, we set up an Average Equation for all 12 scores. We plugged in the numbers we had available, which set us up to algebraically solve for the *sum* of the first 12 scores.

Next, we set up a second Average Equation for the remaining 11 scores after one was removed from the data set. Again we plugged in the numbers for this situation, which allowed us to solve for the sum of the remaining 11 scores. The *difference* between these two sums gave us the missing score.

Maybe you see what I mean now when I say that many students *think* they know everything about Averages, but are often surprised and frustrated by the ways the SAT Math presents this topic on test day.

Pretest Question #2

Let's take a look at another Pretest question. Try it yourself before you look at my explanation below the question. Beware though, this is a tough one. Be patient with it.

If a is the average of 12 and b, b is the average of c and 6, and c is the average of 8 and b, what is the average of a, b, and c in terms of b?

(A) $\frac{5}{12}b+5$

(B) $\frac{5}{6}b+10$

(C) $\frac{12}{5b+60}$

(D) $\frac{5b+12}{60}$

First, write down the essential Average formula as a reference:

$$\text{Avg} = \frac{\text{sum}}{\#}$$

Now let's start setting up some averages. Based on the question, we'll need three to begin with:

$$\text{Avg}_1 = \frac{\text{sum}_1}{\#_1}$$

$$a = \frac{12+b}{2}$$

$$\text{Avg}_2 = \frac{\text{sum}_2}{\#_2}$$

$$b = \frac{c+6}{2}$$

$$\text{Avg}_3 = \frac{\text{sum}_3}{\#_3}$$

$$c = \frac{8+b}{2}$$

Feeling overwhelmed yet? Sorry - so am I, to be honest. But we can work from here. Consider our position - we've already set up most of the word problem into three useful equations. The question is - how do we make them *talk* to each other?

Well, you're probably not going to love this - but there's one final average equation to set up from the question: the average of a, b, and c. We'll combine our three previous averages into one mega-average of three values, which I'll call "$\text{Avg}_{\text{final}}$". Here we go:

$$\text{Avg}_{\text{final}} = \frac{\text{sum}_{a+b+c}}{\#_{\text{final}}}$$

$$\text{Avg}_{\text{final}} = \frac{(\frac{12+b}{2})+(\frac{c+6}{2})+(\frac{8+b}{2})}{3}$$

I know the equation looks ugly right now, but do you clearly understand how we got to this point? We set up three individual averages, then set up a "final" average of those three averages. If you're not clear, carefully review the previous steps before continuing.

Alright, ready? Let's do this together. We'll clean things up using **Basic Algebra 1**, Combining Like Terms, and our knowledge of **Fractions** before we worry about doing anything else:

$$\text{Avg}_{\text{final}} = \frac{(\frac{12+b}{2})+(\frac{c+6}{2})+(\frac{8+b}{2})}{3}$$

$$\text{Avg}_{\text{final}} = \frac{(\frac{12+b+c+6+8+b}{2})}{3}$$

$$\text{Avg}_{\text{final}} = \frac{(\frac{26+2b+c}{2})}{3}$$

$$\text{Avg}_{\text{final}} = \frac{26+2b+c}{6}$$

I wish we were done - but not quite yet. See, our answer choices can't have any c variables in them; we're required to answer "in terms of b." That means we have to find a way to replace the c in our current equation with a b.

Luckily, we already have a substitution available - remember our third equation for Avg_3? That equation will let us substitute for c with an equation based on b.

Let's make that substitution now. Again, we're using our original Avg_3 equation to make a substitution for c into our $\text{Avg}_{\text{final}}$ equation:

$$\text{Avg}_{\text{final}} = \frac{26+2b+c}{6}$$

$$c = \frac{8+b}{2}$$

$$\text{Avg}_{\text{final}} = \frac{26+2b+(\frac{8+b}{2})}{6}$$

Now time to clean up (again) based on our knowledge of **Fractions:**

$$\text{Avg}_{\text{final}} = \frac{26 + 2b + (\frac{8+b}{2})}{6}(\frac{2}{2})$$

$$\text{Avg}_{\text{final}} = \frac{2(26) + 2(2b) + 2(\frac{8+b}{2})}{2(6)}$$

$$\text{Avg}_{\text{final}} = \frac{52 + 4b + (8+b)}{12}$$

$$\text{Avg}_{\text{final}} = \frac{60 + 5b}{12}$$

We can take this final fraction and reduce and rewrite it one final time to match the answer choices:

$$\text{Avg}_{\text{final}} = \frac{5}{12}b + 5$$

Our final answer is **Choice A**.

This goes to show how difficult and time-consuming a difficult Averages with Algebra question can be. But, to be fair, most of the difficulty on this question came from the **Algebra 1** techniques, not from the Averages themselves.

Also, this particular question was substantially more difficult and time-consuming than the vast majority of these questions that you'll see on the SAT test. If you can solve it confidently, you're in great shape for the test.

Review & Encouragement

Although some Averages with Algebra problems can involve significant complexity, multiple equation setups and solution steps, that complexity can all be reduced to a few simple concepts. We've got the Average Equation itself, the Algebra 1 techniques, and the Fraction skills to handle every situation.

Remember your training. Use the Average Equation to set up Algebra-based problems that let you solve for any component of an average. Be ready to set up multiple averages within the same question. Don't be surprised if you have to work through a series of Algebra equations, including substitutions, before you finish.

Now that you know what you're up against, you'll be light-years ahead of the rest of the testing room. Practice these techniques on the following question set!

Averages with Algebra Practice Questions

YOU MAY USE A CALCULATOR FOR ALL OF THE FOLLOWING PRACTICE PROBLEMS.

$200, 300, 1100, 1300, x$

1. FREE RESPONSE: If the mean of the five numbers above is 900, what is the value of x?

1600

$\{6, a, 11, 14, 14, b, 37\}$

2. FREE RESPONSE: If the mean of the data set above is 15, what is the value of $a+b$?

23

3. A car club holds a restoration contest for classic cars. Each car is scored on a scale of 1 to 10. There are six cars entered into the competition. If these six cars received scores of 1, 2, 3, 5, 9, and 10, how many cars received a score that exceeded the mean score?

(A) Two

(B) Three

(C) Four

(D) Five

4. Christian is preparing to run a marathon. His goal is to run an average of at least 85 miles per week for 4 weeks. He ran 65 miles the first week, 75 miles the second week, and 85 miles the third week. Which inequality can be used to represent the number of miles, n, that Christian could run on the fourth week to meet his goal?

(A) $\frac{65+75+85}{3}+n \geq 85$

(B) $\frac{65}{4}+\frac{75}{4}+\frac{85}{4}+n \geq 85$

(C) $65+75+85+n \geq 340$

(D) $65+75+85+4n \geq 340$

Masses of Pyrite (grams)

Tim	5	7	3	9	4
Ellen	2	12	8	4	6
Jerome	x	13	11	1	3

5. FREE RESPONSE: Tim, Ellen, and Jerome are collecting the mineral pyrite in the river behind their school. Each student collects five chunks of the mineral. The masses of these chunks are shown in the table above. The mean of the masses of the chunks that Jerome collected is 3 grams greater than the average of the mean of the masses of the chunks that Ellen collected and the mean of the chunks that Tim collected. What is the value of x?

6. FREE RESPONSE: The mean lap time of sixteen motorcycle riders in a race was 150 seconds. If the fastest average lap time is removed, the mean lap time of the remaining fifteen racers becomes 153 seconds. What was the fastest lap time, in seconds?

7. If n is the average of t and 15, m is the average of $2n$ and 21, and x is the average of $5n$ and 32, what is the average of m, n, and x in terms of n?

(A) $\frac{3n+53}{9}$

(B) $\frac{9n+83}{3}$

(C) $\frac{9n+53}{3}$

(D) $\frac{9n+53}{6}$

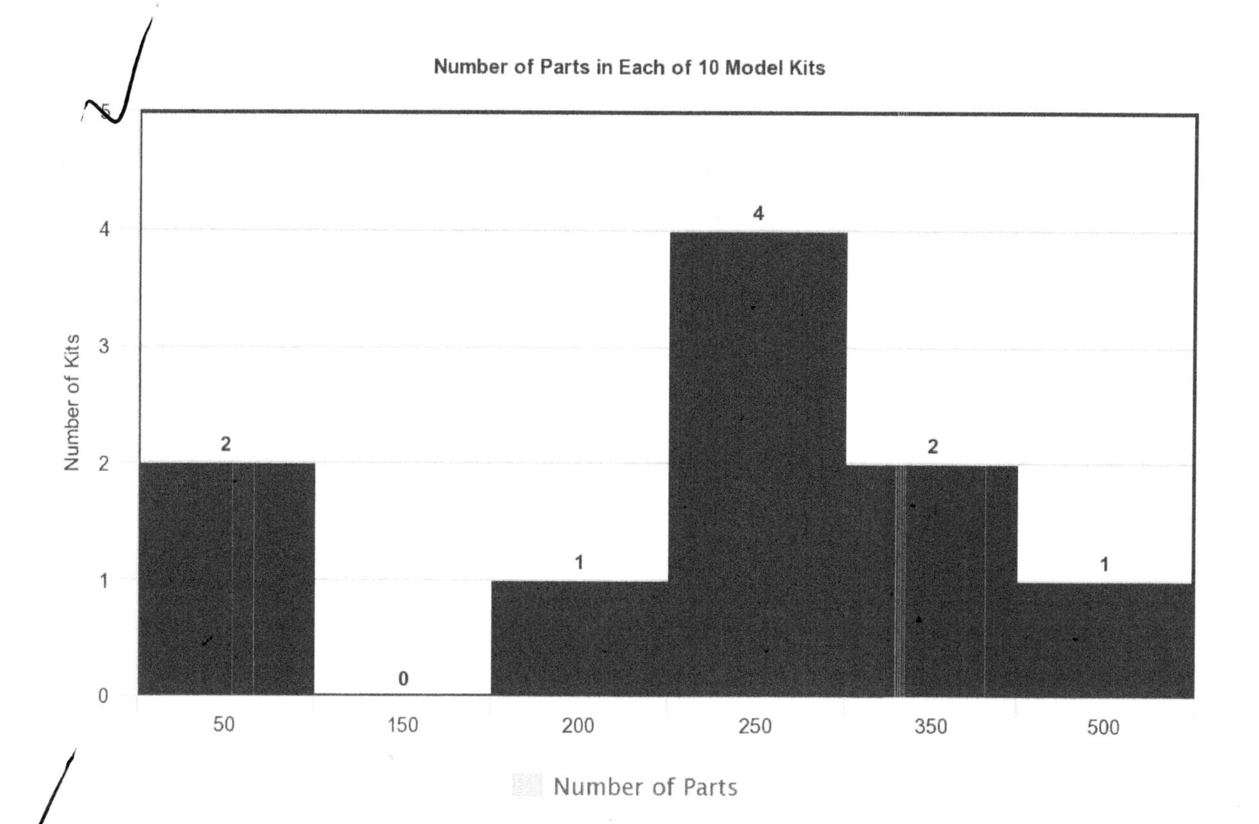

8. Based on the histogram above, of the following, which is closest to the average number of parts per model kit?

(A) 200

(B) 225

(C) 250

(D) 417

9. FREE RESPONSE: Two basketball teams, the City Slickers and the Country Captains, are competing in the playoffs. The City Slickers have 12 players and their average points per player is 6. The Country Captains have 10 players and their average points per player is 17. What are the average points per player for both teams combined?

10. FREE RESPONSE: A new computer game receives critical reviews between 0 and 100, inclusive. In the first 12 ratings, the average of the ratings was 82. What is the least value the game can receive for the 17th rating and still be able to have an average of at least 85 for the first 20 ratings?

Averages with Algebra Answers

1. 1600
2. 23
3. A
4. C
5. 17
6. 105
7. D
8. C
9. 11
10. 16

Averages with Algebra Explanations

1. **1600.** The first thing we should do is pull out our secret weapon - the Average Equation - and set it up with all the information we have available from the word problem. We can also clean things up by Combining Like Terms once the setup is complete:

$$\text{Avg} = \frac{\text{sum}}{\#}$$

$$900 = \frac{200 + 300 + 1100 + 1300 + x}{5}$$

$$900 = \frac{2900 + x}{5}$$

Now just solve for x using **Basic Algebra 1**.

$$(5)900 = \frac{2900 + x}{5}(5)$$

$$4500 = 2900 + x$$

$$-2900 \quad -2900$$

$$1600 = x$$

2. **23.** This is an everyday basic Averages with Algebra question, so let's start with the Average Equation and set it up with all the information we have from the word problem:

$$\text{Avg} = \frac{\text{sum}}{\#}$$

$$15 = \frac{6 + a + 11 + 14 + 14 + b + 37}{7}$$

$$15 = \frac{82 + a + b}{7}$$

Now let's use **Basic Algebra 1** to solve for the sum of $a + b$, as the question requires:

$$(7)15 = \frac{82 + a + b}{7}(7)$$

$$105 = 82 + a + b$$

$$-82 \quad -82$$

$$23 = a + b$$

Boom! We have found that $a + b = 23$. Notice that there's no need to go any further; this is exactly what the question asked us for.

3. **A.** As usual with these simpler Average questions, we'll start by setting up the Average Equation and filling it in with whatever key data we have from the question:

$$\text{Avg} = \frac{\text{sum}}{\#}$$

$$\text{Avg} = \frac{1 + 2 + 3 + 5 + 9 + 10}{6}$$

$$\text{Avg} = \frac{30}{6}$$

$$\text{Avg} = 5$$

We have found that the average of the scores is 5, but the question asked "how many cars scored higher than the mean?" So, just count how many scores were higher than 5. Note that does not include the car with a score of 5. There were only two cars (scores of 9 and 10) that scored higher than the average.

4. **C.** As usual, we'll start this question by setting up the Average Equation and plugging in all the numbers we have available. However, this time we'll need an "less than or equal" inequality symbol (covered in **Basic Algebra 1**) to represent the fact that Christian must meet or exceed his target number of miles.

$$\text{Avg} = \frac{\text{sum}}{\#}$$

$$85 \le \frac{65 + 75 + 85 + n}{4}$$

Now that our inequality is set up, we need to take a look at the answer choices and use **Basic Algebra 1** to make any final moves so it appears in the same form as one of the answer choices. Notice that the answer choices did not simplify by Combining Like Terms. That's a little strange, but we can easily play along. All we have to do for our final answer is multiply both sides by 4:

$$(4)85 \le \frac{65 + 75 + 85 + n}{4}(4)$$

$$340 \le 65 + 75 + 85 + n$$

5. **17.** This Averages with Algebra problem presents us with a table (missing a value x for one of Jerome's data points) and asks us to calculate three separate averages before moving on - one for each student.

It makes sense, then, to begin with the easier students: both Tim and Ellen have complete data sets and we can easily calculate their averages to begin the question. Set both Tim & Ellen up with Average Equations, using sub-labels to keep track of the different averages:

$$\text{Avg}_{\text{Tim}} = \frac{\text{sum}_{\text{Tim}}}{\#}$$

$$\text{Avg}_{\text{Tim}} = \frac{5+7+3+9+4}{5}$$

$$\text{Avg}_{\text{Tim}} = \frac{28}{5}$$

$$\text{Avg}_{\text{Tim}} = 5.6$$

$$\text{Avg}_{\text{Ellen}} = \frac{\text{sum}_{\text{Ellen}}}{\#}$$

$$\text{Avg}_{\text{Ellen}} = \frac{2+12+8+4+6}{5}$$

$$\text{Avg}_{\text{Ellen}} = \frac{32}{5}$$

$$\text{Avg}_{\text{Ellen}} = 6.4$$

Now we've calculated that Tim's average was 5.6 and Ellen's average was 6.4.

However, the question gets a little twisted when it asks us to compare Jerome's average to "the average of the mean of Ellen's chunks and Tim's chunks." In other words, we still have to do *another* average - this time, the average of Ellen's average and Tim's average. Let's set that up:

$$\text{Avg} = \frac{\text{sum}}{\#}$$

$$\text{Avg}_{\text{Tim \&Ellen}} = \frac{\text{Avg}_{\text{Tim}} + \text{Avg}_{\text{Ellen}}}{2}$$

$$\text{Avg}_{\text{Tim \&Ellen}} = \frac{5.6+6.4}{2}$$

$$\text{Avg}_{\text{Tim \&Ellen}} = \frac{12}{2}$$

$$\text{Avg}_{\text{Tim \&Ellen}} = 6$$

Not done yet. Now it's time to focus on Jerome. The word problem tells us that Jerome's average is "3 grams greater" than $\text{Avg}_{\text{Tim \&Ellen}}$, which we know is 6.

So, Jerome's average must be:

$$\text{Avg}_{\text{Tim \&Ellen}} + 3$$

$$= 6 + 3 = 9$$

Now that we know Jerome's average must be 9, let's set up an Average Equation for Jerome, including the missing x data point:

$$\text{Avg}_{\text{Jerome}} = \frac{\text{sum}_{\text{Jerome}}}{\#}$$

$$9 = \frac{x+13+11+1+3}{5}$$

$$9 = \frac{x+28}{5}$$

Now let's just use **Basic Algebra 1** to finish it off and solve for x:

$$(5)9 = \frac{x+28}{5}(5)$$

$$45 = x + 28$$

$$-28 \quad\quad -28$$

$$17 = x$$

Great! Finally, we've finished the question and found that Jerome's missing data value was 17.

6. **105.** This question is nearly identical to Pretest Question #1 in this lesson.

Like all the other questions in this topic, we'll start by setting up Average Equations with the known data from the word problem.

Note my use of sub-labels to keep track of the group of 16 riders vs the group of 15 riders:

$$\text{Avg}_{16} = \frac{\text{sum}_{16}}{16}$$
$$150 = \frac{\text{sum}_{16}}{16}$$
$$(16)150 = \frac{\text{sum}_{16}}{16}(16)$$
$$2400 = \text{sum}_{16}$$

$$\text{Avg}_{15} = \frac{\text{sum}_{15}}{15}$$
$$153 = \frac{\text{sum}_{15}}{15}$$
$$(15)153 = \frac{\text{sum}_{15}}{15}(15)$$
$$2295 = \text{sum}_{15}$$

Just as in Pretest Question 1, we'll use subtraction to find the *difference* between the sums of the times for 16 riders and the sums of the 15 slower riders, which will reveal the lap time of the fastest rider:

$$\text{sum}_{16} - \text{sum}_{16} = \text{fastest rider's time}$$
$$2400 - 2295 = 105$$

Just do a quick "reality check" - is 105 seconds faster than the average time of 150 seconds? It should be, since we've calculated the lap time of the fastest rider.

Yes, 105 seconds is significantly faster than the average lap time of the 16 riders. That doesn't *guarantee* that we've got the correct number, but it's a good, quick "common sense" double-check to do before giving your final answer. If our final time was *slower* than the average, we'd know we had made a mistake.

7. **D.** This question is going to be a rather complex Averages with Algebra question that has a lot in common with Pretest Question #2. First, let's set up a series of averages directly from the word problem:

$$n = \frac{t+15}{2}$$

$$m = \frac{2n+21}{2}$$

$$x = \frac{5n+32}{2}$$

Now we're asked to find the average of m, n, and x. Let's set that up:

$$\text{Avg}_{m,n,x} = \frac{n+m+x}{3}$$

$$\text{Avg}_{m,n,x} = \frac{(\frac{t+15}{2}) + (\frac{2n+21}{2}) + (\frac{5n+32}{2})}{3}$$

Now calling upon our knowledge of **Fractions** and **Basic Algebra 1**, we can clean this equation up and see what happens.

$$\text{Avg}_{m,n,x} = \frac{(\frac{t+15}{2}) + (\frac{2n+21}{2}) + (\frac{5n+32}{2})}{3}$$

$$\text{Avg}_{m,n,x} = \frac{(\frac{t+15+2n+21+5n+32}{2})}{3}$$

$$\text{Avg}_{m,n,x} = \frac{(\frac{t+7n+68}{2})}{3}$$

$$\text{Avg}_{m,n,x} = \frac{t+15+2n+21+5n+32}{3(2)}$$

$$\text{Avg}_{m,n,x} = \frac{t+7n+68}{6}$$

Alright, we're getting close, but we're momentarily stuck. The question asked for us to solve "in terms of n", which means that t variable won't be allowed to remain in our equation - notice that all of our answer choices only have n variables in them. Somehow we'll have to replace that t with something based on n.

Do we have anything to work with from earlier in the problem that would allow us to do an algebraic substitution for t with something based on n? Yes we do - the very first average we set up was:

$$n = \frac{t+15}{2}$$

We can manipulate this equation with **Basic Algebra 1** to isolate t and then plug it into our $\text{Avg}_{m,n,x}$ equation:

$$n = \frac{t+15}{2}$$
$$(2)n = \frac{t+15}{2}(2)$$
$$2n = t+15$$
$$-15 \quad -15$$
$$2n - 15 = t$$

Now that we've isolated t in terms of n, we can substitute into the equation for $\text{Avg}_{m,n,x}$, like so:

$$\text{Avg}_{m,n,x} = \frac{t+7n+68}{6}$$
$$\text{Avg}_{m,n,x} = \frac{(2n-15)+7n+68}{6}$$

Now clean things up and we'll be finished:

$$\text{Avg}_{m,n,x} = \frac{(2n-15)+7n+68}{6}$$
$$\text{Avg}_{m,n,x} = \frac{2n-15+7n+68}{6}$$
$$\text{Avg}_{m,n,x} = \frac{9n+53}{6}$$

It's been a lot of work, with multiple Average setups, and steps involving Algebra and Fractions. But we're finished!

8. **C.** This Averages question asks us to use a histogram chart to calculate an average. Note that the title tells us there are 10 model kits, so we should have 10 data values. This is something many students overlook. Let's set it up:

$$\text{Avg} = \frac{\text{sum}}{\#}$$
$$\text{Avg} = \frac{2(50)+0(150)+1(200)+4(250)+2(350)+1(500)}{10}$$

Notice how I set up the top of my fraction. I use pairs of numbers for each bar of the histogram: the first number in each pair of multiples is how many model kits are in the bar, and the second number is the number of parts in that type of kit. This helps me quickly set up and organize my data.

Now just clean things up and finish the calculation.

$$\text{Avg} = \frac{2(50)+0(150)+1(200)+4(250)+2(350)+1(500)}{10}$$
$$\text{Avg} = \frac{100+0+200+1000+700+500}{10}$$
$$\text{Avg} = \frac{2500}{10}$$
$$\text{Avg} = 250$$

9. **11.** Like so many other questions in this lesson, we'll start by setting up a pair of average equations, one for each team, using sub-labels to keep track of which is which:

$$\text{Avg}_{\text{City Slickers}} = \frac{\text{sum}_{\text{City Slickers}}}{\#_{\text{City Slickers}}}$$
$$6 = \frac{\text{sum}_{\text{City Slickers}}}{12}$$
$$(12)6 = \frac{\text{sum}_{\text{City Slickers}}}{12}(12)$$
$$72 = \text{sum}_{\text{City Slickers}}$$

$$\text{Avg}_{\text{Country Captains}} = \frac{\text{sum}_{\text{Country Captains}}}{\#_{\text{Country Captains}}}$$
$$17 = \frac{\text{sum}_{\text{Country Captains}}}{10}$$
$$(10)17 = \frac{\text{sum}_{\text{Country Captains}}}{10}(10)$$
$$170 = \text{sum}_{\text{Country Captains}}$$

Now that we know the sums of the two teams' scores, we can total them up and divide by the total number of players. Be very careful not to divide by 2! I know there are two teams, but we're calculating the average score *per player*, which means we must divide by the total number of players on the two teams combined.

Let's set this up and finish:

$$\text{Avg}_{\text{combined}} = \frac{\text{sum}_{\text{City Slickers}} + \text{sum}_{\text{Country Captains}}}{\#_{\text{City Slickers}} + \#_{\text{Country Captains}}}$$

$$\text{Avg}_{\text{combined}} = \frac{72+170}{12+10}$$

$$\text{Avg}_{\text{combined}} = \frac{242}{22}$$

$$\text{Avg}_{\text{combined}} = 11$$

The combined average score of all the players on both teams is 11 points per player.

10. **16.** We'll start by setting up an average for the first 12 ratings. We'll figure out where to go after that.

$$\text{Avg}_{1\text{-}12} = \frac{\text{sum}_{1\text{-}12}}{12}$$

$$82 = \frac{\text{sum}_{1\text{-}12}}{12}$$

$$(12)82 = \frac{\text{sum}_{1\text{-}12}}{12}(12)$$

$$984 = \text{sum}_{1\text{-}12}$$

OK, at least now we know something new: the sum of all scores for the first 12 ratings was 984 points.

Now let's set up an average for the first 20 ratings:

$$\text{Avg}_{1\text{-}20} = \frac{\text{sum}_{1\text{-}20}}{20}$$

The word question tells us that this average must be "at least 85", so we can continue setting that up as an inequality:

$$85 \le \frac{\text{sum}_{1\text{-}20}}{20}$$

$$(20)85 \le \frac{\text{sum}_{1\text{-}20}}{20}(20)$$

$$1700 \le \text{sum}_{1\text{-}20}$$

But where to go from here? A good next step could be to find the difference (subtraction) between the sum of the first 12 scores and the sum of all 20 scores. Think of this as telling us the difference between "where we are now" and "where we need to go."

$$\text{sum}_{1\text{-}20} - \text{sum}_{1\text{-}12}$$
$$= 1700 - 984$$
$$= 716$$

This tells us that the 13th through 20th ratings must add to a total of 716 additional points. This is very useful.

Here's the final process: we know from the lesson on **Algebra 1 Word Problems** that if we want to find the *minimum of one value*, we should use the *maximum for all other values*. So, since we want the *minimum* possible score for one rating, we should use the *maximum possible score for all the other ratings*.

Remember, we are focusing now on the unknown points from the 13th through the 20th score. That's 8 total scores (count on your fingers if you must). We want one of those eight scores to be as *low* as possible - that means the other 7 scores should be as *high* as possible. Seven scores of a perfect 100 would give us 700 total points.

Take these "maximized" 700 points out of the total of 716 points that we require to hit our target average:

$$716 - 700 = 16$$

This is how we know that our 17th rating could be as low as just 16 points! If all the other ratings, from the 13th to the 20th, were perfect 100s, we could still hit our target average of 85.

Lesson 7: Ratios & Proportions

Percentages

- 2.4% of Whole Test
- 0.5% of No-Calculator Section
- 3.4% of Calculator Section

Prerequisites

- Basic Algebra 1
- Fractions
- Algebra 1 Word Problems
- Advanced Algebra 1 (Recommended)

In this lesson we'll be exploring the concept of **Ratios & Proportions**. This topic involves relationships of smaller numbers to bigger numbers - for example, scaled-down models of larger objects.

A "Ratio" is simply a fraction. A "Proportion" is simply two fractions set equal to each other.

Almost all Ratio & Proportion questions on the SAT Math are based on **Algebra 1 Word Problems**. Most are not very hard. It's unusual to see difficult questions based on Ratios & Proportions, because the topic of Ratios & Proportions simply isn't very complicated.

This topic is most common on the Calculator section, where it appears frequently. Occasionally, you may encounter it on the No-Calculator section, but this is rare.

Ratios & Proportions Quick Reference

- "Ratio," in simplest terms, just means a fraction setup.
- A "Proportion" is two Ratios that are set equal to each other.
- Use a "Word Fraction" setup to help write out any fractions you need to create, *before* plugging values and variables into your setup.
- Word Problems, Cross-Multiplying and **Basic Algebra 1** are very common when solving these questions.

Introduction to Ratios

After 10 years of SAT tutoring, I can say that "ratio" is a very misunderstood word.

Ratio just means "fraction." A ratio of 4 to 3 (often written "$4:3$") is just the fraction $\frac{4}{3}$.

Notice that the *order matters* in the examples above. The ratio "4 to 3" *must* have 4 on top and 3 on the bottom.

This is always true. A ratio of x to y means $\frac{x}{y}$.

Introduction to Proportions

A "Proportion" is simply setting two Ratios equal to each other - in other words, setting one fraction equal to another fraction.

For example, if a full-sized boat is 20 feet long, and a scale model of the boat is 4 feet long, we can set up this ratio and then reduce the fraction:

$$\frac{\text{full size length}}{\text{scale model length}} = \frac{20}{4} = \frac{5}{1}$$

Notice that on the left, I've used what I call a "Word Fraction" to help myself understand what values go on top and on bottom of my ratio (fraction). Also notice that it's acceptable (and usually smart) to reduce the resulting fraction, if possible.

Now, since the scale model should follow the same proportions of the boat, we should have the *same* ratio between the full sized boat's *width* and the scale model width. Therefore, if the real boat is 10 feet wide, the scale model should be 2 feet wide.

This could be expressed as a Proportion, like so:

$$\frac{\text{20 feet full size length}}{\text{4 feet model length}} = \frac{\text{10 feet fullsized width}}{\text{2 feet model width}}$$

Notice that both of these fractions are equal to each other. They both reduce to the fraction $\frac{5}{1}$. In other words, this equation is made of two fractions ("ratios") that are equal to each other. This equation made of equal fraction is exactly what a "Proportion" is.

On the SAT, your Proportions will usually include an unknown value. For example, if we didn't know the length of the original full-sized boat (but we were given all the the other numbers), then we could put an unknown such as x into our Proportion in place of "20 feet long full-sized boat" and solve for x.

Once you've set up a Proportion correctly, it's easy to solve with Cross-Multiplication and **Basic Algebra 1**.

Now let's take a look at our first Pretest question on this topic. Try it yourself if you got it wrong the first time.

Pretest Question #1

(CALCULATOR) Tree A is currently 18 feet tall, and Tree B is currently 24 feet tall. The ratio of heights of Tree A to Tree B is equal to the ratio of the heights of Tree C to Tree D. If Tree C is 28 feet tall, what is the height of Tree D in feet?

(A) $13\frac{1}{2}$

(B) 21

(C) $33\frac{1}{3}$

(D) $37\frac{1}{3}$

One of my favorite tricks with Ratio & Proportion questions is to first write a "word fraction": a descriptive fraction that puts the right things in the right places, but without getting too involved in the numbers themselves. For example, in this problem I can set up the "word fraction" ratio of Tree A to Tree B:

$$\frac{\text{Tree A}}{\text{Tree B}}$$

I also know that this ratio will be equal to another ratio, given in the problem as "Tree C to Tree D":

$$\frac{\text{Tree A}}{\text{Tree B}} = \frac{\text{Tree C}}{\text{Tree D}}$$

This is a perfect example of my "Word Fractions" in action. Only now do I move on to the values themselves and start inputting actual numbers from the question:

$$\frac{18}{24} = \frac{28}{\text{Tree D}}$$

Notice that I continue to include "Tree D" in my fraction, since I don't know the value for it yet. It might have a cleaner appearance if I replace "Tree D" with a simple variable like D:

$$\frac{18}{24} = \frac{28}{D}$$

My setup is complete. Now it's time to solve using cross-multiplication:

$$\frac{18}{24} = \frac{28}{D}$$
$$18D = (24)(28)$$
$$18D = 672$$
$$\frac{18D}{18} = \frac{672}{18}$$
$$D = 37\frac{1}{3}$$

The correct answer is **Choice D**! Easy!

Trace my steps again: I noticed the question was about Ratios & Proportions through the language of the word problem. Then I set up two "Word Fractions" that were equal to each other.

Next I plugged in the provided values and replaced my unknown with a variable. Finally, I solved the equation through the use of Cross-Multiplying and Basic Algebra 1.

Multiple Ratios

Most SAT questions involving Ratios & Proportions can be easily solved through nothing more than an accurate application of the concepts demonstrated in Pretest Question #1.

However, there is another Ratio topic that shows up occasionally on SAT tests. I call this "Multiple Ratios."

For example, the question might give us a mixture made of of three liquids in a ratio of "four parts water to three parts chlorine to two parts lemon juice" by volume. (This could also be expressed as $4:3:2$).

These questions are easy if you know how to work them. The first step is to add the entire ratio together, which I call getting the "Total Parts" of the ratio (because we're literally totaling up the sum of all the parts.)

In the $4:3:2$ ratio mentioned above, the total parts would simply be $4+3+2=9$.

How does this help us? Well, it means that our mixture can be divided into 9 equal parts, and then those parts will be "redistributed" to each of the three liquids, depending on their ratio values.

This sounds complicated in word form, but it's really easy. Here's what I mean. The "Total Parts" value was 9. Our water was "4 parts" of that. Therefore, the mixture is $\frac{4}{9}$ water by volume.

Again, the Total Parts value was 9. Our chlorine was 3 parts of that. Therefore, the mixture is $\frac{3}{9}$ or $\frac{1}{3}$ chlorine by volume (remember, we should usually reduce our Ratios to their simplest fraction forms).

And finally, Lemonade was 2 parts of the mixture. Again, the Total Parts value was 9. Therefore, the mixture is $\frac{2}{9}$ lemonade by volume.

Just to confirm this, if we add our three fractions together, we should return to $\frac{9}{9}$, which reduces to "1", and tells us that if we add all our parts together, we return to the original whole amount of mixture.

Let's just confirm by adding the fractions of all three liquids:

$$\frac{4}{9}+\frac{3}{9}+\frac{2}{9}=\frac{9}{9}=1$$

We do in fact return to a value of $\frac{9}{9}$ or just "1". This simply confirms that we accurately divided the ratios of our total mixture into its component parts by applying the "Total Parts" technique.

Pretest Question #2

Let's take a look at another Pretest question. Try it yourself before you look at my explanation below the question.

(CALCULATOR) FREE RESPONSE: A construction worker is mixing sand, gravel, cement, and earth in a ratio of $2:3:5:9$ by volume, respectively. If the total amount of mixture is 95 liters in volume, what was the total volume of the sand and earth combined, in liters?

In this question, we'll use the technique described above to get our "Total Parts." Add up all the parts of the mixture:

$$2+3+5+9=19$$

We now can divide these into our different part fractions:

$$\frac{2}{19} \text{ of the mixture is sand}$$
$$\frac{3}{19} \text{ of the mixture is gravel}$$
$$\frac{5}{19} \text{ of the mixture is cement}$$
$$\frac{9}{19} \text{ of the mixture is earth}$$

Note that adding all four fractions will result in a total of exactly "1," as it always should (this tells us that the sum of all our *parts* is equal to the total):

$$\frac{2}{19}+\frac{3}{19}+\frac{5}{19}+\frac{9}{19}=\frac{19}{19}=1$$

Now, let's focus on the fractions for *sand* and *earth*, just as the word problem asked us for. Remember:

$$\frac{2}{19} \text{ of the mixture is sand}$$
$$\frac{9}{19} \text{ of the mixture is earth}$$

So, if we add those together, we'll get:

$$\frac{2}{19}+\frac{9}{19}=\frac{11}{19}$$

Now we know something very useful: $\frac{11}{19}$ of the mixture is "sand and earth combined." All that's left is to set up a final Proportion:

$$\frac{11}{19}=\frac{\text{sand \& earth combined}}{\text{all mixture}}$$

We can finalize this question by plugging in the value we know (the whole mixture was 95 liters) and solving for the unknown value we want (the volume of sand & earth combined, for which I will use the variable C).

$$\frac{11}{19} = \frac{\text{sand \& earth combined}}{95}$$
$$\frac{11}{19} = \frac{C}{95}$$
$$(11)(95) = 19C$$
$$1045 = 19C$$
$$\frac{1045}{19} = \frac{19C}{19}$$
$$55 = C$$

Finished! The total volume of sand and earth combined was **55 liters**.

Review & Encouragement

Ratios and Proportions are a relatively simple topic that many students either never learned, got rusty on, or were poorly taught by their teachers.

As a tutor, I wince when I see students get stuck on these questions or make simple mistakes - only because I know those same students will find the question *easy* when they understand that "Ratios are fractions" and "Proportions are two equal Ratios."

Be sure to recognize giveaway keywords like "ratio," "proportional," "scale model," "$x : y$", "x parts per," "at this rate" (although a keyword like "constant rate" might also signify a **d=rt** or **Linear Equation** question) and so forth.

Use "Word Fractions" to clearly set up your fractions before you move onto using actual numbers and variables.

Now complete the Practice Questions to lock this concept in and get more points on your next SAT test!

Ratios & Proportions Practice Questions

YOU MAY USE A CALCULATOR FOR ALL OF THE FOLLOWING PRACTICE PROBLEMS.

1. James swam a lap in 2 minutes. Sarah swam the same distance in 90 seconds. What is the ratio of Sarah's lap time to James's lap time?

   (A) 1 to 2

   (B) 3 to 4

   (C) 5 to 9

   (D) 1 to 7

2. A random sample of a bee farm 400 bees produces 8 bees with unusual coloration. At this rate, how many of the 200,000 bees on the farm will have this unusual coloration?

   (A) 400

   (B) 4,000

   (C) 100,000

   (D) 10,000,000

3. FREE RESPONSE: A breakdancer earns money by performing on Main Street for the passing crowds. Her income is directly proportional to the number of people walking on Main Street. If she makes $30 on a Tuesday evening when 300 people are walking on Main street, how much does she make (in dollars) on a Saturday afternoon when 700 people are walking on Main Street?

$70

4. FREE RESPONSE: A factory produces 40,000 motorcycle throttle bodies per day. A random quality-control sampling of 200 units produces 3 defective units. Assuming this rate of defective units holds true for the day's production, how many defective throttle bodies does the factory produce per day?

5. The superyacht *Sheladia* is approximately 132 meters long, 26 meters wide across the beam (the widest point of the ship), and 22 meters high from the lowest point of its rudder to the tip of its exhaust pipes, which are the highest point of the yacht. If a scale model of the *Sheladia* is built at a $\frac{1}{40}$th scale, how wide, in centimeters, would the model be at its widest point?

(A) .65

(B) 3.3

(C) 65

(D) 330

6. If an artist is mixing blue, teal, yellow, and green paints in a ratio of $1:2:2:3$ by volume respectively, what is the total volume of paint in milliliters if she uses 6 milliliters of green paint?

(A) 16

(B) 18

(C) 24

(D) 48

7. A certain paint is used for painting the exterior of large airplanes. This paint is so effective that a single gallon can up to 4 football fields. If a football field has an area of approximately $1\frac{2}{3}$ acres, about how many acres could 36 gallons of this paint cover? (Round your answer to the nearest acre)

(A) 9

(B) 86

(C) 240

(D) 792

8. FREE RESPONSE: There are two atoms of oxygen and one atom of carbon in one molecule of carbon dioxide. How many atoms of oxygen are there in 73 molecules of carbon dioxide?

Animal	1999	2000	2001	2002
Pheasants	56	72	86	101
Sea Otters	14	18	26	38
Polar Bears	2	2	3	3
African Elephants	6	7	7	9
Bald Eagles	6	9	11	14
Hyenas	43	48	53	57

9. The table above shows the population of four animal species in a wild game reserve for four years in the period 1999 through 2002. Which animal's ratio of its 1999 population to its 2002 population is closest to the sea otters' ratio of its population in 1999 to its population in 2000?

(A) Pheasants

(B) Polar Bears

(C) Bald Eagles

(D) Hyenas

10. A certain line in a coordinate plane passes through the origin and the point (4,12). If Point B lies on the graph of this line at coordinates (n, m), what is the ratio of m to n?

(A) 0

(B) $\frac{1}{3}$

(C) 1

(D) 3

11. A gear ratio $r:s$ is the ratio of the number of teeth of two connected gears. The ratio of the number of revolutions per minute (rpm) of the two gear wheels is $s:r$. Gear X, with 80 teeth, is driven by a motor. Gear X turns Gear Y, which has 400 teeth. Gear Y turns Gear Z, which has 20 teeth. If Gear X turns at 40 rpm, what is the number of revolutions per minute for Gear Z?

(A) 4

(B) 20

(C) 160

(D) 200

12. FREE RESPONSE: The weight of an object on Mars is approximately $\frac{1}{3}$ of its weight on Earth. The weight of an object on Neptune is approximately $\frac{11}{10}$ of its weight on Earth. If an object weighs 66 pounds on Earth, how many more pounds does it weigh on Neptune than it does on Mars?

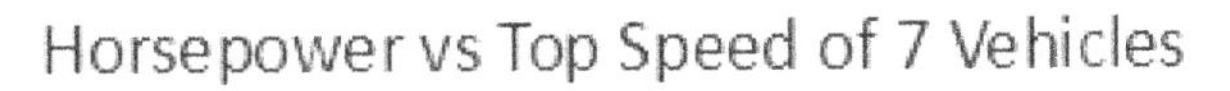

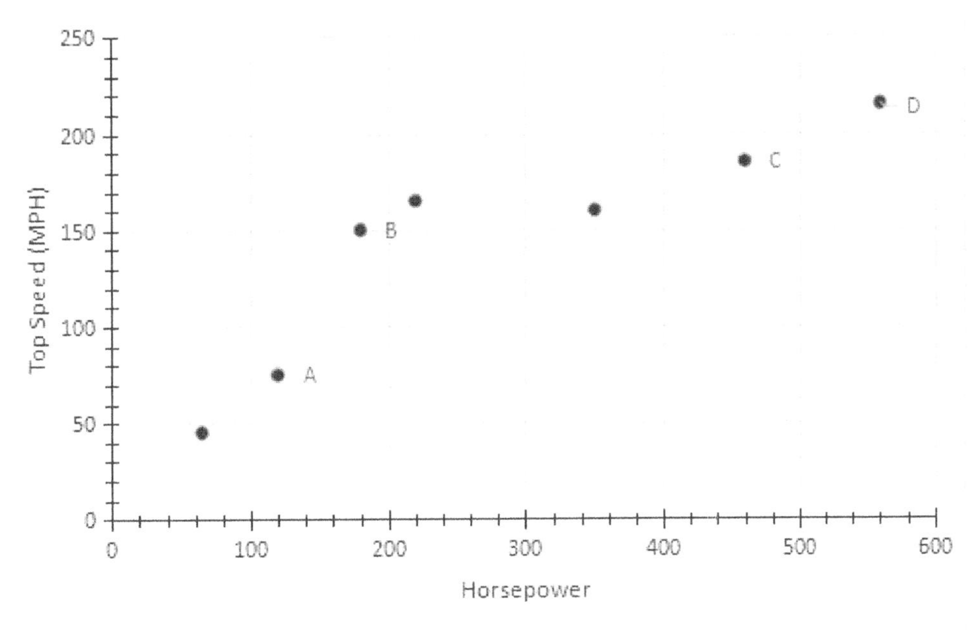

13. The scatterplot above charts the relationship of horsepower to top speed for seven different vehicles. Of the labeled points, which represents the vehicle for which the ratio of top speed to horsepower is greatest?

(A) A

(B) B

(C) C

(D) D

Ratios & Proportions Answers

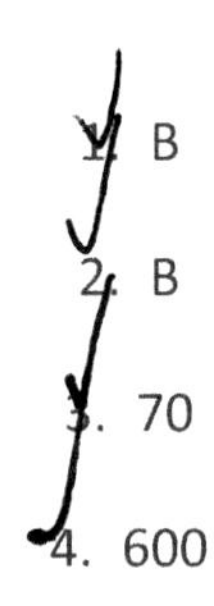

1. B
2. B
3. 70
4. 600
5. C
6. A
7. C
8. 146
9. D
10. D
11. C
12. 50.6
13. B

Ratios & Proportions Explanations

1. **B.** This is an easy question. First, notice that it's definitely a Ratios & Proportions question because of the language ("the ratio of Sarah's lap time to James' lap time"). This will be set up as a fraction, like all Ratios. To be safe, make a Word Fraction first:

$$\frac{\text{Sarah's Lap Time}}{\text{James' Lap Time}}$$

Then plug in the values you've been given. But wait! The units are different. You can either work in units of minutes or seconds. I'll choose seconds, because it's very easy to convert "2 minutes" into "120 seconds":

$$\frac{90}{120}$$
$$=\frac{3}{4}$$

The ratio must be 3 to 4. And we're done!

2. **B.** We can tell this is a Ratios & Proportions question, although it's a bit more subtle than Question #1: there's a comparison of "8 out of 400" to "*how many* out of 200,000", a classic Proportion-based word problem.

$$\frac{\text{weird bees}}{\text{all bees}}=\frac{8}{400}=\frac{\text{how many}}{\text{whole farm}}$$

Now use an x in place of the words "how many":

$$\frac{8}{400}=\frac{x}{200000}$$

Now solve using Cross-Multiplying and **Basic Algebra 1**:

$$\frac{8}{400}=\frac{x}{200000}$$
$$400x=8(200000)$$
$$400x=1600000$$
$$\frac{400x}{400}=\frac{1600000}{400}$$
$$x=4000$$

3. **70.** We can tell this is a Ratios & Proportions question because of the wording ("directly proportional"). Make a fraction for Tuesday and set it equal to the fraction for Saturday, using x for her income on Saturday:

$$\frac{\$}{\text{people}}=\frac{\$30}{300\text{ people}}=\frac{\$x}{700\text{ people}}$$

Now cross-multiply and solve using **Basic Algebra 1**.

$$\frac{30}{300}=\frac{x}{700}$$
$$300x=30(700)$$
$$300x=21000$$
$$x=70$$

Her income on Saturday must be \$70.

4. **600.** This question is similar to Question 2. We can set up a Proportion like this:

$$\frac{\text{defective units}}{\text{all units}}=\frac{3\text{ defective units}}{200\text{ all units}}=\frac{x\text{ defective units}}{40{,}000\text{ all units}}$$

Now solve using Cross Multiplication and **Basic Algebra 1**:

$$\frac{x}{40{,}000}=\frac{3}{200}$$
$$200x=3(40{,}000)$$
$$200x=120{,}000$$
$$\frac{200x}{200}=\frac{120{,}000}{200}$$
$$x=600$$

5. **C.** Any question about "scale models" or shapes "scaled up" or "scaled down" is likely to involve Ratios & Proportions. Unfortunately, this word problem has a lot of distracting details (we practiced this in **Algebra 1 Word Problems**).

Luckily, the only part that matters is the final question: "How wide would the model be at its widest point?" The details say that the yacht is "26 meters wide across the beam" which is "the widest point of the ship."

Therefore, we can start setting up a proportion like this:

$$\frac{\text{scale model}}{\text{real yacht}}=\frac{1}{40}=\frac{x\text{ meters}}{26\text{ meters}}$$

This states that for every 40 meters in our real yacht, the scale model will be 1 meter. Obviously the scale model will be smaller than the real thing! x represents the unknown dimensions of our scale model at its widest point.

Now go ahead and use Cross Multiplication and solve the algebra.

$$\frac{1}{40} = \frac{x \text{ meters}}{26 \text{ meters}}$$
$$1(26 \text{ meters}) = 40x \text{ meters}$$
$$\frac{26 \text{ meters}}{40} = \frac{40x \text{ meters}}{40}$$
$$.65 \text{ meters} = x \text{ meters}$$

However, notice that the Word Problem asks for the width in *centimeters* and we're currently working in meters. Do a quick, simple **Unit Conversion**:

$$.65 \text{ meters}\left(\frac{100 \text{ centimeters}}{1 \text{ meter}}\right)$$
$$= 65 \text{ centimeters}$$

And we're done!

6. **A.** This question is very similar to Pretest Question #2. Remember to get our "Total Parts" by adding up all the separate parts of the ratio:

$$1+2+2+3=8$$

Now we can determine the fraction of the whole mixture for each paint:

$$\frac{1}{8} \text{ of the paint is blue}$$
$$\frac{2}{8} \text{ of the paint is teal}$$
$$\frac{2}{8} \text{ of the paint is yellow}$$
$$\frac{3}{8} \text{ of the paint is green}$$

Notice that if you add all four fractions, you will end up back at $\frac{8}{8}$, which simply indicates that we've correctly divided our entire mixture of paint into fractions of four colors.

Now, we know that we have 6 milliliters of green paint (from the question) and we're asked to find the total volume of paint. This can be set up into a Proportion:

$$\frac{3 \text{ parts green}}{8 \text{ all paint colors}} = \frac{6 \text{ mL of green}}{x \text{ mL of all paint colors}}$$

This can be solved easily with Cross Multiplication and **Basic Algebra 1**.

$$\frac{3}{8} = \frac{6}{x}$$
$$3x = (6)(8)$$
$$3x = 48$$
$$\frac{3x}{3} = \frac{48}{3}$$
$$x = 16$$

Now we know that the total volume of all paint colors is 16 milliliters.

7. **C.** This is another standard Ratios & Proportions question. There's a $1:4$ ratio of paint gallons to football fields. It will also involve a **Unit Conversion**, but I think I'll save that step until last.

Let's set up the initial proportion:

$$\frac{1 \text{ gallon}}{4 \text{ football fields}} = \frac{36 \text{ gallons}}{x \text{ football fields}}$$

Now use Cross Multiplication and **Basic Algebra 1** to solve for the x football fields we can cover:

$$\frac{1}{4} = \frac{36}{x}$$
$$1x = (4)(36)$$
$$x = 144$$

Now we know that our 36 gallons of paint could cover 144 football fields. All that's left is to do a **Unit Conversion** from football fields to acres:

$$144 \text{ football fields}\left(\frac{1\frac{2}{3} \text{ acres}}{1 \text{ football field}}\right)$$
$$= 144\left(1\frac{2}{3} \text{ acres}\right)$$
$$= 144\left(\frac{5}{3} \text{ acres}\right)$$
$$= \frac{720}{3} \text{ acres}$$
$$= 240 \text{ acres}$$

8. **146.** Let's set up a Word Fraction proportion for oxygen atoms to molecules:

$$\frac{2 \text{ oxygen atoms}}{1 \text{ molecule}} = \frac{x \text{ oxygen atoms}}{73 \text{ molecules}}$$

Notice that we just disregard the carbon atoms, since the question doesn't ask about them and they don't help us create a proportion between oxygen atoms and molecules of carbon dioxide.

Now we can just use cross-multiplication and solve for x with **Basic Algebra 1**.

$$\frac{2}{1} = \frac{x}{73}$$
$$(73)(2) = 1x$$
$$146 = x$$

Pretty easy! There must be 146 atoms of oxygen in the 73 molecules of carbon dioxide.

9. **D.** To be honest, this question is a bit of a pain in the butt, mainly because of information overwhelm and the fact that there's no super-quick trick to solve it. First, we'll need to calculate the sea otter ratio of 1999 to 2000, then try each of the four answer choices to see which produces the closest value.

First, the sea otter ratio:

$$\frac{\text{sea otters in 1999}}{\text{sea otters in 2000}} = \frac{14}{18}$$
$$= .777....$$

In this case, it will be easiest to leave the result as a decimal (.777 repeating). This will make it easier to compare values with our four answer choices than staying in fraction form.

Now let's test each of the four answer possibilities. Remember, we're looking for the number that is closest to .777. Also, we have to use the years 1999 and 2002, as noted in the question. Word Fraction setups are essential to avoid mix-ups and Careless Mistakes:

$$\frac{\text{Pheasants in 1999}}{\text{Pheasants in 2002}} = \frac{56}{101}$$
$$= .554...$$

$$\frac{\text{Polar Bears in 1999}}{\text{Polar Bears in 2002}} = \frac{2}{3}$$
$$= .666...$$

$$\frac{\text{Bald Eagles in 1999}}{\text{Bald Eagles in 2002}} = \frac{6}{14}$$
$$= .429...$$

$$\frac{\text{Hyenas in 1999}}{\text{Hyenas in 2002}} = \frac{43}{57}$$
$$= .754...$$

Now we know that Hyenas are the animal with the closest ratio value to our original .777 Sea Otters ratio.

10. **D.** For this question, remember (as always) that the word "ratio" simply means "fraction". In this question, we need to create a fraction for the ratio of m to n, like so:

$$\frac{m}{n}$$

One problem, though - we don't know the coordinates of point (n,m) and so we can't complete the fraction with any actual values.

However, we do know quite a bit about the line itself. I'm going to use some techniques from the upcoming lessons on **Linear Equations (Algebraic)**. Basically, my plan is to create an equation for the line, then use it to find some sample coordinates for n and m.

We'll use the $y = mx + b$ equation for our line. The line passes through the origin, so it has a *y*-intercept (*b*-value) of 0. Now we need the slope:

$$\frac{\text{rise}}{\text{run}} = \frac{y_2 - y_1}{x_2 - x_1} = \frac{12-0}{4-0}$$
$$= \frac{12}{4} = \frac{3}{1} = 3$$

The slope is 3. You know what's interesting? "Rise over run" is a fraction - a ratio! In fact, the very concept of *slope* itself is simply a ratio: the relationship of rise to run, written as a fraction. Hmm... could be worth remembering.

But for now, we know the slope of this line is 3 and the *y*-intercept is 0. Now we can finish our Linear Equation:

$$y = 3x + 0$$

The next part of my plan is to use any *x*-value I want, and simply plug it in. The question doesn't give any details about coordinate (n,m) other than that "it's on the graph of the line."

I'll use "5" for my *x*-value, just because I like the number 5. It really doesn't matter what I use. Plug it into my line equation for x, and what do I get for my *y*-coordinate?

$$y = 3(5) + 0$$
$$y = 15$$

If my x was 5 and I get a y of 15, my coordinate for (n, m) can be written as $(5,15)$.

I'm almost done with the question. Now I can finish the problem. The "ratio of m to n" would be the fraction $\frac{m}{n} = \frac{15}{5}$, which reduces to 3.

Wait a second - wasn't the slope of the line also 3? Is that just a coincidence?

No, it isn't! Remember that slope, which we study more in the two lessons on **Linear Equations**, is a *ratio* of rise over run. And, our ratio of m to n was comparing a *y*-value to an *x*-value. When the line itself passes through the origin, then the ratio of y to x will be identical to the ratio of rise over run - we'll just end up getting the slope value. You don't need to remember this; I just thought I'd point it out.

11. **C.** This question might seem pretty complicated. In fact, it's based on an official SAT practice question.

I'll keep my eyes on the end goal: finding the RPM for Gear Z. But I'll also have to start at the beginning of the gear system: Gear X, with 80 teeth, is turning Gear Y, with 400 teeth.

According to the information in the question, the "gear ratio" of $X:Y$ would be 80:400. Remember that ratios are just fractions, so we can simplify and reduce that ratio to smaller numbers by thinking of it like a fraction:

$$\frac{X_{\text{teeth}}}{Y_{\text{teeth}}} = \frac{80}{400} = \frac{1}{5} \text{ gear ratio of X : Y}$$

The question tells us the ratio of RPM is the reverse of the gear ratio (or "reciprocal" fraction, if you prefer) . That means that the RPM ratio of Gear X to Gear Y is $5:1$.

In other words, for every five times Gear X completes a revolution, Gear Y will complete a single revolution. This can also be expressed as a proportion when we include the 40 RPM of Gear X (given to us in the question):

$$\frac{5_{X\text{ rpm}}}{1_{Y\text{ rpm}}} = \frac{40_{X\text{ rpm}}}{Y_{\text{rpm}}}$$

We can now solve for Gear Y's RPM:

$$\frac{5}{1} = \frac{40}{Y_{\text{rpm}}}$$
$$40(1) = 5Y_{\text{rpm}}$$
$$\frac{40}{5} = \frac{5Y_{\text{rpm}}}{5}$$
$$8 = Y_{\text{rpm}}$$

Now we've shown that Gear Y is turning at 8 RPM. Not only that - we've also created a workflow to figure out how quickly any two gears are turning, based upon the number of teeth they have.

We can use the same workflow as we move into the next phase: turning Gear Z via the revolutions of Gear Y. First, set up the gear ratio for Gear Y and Gear Z, and reduce the fraction:

$$\frac{Y_{\text{teeth}}}{Z_{\text{teeth}}} = \frac{400}{20} = \frac{20}{1} \text{ gear ratio of Y : Z}$$

Now use the "reverse ratio" or "reciprocal fraction" (different names for the same idea) to find the ratio of RPM between Gear Y and Gear Z. This would be $1:20$; for every single turn of Gear Y, the Gear Z will make 20 revolutions (this makes logical sense if you think about it, because Gear Y with 400 teeth is much larger than Gear Z with 20 teeth; for every single turn of Gear Y, Gear Z will turn many times.)

Using the 8 RPM for Gear Y that we found in the first stage and the ratio we just created for RPMs of Gears $Y:Z$, now we can set up a final proportion to solve for Gear Z's RPMs, cross-multiply, and solve:

$$\frac{1_{Y\text{ rpm}}}{20_{Z\text{ rpm}}} = \frac{8_{Y\text{ rpm}}}{Z_{\text{rpm}}}$$
$$(20)(8) = 1Z_{\text{rpm}}$$
$$160 = Z_{\text{rpm}}$$

And we're done! Gear Z will turn at 160 RPM.

12. **50.6.** We'll need to set up two completely separate proportions for this equation. One will calculate the weight of the object on Neptune; the other will calculate the weight of the object on Mars. Then we'll find the difference (subtraction) between those values.

First, Neptune:

$$\frac{\text{Weight on Neptune}}{\text{Weight on Earth}} = \frac{\frac{11}{10}}{1}$$

Notice that we've used a "Word Fraction" to get our ideas organized before plugging in the actual weight of the object.

Now let's plug in values, then solve for the weight on Neptune (which I'll label " N ").

$$\frac{N}{66} = \frac{\frac{11}{10}}{1}$$
$$1N = (\tfrac{11}{10})(66)$$
$$N = 72.6$$

We've found that the object will weigh 72.6 pounds on Neptune.

Now repeat an identical workflow to find the weight on Mars. We'll use the variable " M " for the weight on Mars:

$$\frac{\text{Weight on Mars}}{\text{Weight on Earth}} = \frac{\frac{1}{3}}{1}$$
$$\frac{M}{66} = \frac{\frac{1}{3}}{1}$$
$$1M = (\tfrac{1}{3})(66)$$
$$M = 22$$

Now we know the object will weigh 22 pounds on Mars.

To finish the question, simply take the difference ("subtraction") between Neptune and Mars:

21. $72.6 - 22 = 50.6$

13. **B.** Despite the unusual presentation of a Ratio question involving a scatterplot, this is quite simple to answer. We'll just need to calculate the ratio for each of the 4 points. To do so, we'll also have to make our best estimates for the coordinates of each point.

The question asked for the ratio of top speed to horsepower. Set up a Word Fraction to remind yourself of this:

$$\frac{\text{Top Speed}}{\text{Horsepower}}$$

Now make your estimates for the coordinates of each of the four labeled points. Don't trust your eyes! Use your pencil and mark vertical and horizontal lines from each point to the axes. For a question like this, we'll need to be exact as possible (one way you can tell the importance of using accurate coordinates is because the graphs have been thoroughly labeled with precise tick marks and values).

Here are my best estimates:

Point A is at (120 HP, 75 MPH). Point B is at (180 HP, 150 MPH). Point C is at (460 HP, 180 MPH). Point D is at (560 HP, 215 MPH).

Keep in mind that these are only my estimates; several points fall between tick marks. That's OK - I've been as accurate as I possibly can be. The numbers should work out correctly, as long as I was as close as possible.

Now let's plug in the coordinates of each point one by one to calculate the ratio values for each of the four points. I'm going to reduce the ratios to decimal form, which will be easier to directly compare than fractions would be.

Point A: $\frac{\text{Top Speed}}{\text{Horsepower}} = \frac{75}{120} = 0.625$

Point B: $\frac{\text{Top Speed}}{\text{Horsepower}} = \frac{150}{180} = 0.8333...$

Point C: $\frac{\text{Top Speed}}{\text{Horsepower}} = \frac{180}{460} = 0.3913...$

Point D: $\frac{\text{Top Speed}}{\text{Horsepower}} = \frac{215}{560} = 0.3839...$

From our estimated values, taken from the scatterplot and plugged into our ratios, we can determine that Point B has the highest value for the ratio of Top Speed to Horsepower. Even if our estimated coordinates are slightly off, it's still ahead by a wide margin.

Lesson 8: Unit Conversions

Percentages

- 1.3% of Whole Test
- 0% of No-Calculator Section
- 2% of Calculator Section

Prerequisites

- Basic Algebra 1
- Algebra 1 Word Problems
- Ratios & Proportions

Unit Conversions are a common basic math concept typically found within the Calculator section of the SAT.

This topic can be recognized by Word Questions that mix multiple units, such as:

- Time units like "hours," "minutes," and "seconds"
- Distance units like "miles," "feet," and "inches"
- Volume units like "milliliters" and "liters"
- Weight units like "kilograms" and "grams" or "pounds" and "ounces"
- Exotic or Obsolete units like "cubits" and "palms"

Unit Conversions can also be found mixed into other types of questions on the SAT Math. From **Algebra 1 Word Problems** to Geometry topics like **Mixed Shapes**, we will sometimes find Unit Conversions built into other questions, where they add an extra layer of complexity.

Therefore, it's important to feel confident about your Unit Conversion techniques. By the end of this lesson, you will!

Unit Conversions Quick Reference

- Unit Conversions always involve a word question.
- Recognize this topic by keywords that involve multiple different units of measurements, such as in the bullet-point list above.
- Unit Conversions are also found mixed into many other types of math questions.
- The Fencepost Method is the most versatile and effective method to use on Unit Conversions.
- Most SAT Math questions provide you any necessary conversion values (e.g. "16 ounces = 1 pound" will be given in the question), so memorizing the various units is not important.
- Unit Conversions are vulnerable to **Careless Mistakes**. It's easy to divide when you should multiply, and vice versa. Misreading the word problem is also an ever-present danger.

The Fencepost Method for Unit Conversions

The bad news first: students are often sloppy with their Unit Conversions, and the SAT knows this. Therefore, questions will often have "trap" answer choices that will match up perfectly with the common errors students make in their Unit Conversions.

The good news is that there's only *one* technique needed to master Unit Conversions. Regardless of whether you're working in time, distance, volume, or weight, this technique will work every time.

It's called the "Fencepost Method," and we're going to learn it right now (and use it throughout this lesson).

The Fencepost Method uses a series of one or more fractions to accomplish any Unit Conversion, and gets its name from a visual image that I'll explain momentarily.

For example, what if we wanted to make a simple, familiar Unit Conversion, such as converting 45 seconds into an equivalent number of minutes? Here's what the Fencepost setup would look like:

$$(45 \text{ seconds})\left(\frac{1 \text{ minute}}{60 \text{ seconds}}\right)$$

So why do we call it the "Fencepost Method"? Well, take your pencil or pen and do something for me: Draw a short vertical line through each of the two terms in parentheses. Then draw a diagonal line through the two times we wrote "seconds" in the equation.

With a little bit of imagination, you can see how this resembles the image of a wooden fence: two vertical posts and a diagonal crossbeam:

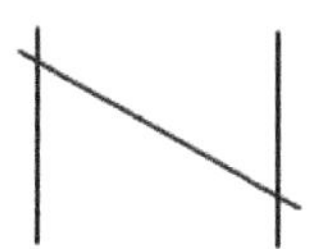

Visualize the "45 seconds" as the left vertical line, and the fraction $\frac{1 \text{ minute}}{60 \text{ seconds}}$ as the right vertical line. Then the diagonal slash would be the cancellation of "seconds" on the left with "seconds" on the right.

This might make sense to you already, but if it doesn't - don't worry. I'll clarify it more throughout the lesson and you'll get it. Just keep trying to see the image of a series of vertical fenceposts with diagonal crossbars of unit cancellations. You'll see it soon!

Back to our current example. Now that our Fencepost Setup is completed, we'll multiply the two fractions, cancel the units of "seconds" - which leaves only units of "minutes" - and then simplify the fraction.

$$(45 \text{ seconds})\left(\frac{1 \text{ minute}}{60 \text{ seconds}}\right)$$
$$= \frac{45}{60} \text{ minutes}$$
$$= \frac{3}{4} \text{ minutes}$$

This is how we can use the Fencepost method to convert 45 seconds into its equivalent, $\frac{3}{4}$ of a minute.

Now let's try a longer Fencepost Method conversion.

How many inches are in 3 miles? (Note: there are 5,280 feet in one mile)

Let's do a Fencepost setup first:

$$3 \text{ miles}\left(\frac{5280 \text{ feet}}{1 \text{ mile}}\right)\left(\frac{12 \text{ inches}}{1 \text{ foot}}\right)$$

Can you see the "Fencepost" visual image in your mind?

We have three "vertical posts" and two "diagonals". The three vertical posts are "3 miles," $\left(\frac{5280 \text{ feet}}{1 \text{ mile}}\right)$, and $\left(\frac{12 \text{ inches}}{1 \text{ foot}}\right)$. The two diagonals are the cancellations of "miles" with "mile", and "feet" with "foot".

Now let's finish the work by multiplying and simplifying:

$$\begin{aligned} &3 \text{ miles}\left(\frac{5280 \text{ feet}}{1 \text{ mile}}\right)\left(\frac{12 \text{ inches}}{1 \text{ foot}}\right) \\ &= \frac{(3)(5280)(12)}{(1)(1)} \text{inches} \\ &= 190{,}080 \text{ inches} \end{aligned}$$

Notice that after our two cancellations of miles with miles and feet with feet, the only unit of measurement that remains is "inches" - the final unit that we were trying to convert into.

Pretest Question #1

Let's take a look at our first Pretest question on this topic. Try it yourself if you got it wrong the first time.

Jeffery is practicing for an archery competition. While practicing, he shoots an average of n arrows every 12 seconds. Which of the following can be used to solve for h, the average number of arrows he shoots per hour?

(A) $h = 12n$

(B) $h = 300n$

(C) $h = \frac{300}{n}$

(D) $h = \frac{n}{12}$

There are only a few small differences in this question than in our previous Fencepost examples: this time we're using *variables* in the place of specific numbers. This time, we need to set up a balanced *equation*.

My setup is below. "1 arrow every 12 seconds" can be set up as the **Ratio** $\frac{n \text{ arrows}}{12 \text{ seconds}}$ and "h arrows per hour" can be set up as $\frac{h \text{ arrows}}{1 \text{ hour}}$.

$$\frac{n \text{ arrows}}{12 \text{ seconds}}\left(\frac{60 \text{ seconds}}{1 \text{ minute}}\right)\left(\frac{60 \text{ minutes}}{1 \text{ hour}}\right) = \frac{h \text{ arrows}}{1 \text{ hour}}$$

Also notice that almost all the units cancel with Fencepost diagonals - seconds cancel diagonally with seconds, and minutes cancel with minutes. The only units that don't cancel are "arrows" on top, and "hours" on bottom, leaving me with $\frac{\text{arrows}}{\text{hour}}$ or "arrows per hour" as my final unit.

Now let's multiply and clean up the fractions:

$$\frac{n \text{ arrows}}{12 \text{ seconds}}\left(\frac{60 \text{ seconds}}{1 \text{ minute}}\right)\left(\frac{60 \text{ minutes}}{1 \text{ hour}}\right) = \frac{h \text{ arrows}}{1 \text{ hour}}$$

$$\frac{(n)(60)(60) \text{ arrows}}{(12)(1)(1) \text{ hours}} = h \text{ arrows per hour}$$

$$\frac{3600n \text{ arrows}}{12 \text{ hours}} = h$$

$$300n = h$$

Voila! We have our answer: $300n = h$, or **Choice B**.

Advanced Unit Conversions: The Fencepost Method Expanded

The great thing about the Fencepost Method is it works no matter how many units you need to convert between.

We've already made small hops (from seconds to minutes, or from miles to feet). But even for larger leaps of Unit Conversions, the Fencepost Method just keeps working!

For example, let's try converting years into seconds. Try setting up and solving with the Fencepost Method up for this question before looking at my explanation below.

How many seconds are in 2 years?

Try it yourself before you look at my work!

Here's my Fencepost Setup. Notice that all the units will cancel diagonally except for seconds, which is the final unit that I want to end up with:

$$2 \text{ years}\left(\frac{365 \text{ days}}{1 \text{ year}}\right)\left(\frac{24 \text{ hours}}{1 \text{ day}}\right)\left(\frac{60 \text{ minutes}}{1 \text{ hour}}\right)\left(\frac{60 \text{ seconds}}{1 \text{ minute}}\right)$$

And here's the finished work as I clean up the fractions:

$$2 \text{ years}\left(\frac{365 \text{ days}}{1 \text{ year}}\right)\left(\frac{24 \text{ hours}}{1 \text{ day}}\right)\left(\frac{60 \text{ minutes}}{1 \text{ hour}}\right)\left(\frac{60 \text{ seconds}}{1 \text{ minute}}\right)$$
$$= \frac{(2)(365)(24)(60)(60)}{(1)(1)(1)(1)} \text{ seconds}$$
$$= 63{,}072{,}000 \text{ seconds}$$

Now we know that there are 63,072,000 seconds in two years. That's a lot of time!

Pretest Question #2

Let's take a look at another Pretest question. Try it yourself before you look at my explanation after the question.

FREE RESPONSE: A sarpler was a historical unit used to measure the weight of wool. A sarpler is equivalent to 80 tods, and three tods are equivalent to 84 pounds. A weight of 160 stone is equivalent to 2,240 pounds. How many stone are equivalent to a 40-sarpler weight of wool?

Guess what? The Fencepost Method will empower you to accomplish *any* Unit Conversion - no matter how weird or unfamiliar the units of measurement are (for example in this question - what the heck are "sarplers" or "tods"? It doesn't matter to the Fenceposts!)

We'll just create a Fencepost Setup, as always - and focus on diagonally canceling units, with no need to deeply understand what each unit means.

$$40 \text{ sarpler}\left(\frac{80 \text{ tods}}{1 \text{ sarpler}}\right)\left(\frac{84 \text{ pounds}}{3 \text{ tods}}\right)\left(\frac{160 \text{ stone}}{2240 \text{ pounds}}\right)$$

Hint: If you're wondering where to even *start*, the clue is in the final sentence of the question - we want to start with a "40-sarpler weight of wool".

From there I looked for *any* Unit Conversion of sarplers into "some other unit," and I found it with a "sarpler to tods" conversion. Then I just keep stepping forward, one fraction at a time: I found a conversion of "tods to pounds." Finally, there was some unused info about converting "pounds to stone" that enabled me to create the final Fencepost fraction.

Notice how the diagonal Fencepost cancellations in my setup above will eliminate sarplers, tods, and pounds, leaving me only with stone as my final unit, as the question asked.

Now let's finish the fraction multiplication and simplification:

$$\begin{aligned} &40 \text{ sarpler}\left(\frac{80 \text{ tods}}{1 \text{ sarpler}}\right)\left(\frac{84 \text{ pounds}}{3 \text{ tods}}\right)\left(\frac{160 \text{ stone}}{2240 \text{ pounds}}\right) \\ &= \frac{(40)(80)(84)(160)}{(1)(3)(2240)} \text{ stone} \\ &= \frac{43{,}008{,}000}{6{,}720} \text{ stone} \\ &= 6{,}400 \text{ stone} \end{aligned}$$

Our final answer is **6400 stone.**

Review & Encouragement

Unit Conversions may not be the *most* common topic on the SAT Math, but they definitely are on the test - and they can show up when you're least expecting them.

Whether the entire question is focused on units of measurement, or if units are simply one small piece of the puzzle, you'll be rewarded with more points if you can confidently, quickly, and accurately convert from one unit to another using the Fencepost Method.

Set up your fractions so that all the units cancel each other diagonally *except* for the final units you want to end with. Be sure to use all the information in the question if you get stuck.

Don't "improvise" on your Unit Conversions - that's when common mistakes happen the most. Instead, remember the Fencepost Method and be sure to practice it thoroughly and rigorously on the following set of practice problems.

Unit Conversions Practice Questions

YOU MAY USE YOUR CALCULATOR ON ALL OF THE FOLLOWING QUESTIONS.

1. FREE RESPONSE: Theodore can ride his electric scooter up to 15 kilometers on a single charge before it runs out of power. Rounded to the nearest tenth of a mile, approximately how many miles can Theodore ride his scooter on a single charge? (1 mile = 1.6093 kilometers)

2. Horatio is planning a party. He has a punch bowl that can hold three gallons of punch. However, the punch bowl doesn't fit in his refrigerator, so he had to divide the three gallons of punch into twelve equally-sized jugs. To the nearest fluid ounce, how many fluid ounces of punch is each jug holding? (1 gallon = 128 fluid ounces).

(A) 6

(B) 11

(C) 32

(D) 512

3. FREE RESPONSE: Samantha is mailing donation request letters for a grassroots political campaign. Each envelope requires 4 centimeters of tape to be sealed securely. If the rolls of tape she is using each contain 5 meters of tape, what is the maximum number of envelopes that can be sealed with two rolls of tape? (1 meter = 100 centimeters)

4. FREE RESPONSE: Two obsolete units of length are furlongs and rods. A furlong is equivalent to 40 rods. A rod is equivalent to $5\frac{1}{2}$ yards. A yard is equivalent to 3 feet. How many feet is equivalent to 2 furlongs?

5. FREE RESPONSE: Historically used to measure the weight of precious metals, a pennyweight is a unit of mass that is equal to $\frac{1}{240}$ of a troy pound. There are 12 troy ounces in a troy pound. A certain bank holds stock in silver bars that each weigh 600 pennyweights. If there are five of these bars in the bank's vault, representing the bank's entire stock of silver, how many troy ounces of silver are in the bank's vault?

6. FREE RESPONSE: On a certain date in 2019, platinum was worth $925 per ounce, and a large metals company spent $70,300 purchasing platinum at that rate. What was the weight,in pounds, of this stock of platinum? (16 ounces = 1 pound)

7. Watts and horsepower are both units of measure of power. They are directly proportional to each other, and 15 horsepower is equivalent to 11,190 watts. If a motorcycle's engine produces 195 horsepower, how many watts is this equivalent to, rounded to the nearest watt?

(A) 4

(B) 145,470

(C) 2,182,050

(D) 32,730,750

8. Which of the following equations could be used to solve for the number of seconds, s, equivalent in duration to y years?

(A) $s = 31{,}536{,}000y$

(B) $s = \dfrac{54{,}750}{y}$

(C) $s = \dfrac{y}{54{,}750}$

(D) $s = \dfrac{y}{31{,}536{,}000}$

9. The pood is an obsolete unit of mass once used in Russia. One pood is equivalent to 40 funt. A zolotnik is equivalent to $\frac{1}{96}$ of a funt. Which of the following equations would allow you to solve for the number of zolotniks, z, equivalent to p poods?

(A) $z = \dfrac{5p}{12}$

(B) $z = \dfrac{12p}{5}$

(C) $z = \dfrac{p}{3840}$

(D) $z = 3840p$

10. FREE RESPONSE: The daktylos (plural: "daktyloi"), orthodoron, and pygon were three distance units of Ancient Greek measurement. An orthodoron was equivalent to 11 daktyloi. A pygon was equivalent to 20 daktyloi. What is the equivalent of 440 pygons in orthodorons?

Unit Conversions Answers

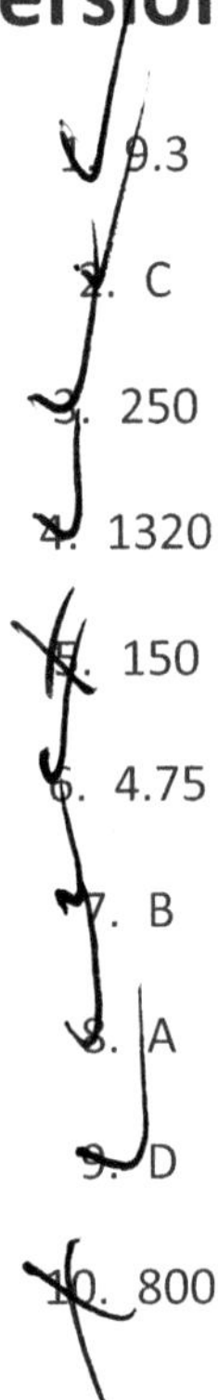

1. 9.3
2. C
3. 250
4. 1320
5. 150
6. 4.75
7. B
8. A
9. D
10. 800

Unit Conversions Explanations

1. **9.3.** Let's make a Fencepost Method setup for this problem:

$$15\text{ km}\left(\frac{1\text{ mile}}{1.6093\text{ km}}\right)$$

And now complete the multiplication and simplify the resulting fraction into a decimal.

$$\begin{aligned}&15\text{ km}\left(\frac{1\text{ mile}}{1.6093\text{ km}}\right)\\&=\frac{15}{1.6093}\text{ miles}\\&=9.3208\text{ miles}\end{aligned}$$

Don't forget to round to the nearest tenth of a mile, like the question asked!

2. **C.** First, let's convert gallons into ounces using a Fencepost Setup. I usually think it's easier to complete my unit conversions early in the problem.

$$\begin{aligned}&3\text{ gallons}\left(\frac{128\text{ fluid ounces}}{1\text{ gallon}}\right)\\&=3(128)\text{ fluid ounces}\\&=384\text{ fluid ounces}\end{aligned}$$

Now we know that Horatio originally has 384 fluid ounces of punch, and we can divide this equally among the 12 jugs by using simple division:

$$\begin{aligned}&\frac{384\text{ fluid ounces}}{12\text{ jugs}}\\&=32\text{ fluid ounces per jug}\end{aligned}$$

3. **250.** The first thing to do is create our Fencepost Setup.

$$2\text{ rolls of tape}\left(\frac{5\text{ meters}}{1\text{ roll of tape}}\right)\left(\frac{100\text{ cm}}{1\text{ meter}}\right)\left(\frac{1\text{ envelope}}{4\text{ cm}}\right)$$

Note that I'm even able to use "rolls of tape" and "envelopes" as units of measurement, since I know the exact conversions as given in the Word Problem (5 meters per 1 roll of tape and 4 cm per 1 envelope).

According to my Fencepost Setup, all of my units but one will cancel each other diagonally, as they should. The unit I'm left with will be "envelopes," which fits the word problem I'm being asked for.

Now cancel the units, complete the multiplication and clean up the result:

$$\begin{aligned}&2\text{ rolls of tape}\left(\frac{5\text{ meters}}{1\text{ roll of tape}}\right)\left(\frac{100\text{ cm}}{1\text{ meter}}\right)\left(\frac{1\text{ envelope}}{4\text{ cm}}\right)\\&=\frac{(2)(5)(100)(1)}{(1)(1)(4)}\text{ envelopes}\\&=\frac{1000}{4}\text{ envelopes}\\&=250\text{ envelopes}\end{aligned}$$

4. **1320.** The beauty of setting up with the Fencepost Method is that it works regardless of how obsolete, complicated, or unfamiliar the units of measurement are. All you need is the **Ratio** or conversion value of one unit to another, allowing you to set up a chain of conversions using the Fencepost Method.

Note below that all my units will cancel each other diagonally except for one - feet - which fits to the question I'm being asked. Also, don't forget to use 2 furlongs at the beginning, as indicated in the question.:

$$2\text{ furlongs}\left(\frac{40\text{ rods}}{1\text{ furlong}}\right)\left(\frac{5\frac{1}{2}\text{ yards}}{1\text{ rod}}\right)\left(\frac{3\text{ feet}}{1\text{ yard}}\right)$$

Now finish the fraction multiplication and clean up, as usual:

$$\begin{aligned}&2\text{ furlongs}\left(\frac{40\text{ rods}}{1\text{ furlong}}\right)\left(\frac{5\frac{1}{2}\text{ yards}}{1\text{ rod}}\right)\left(\frac{3\text{ feet}}{1\text{ yard}}\right)\\&=\frac{(2)(40)(5\frac{1}{2})(3)}{(1)(1)(1)}\text{ feet}\\&=\frac{1{,}320}{1}\text{ feet}\\&=1{,}320\text{ feet}\end{aligned}$$

5. **150.** As always with Unit Conversions, just start by creating a Fencepost Setup. Regardless of how weird the units of measurement are, this method will work - as long as you set up your fractions so that all the units (except one) cancel each other out:

$$5 \text{ silver bars}\left(\frac{600 \text{ pennyweights}}{1 \text{ silver bar}}\right)\left(\frac{\frac{1}{240} \text{ troy pound}}{1 \text{ pennyweight}}\right)\left(\frac{12 \text{ troy ounces}}{1 \text{ troy pound}}\right)$$

Notice that all my units will cancel each other diagonally *except* for "troy ounces," which is the unit I want for my final answer.

Now complete the fraction multiplication and clean up:

$$\begin{aligned} &5 \text{ silver bars}\left(\frac{600 \text{ pennyweights}}{1 \text{ silver bar}}\right)\left(\frac{\frac{1}{240} \text{ troy pound}}{1 \text{ pennyweight}}\right)\left(\frac{12 \text{ troy ounces}}{1 \text{ troy pound}}\right) \\ &= \frac{(5)(600)(\frac{1}{240})(12)}{(1)(1)(1)} \text{ troy ounces} \\ &= \frac{150}{1} \text{ troy ounces} \\ &= 150 \text{ troy ounces} \end{aligned}$$

6. **4.75.** Let's make our Fencepost Setup. "Dollars" are a unit of measurement like any other. Make sure all the units cancel each other diagonally, except for "pounds," which are our intended final unit.

$$\$70{,}300\left(\frac{1 \text{ ounce}}{\$925}\right)\left(\frac{1 \text{ pound}}{16 \text{ ounces}}\right)$$

Now just cancel all units except pounds and multiply and simplify the result:

$$\begin{aligned} &\$70{,}300\left(\frac{1 \text{ ounce}}{\$925}\right)\left(\frac{1 \text{ pound}}{16 \text{ ounces}}\right) \\ &= \frac{(70{,}300)(1)(1)}{(925)(16)} \text{ pounds} \\ &= \frac{70{,}300}{14{,}800} \text{ pounds} \\ &= 4.75 \text{ pounds} \end{aligned}$$

7. **B.** First, make a Fencepost Setup. Then multiply and simplify:

$$\begin{aligned} &195 \text{ horsepower}\left(\frac{11{,}190 \text{ watts}}{15 \text{ horsepower}}\right) \\ &= \frac{2{,}182{,}050}{15} \text{ watts} \\ &= 145{,}470 \text{ watts} \end{aligned}$$

There's nothing unique about this question. It's just another Fencepost Setup, followed by Fraction Multiplication and simplifying - just like all the other questions in this Practice Set so far! By now I hope you're getting the hang of this powerful and effective (yet simple) method.

8. **A.** Everything about this question follows the typical Fencepost Method setup, with two tiny differences (this question has a lot in common with Pretest Question #1).

First of all, we don't know exactly how many years or seconds we'll be using. That's an easy fix - instead of using a constant number like "4 years," we'll just use a variable of "y years." Likewise, we'll set up with "s seconds".

Second, we have to set this Unit Conversion *equal* to something in order to provide our final answer (all of our choices are equations relating y to s).

What will this Unit Conversion equal when it's finished? Well, we're converting years into seconds. We want to know how many seconds will be in y years. And, if we look at the diagonal unit cancellations in our Fencepost Setup below, it will also confirm that all units of measurement will cancel *except* for "seconds." So we can set this entire Fencepost conversion equal to "s seconds" on the right side of the equal sign.

$$y \text{ years}\left(\frac{365 \text{ days}}{1 \text{ year}}\right)\left(\frac{24 \text{ hours}}{1 \text{ day}}\right)\left(\frac{60 \text{ minutes}}{1 \text{ hour}}\right)\left(\frac{60 \text{ seconds}}{1 \text{ minute}}\right) = s \text{ seconds}$$

Now that the Fencepost Method and the variables y and s have helped us create a setup, we can clean up and simplify:

$$y \text{ years}\left(\frac{365 \text{ days}}{1 \text{ year}}\right)\left(\frac{24 \text{ hours}}{1 \text{ day}}\right)\left(\frac{60 \text{ minutes}}{1 \text{ hour}}\right)\left(\frac{60 \text{ seconds}}{1 \text{ minute}}\right) = s \text{ seconds}$$

$$\frac{y(365)(24)(60)(60)}{(1)(1)(1)(1)} \text{seconds} = s \text{ seconds}$$

$$31{,}536{,}000y = s$$

9. **D.** No matter how unfamiliar, obsolete, or funny-sounding the units of measurement are, the Fencepost Method will allow us to quickly create an accurate Unit Conversion. Check out my setup below. This setup has a lot in common with my work on the previous Question 8:

$$p \text{ pood}\left(\frac{40 \text{ funt}}{1 \text{ pood}}\right)\left(\frac{1 \text{ zolotnik}}{\frac{1}{96} \text{ funt}}\right) = z \text{ zolotniks}$$

Now clean up and simplify:

$$p \text{ pood}\left(\frac{40 \text{ funt}}{1 \text{ pood}}\right)\left(\frac{1 \text{ zolotnik}}{\frac{1}{96} \text{ funt}}\right) = z \text{ zolotniks}$$

$$\frac{p(40)(1)}{(1)(\frac{1}{96})} \text{ zolotniks} = z \text{ zolotniks}$$

$$\frac{40p}{(\frac{1}{96})} \text{ zolotniks} = z \text{ zolotniks}$$

$$40p(\tfrac{96}{1}) = z$$

$$3840p = z$$

Notice that near the end I used my knowledge of **Fractions** to divide $40p$ by $\frac{1}{96}$. I did this by flipping $\frac{1}{96}$ upside down and multiplying it. Remember, to divide by a fraction, you just flip the fraction upside down and multiply.

10. **800.** As always, no matter how strange the units of measurement, we can easily set up an accurate Unit Conversion using the Fencepost Method. As long as all of our units of measurement (except one) cancel each other out diagonally, we'll get the right answer.

Here's my Fencepost setup:

$$440 \text{ pygons}\left(\frac{20 \text{ daktyloi}}{1 \text{ pygon}}\right)\left(\frac{1 \text{ orthodoron}}{11 \text{ daktyloi}}\right)$$

Now multiply and simplify:

$$\begin{aligned} &440 \text{ pygons}\left(\frac{20 \text{ daktyloi}}{1 \text{ pygon}}\right)\left(\frac{1 \text{ orthodoron}}{11 \text{ daktyloi}}\right) \\ &= \frac{(440)(20)(1)}{(1)(11)} \text{ orthodorons} \\ &= \frac{8800}{11} \text{ orthodorons} \\ &= 800 \text{ orthodorons} \end{aligned}$$

Lesson 9: Linear Equations (Algebraic)

Percentages

- 4.7% of Whole Test
- 5.5% of No-Calculator Section
- 4.2% of Calculator Section

Prerequisites

- Basic Algebra 1
- Ratios & Proportions (Recommended)

Linear Equations are the number-one single most common topic on the entire SAT Math test. There's so much to talk about, in fact, that we'll also be exploring Linear Equations again in the next chapter (with particular attention to Word Problems, Charts, and Tables).

For now, though, we're going to focus on the key Algebra 1 concepts underpinning Linear Equations. These concepts include the $y = mx + b$ equation, slope, graphs of basic lines, intersections of two lines, graphs of simple inequalities, and other related topics.

This is our first "Coordinate Geometry" topic. Coordinate Geometry simply means work that we do on a graph (also called the "*xy*-coordinate plane" because it's a flat grid made of an *x*-axis and a *y*-axis). Although it's called "Geometry," Coordinate Geometry actually has everything to do with *Algebra* - and almost nothing to do with what you would typically think of as "Geometry".

Linear Equations (Algebraic) Quick Reference

- Linear Equations are the #1 most-common topic on the SAT Math test.
- A "Linear Equation" is an Algebra equation for a straight line on a graph.
- The secret to mastering Linear Equations lies in understanding the $y = mx + b$ equation.
- Slope (or m in $y = mx + b$) is "Rise over Run", calculated with $\frac{y_2 - y_1}{x_2 - x_1}$. Higher slopes are steeper; lower slopes are shallower. Negative slopes descend from left to right.
- *Parallel* lines have *equal* slopes. *Perpendicular* lines have *negative reciprocal* slopes.
- The *y*-intercept (or b in $y = mx + b$) is the coordinate where a line crosses the *y*-axis.
- To find the intersection point of two lines, set their equations equal to each other and use **Basic Algebra 1** to solve for x.
- Inequalities can be graphed by shading above or below their lines, depending on the direction of the inequality sign.

Intro to Linear Equations

A "Linear Equation" is simply a fancy name for "the equation of a line." By "line," we mean any straight line drawn on the *xy*-plane - in other words, any straight line drawn on a graph.

For example, the picture below shows the graph of a typical Linear Equation:

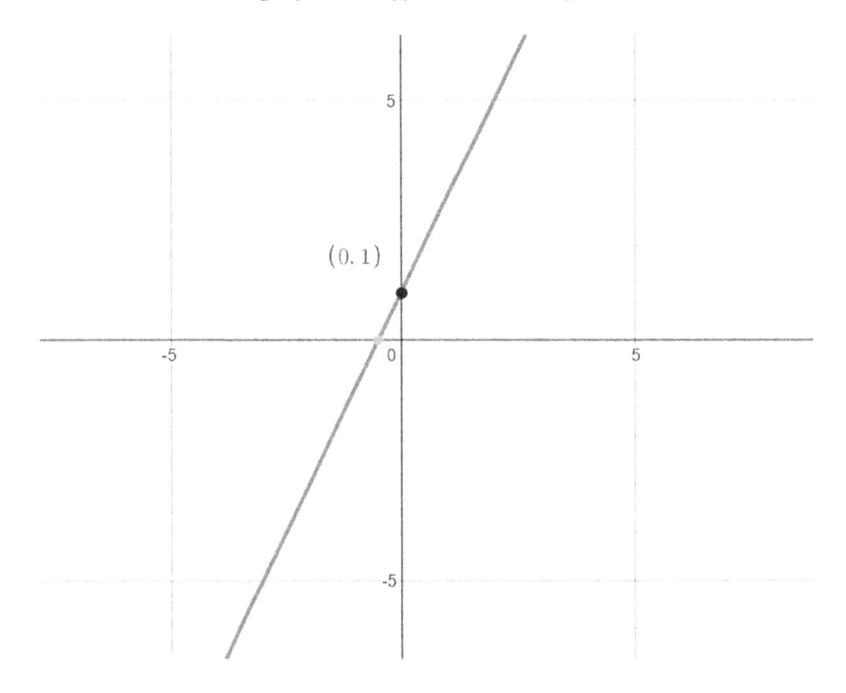

Linear Equations are most commonly described with the equation $y = mx + b$, which we will be exploring throughout this lesson.

In this common equation, m is the *Slope* of the line, and b is the *y-Intercept* of the line (we'll explain both of these ideas in just a moment).

The equation of our line above is $y = 2x + 1$. The coefficient of x is 2, which means the *m*-value or slope is 2. (Note: "coefficient" is just a fancy name for a number that goes in front of a variable).

Put more simply, we would just say that the slope of this line is 2.

The *b*-value of our equation is +1, and in the graph above, the *y*-intercept of the line is 1. That means this line crosses the *y*-axis at 1. This is the coordinate (0, 1) because the value of x is always 0 at the *y*-axis.

The x and y terms in the $y = mx + b$ equation are places to plug in an (x, y) coordinate point that lies on the line. You probably know that all points on an *xy*-plane can be described in terms of their (x, y) coordinates. Any point on a line can be plugged into that line's $y = mx + b$ equation to create a balanced equation. This is useful in a variety of situations.

I'll give a simple demonstration of how we can plug *x* and *y* coordinates into $y = mx + b$. If we take a point on the line above - for example $(2,5)$ - we are able to plug those values into the equation for x and y and return a true equality. Let me show you what happens when I plug in the point $(2,5)$:

$$y = 2x + 1$$
$$5 = 2(2) + 1$$
$$5 = 5$$

Of course, 5 does equal 5, so this equation is true. This is one way of showing that our $y = mx + b$ equation is an accurate Algebraic equation of the line.

Slope of Linear Equations

The word "Slope" means exactly what it sounds like in the real world. For example, a gentle hill has a low slope value, while a steep mountain has a high slope value.

In math terms, "Slope" is the **Ratio** or fraction of "rise over run." In other words, Slope describes how quickly a line goes *up or down* compared to how far it goes *left to righ*t. Higher slopes rise steeper and faster. Lower slopes rise more slowly and shallowly.

A slope of 2 means the line goes "up 2" on the *y*-axis each time it goes "over 1" on the *x*-axis.

There are a few common ways we talk about Slope. All of them are useful to remember:

$$\text{slope is } \frac{\text{rise}}{\text{run}}$$
$$\text{slope is } \frac{y_2 - y_1}{x_2 - x_1}$$
$$\text{slope is } \frac{\text{change in } y}{\text{change in } x}$$

Each of these phrases has the same meaning, but I find it useful to understand all of them.

The first version is the most simplistic and easy to remember. "Slope equals Rise over Run." This is a versatile, all-purpose way of remembering or calculating Slope. "Rise" is how much the line goes up or down in a certain interval. "Run" is how far the line goes from left to right in that same interval.

The second version is the most mathematically useful. It lets us take any two points with the coordinates (x_1, y_1) and (x_2, y_2) and calculate the slope of the line that connects them. We can use this in Algebra setups to calculate a slope from two given points. By the way, it doesn't matter which of the two points you call (x_1, y_1) or (x_2, y_2). Just stay *consistent* with which point is which and you'll be fine.

The third version is just a useful restatement of the first version. "Rise" is the "change in *y*-value" and "run" is the "change in *x*-value."

Negative Slope, Zero Slope, and Undefined Slope

Lines can also have a negative slope. Any descending line has a negative slope. In other words, as you trace the line from left to right, it will fall towards lower and lower y-values.

For example, here is a graph of a line with negative slope. Notice that it *descends* from left to right. That's what negative slope looks like.

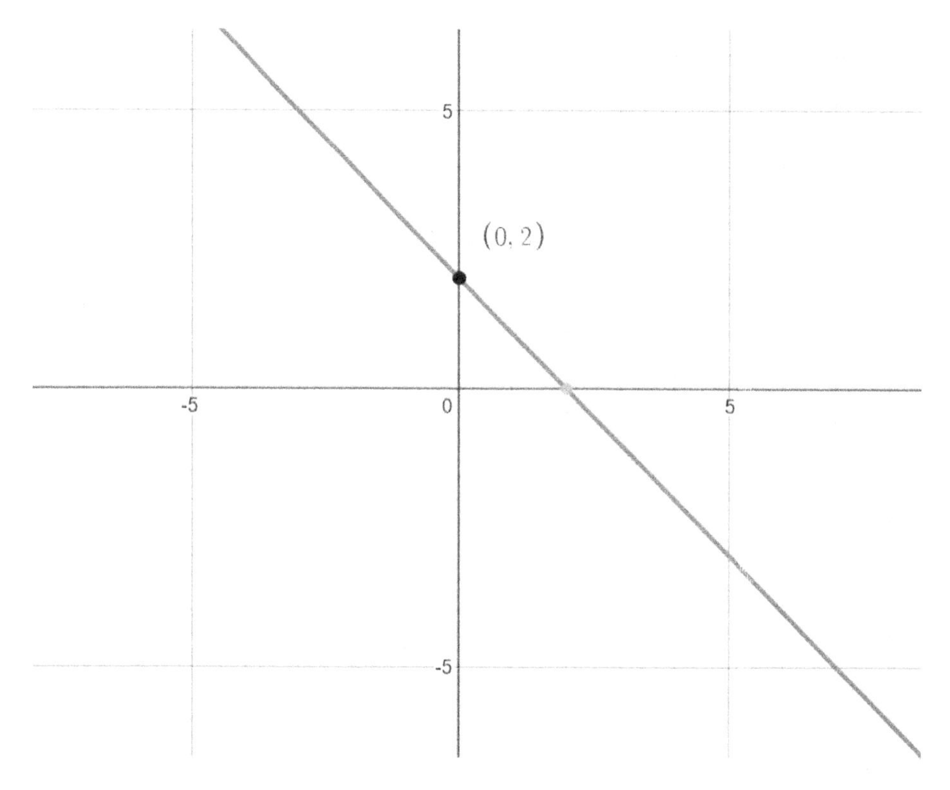

Remember, in the fundamental Linear Equation $y = mx + b$, the slope always plugs in where m is. In other words, in a Linear Equation, the slope is always the "coefficient of x" (remember, "coefficient" is just a fancy name for the number that goes in front of the x).

A line with a slope of zero is a horizontal line. This makes sense: a flat horizontal line has *no* rise, no matter how far it runs from left to right.

A vertical line has an "Undefined" slope. Because a vertical line goes straight up and down, it has infinite vertical rise and zero horizontal run. If you plug that into the Slope Equation, you'll find that it produces a fraction that divides by zero - and we are never allowed to divide by zero, since it produces an "undefined" result.

y=mx+b and *y*-Intercepts

Let's talk more about the $y = mx + b$ format and what the *b*-value means in that equation.

$y = mx + b$ is called "slope-intercept" form, but the name doesn't really matter. There's no benefit on the SAT Math to remembering what it's *called* - only how to *use* it.

As briefly mentioned before, the *b*-value in $y = mx + b$ can also be called the "*y*-intercept." That's where the name "slope-*intercept* form" comes from.

The *y*-intercept is the point where the line crosses the *y*-axis. There's nothing particularly special about this point; it's just used as a common reference point when graphing Linear Equations.

On the graph of the line above (from the section on Negative Slope), the *y*-intercept is 2. You can find it at the point $(0, 2)$. It's right where the line crosses through the *y*-axis.

As always, this *y*-intercept has an *x*-coordinate of zero, because x is always zero whenever you're directly on the *y*-axis.

Graphing Linear Equations

Now that you understand the essentials behind the $y = mx + b$ equation, it's easy to take the next step into sketching graphs of these Linear Equations.

The easiest way is usually to mark the *y*-intercept first (the *b*-value), then use the slope (the *m*-value) to "rise and run" an appropriate amount, mark another point or two, and then connect the points in a straight line.

For practice, try sketching the line $y = x - 3$ on the coordinate plane provided.

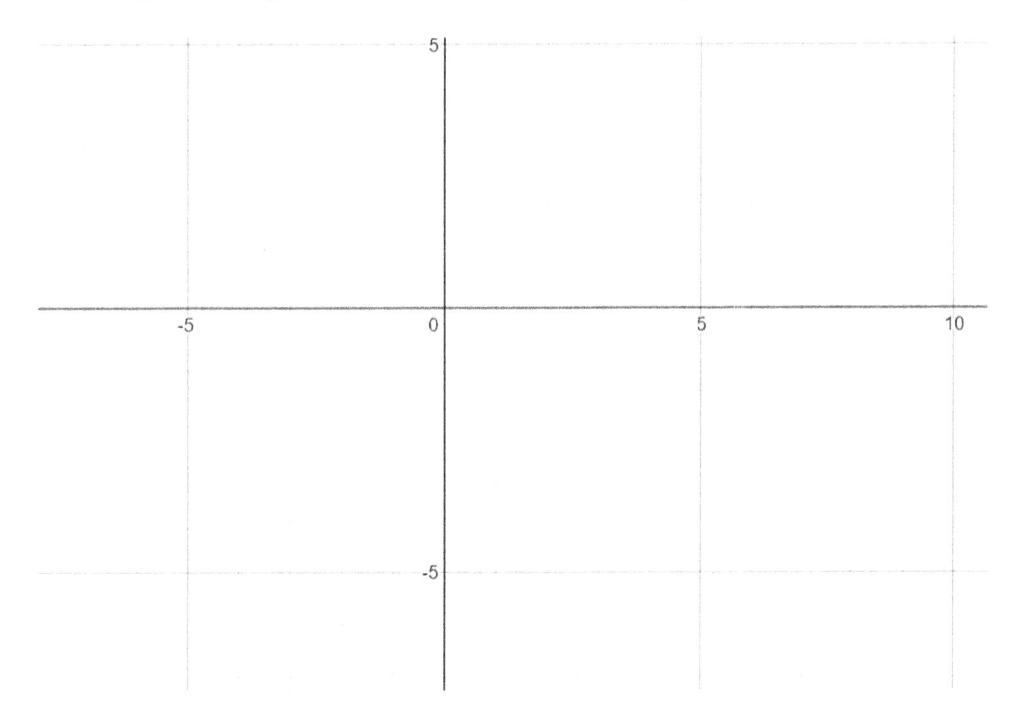

If you're having trouble, first mark the *y*-intercept at -3. Then go "up one, over one" since the slope is 1. Mark this new point at $(1,-2)$.

Just to be safe, you might also mark one more point. Move up one, over one again and mark a third point at $(2,-1)$.

Now connect all three points with a straight line and extend it in both directions. Strictly speaking, you only need two points to create a straight line, but a third point may make you feel more confident.

Finished? Your picture should look something like mine, below. For reference, the line crosses the *x*-axis exactly at the point $(3,0)$.

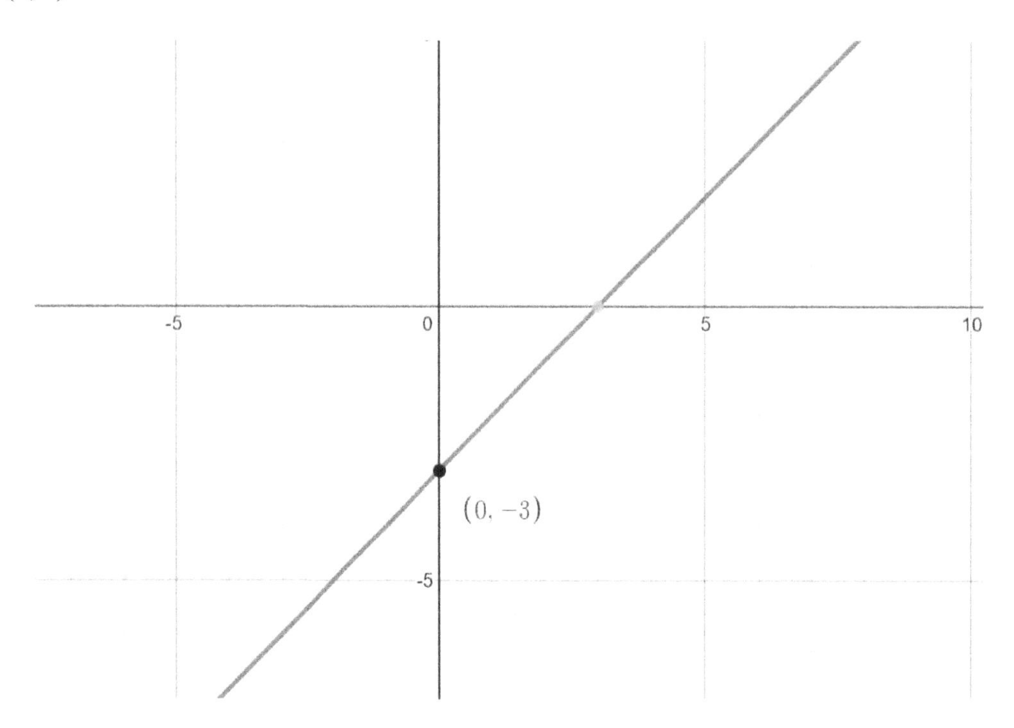

Intersections of Lines

It may seem like there's no end to this lesson - one of the biggest chapters in this book - but we're actually moving into one of the final major concepts of Linear Equations: finding the *intersection point* of two lines.

Here's something worth thinking about - a line is simply a *picture* of an equation, and vice versa: an equation is simply a written-out description of a line. There's *no difference* between a line and its equation. They're just two different ways of representing the exact same thing.

There are many cases on the SAT test when you will need to calculate the point where two different lines cross each other (or "intersect" each other).

It's easy to do this if you already know the equations of the two lines - you simply set the two Linear Equations equal to each other, then solve the resulting Algebra equation.

The point where two lines cross is the point where the two equations have the same x and y coordinates. In other words, two lines intersect when their two equations are equal to each other.

This is true because a line and its equation are the *exact same thing*, just presented in different formats.

Think about it! When two lines intersect, they are "in the same place at the same time." The lines are literally crossing each other in the same place, at a single point of intersection. And, when two equations are equal to each other, that's just another way of mathematically saying "they're the same." That's what *equal* means.

Let's look at an example to make this crystal-clear:

At what coordinate do the lines $y = 2x + 4$ and $y = -4x - 8$ intersect?

Remember, equations and lines are the *same thing*. The point where two lines intersect is also the point where the two equations are equal.

Therefore, we can just set these two equations equal to each other and solve with Algebra to find the point where they intersect.

$$2x + 4 = -4x - 8$$

Make sure you understand exactly how we've come to this point. We took our two $y = mx + b$ equations and set them equal to each other.

Now we can solve this equation for the value of x using **Basic Algebra 1**.

$$\begin{aligned} 2x + 4 &= -4x - 8 \\ -4 \quad & \quad\quad -4 \\ 2x &= -4x - 12 \\ +4x \quad & +4x \\ 6x &= -12 \\ \frac{6x}{6} &= \frac{-12}{6} \\ x &= -2 \end{aligned}$$

OK, great. Now we know that the *x*-coordinate of the point of intersection is $x = -2$. To find the *y*-coordinate, we can plug $x = -2$ into *either* of our original equations.

Note well: it doesn't matter which of the two equations we use. Why? Because both lines are in the *exact same place* right now: they're at the point where they both intersect.

Remember, the lines are *equal* - the same - at the intersect point. So, it doesn't matter which line's equation we use for this next step.

I'll use $y = 2x + 4$, but you could also use $y = -4x - 8$ and get the same result. Now plug in the $x = -2$ that we found in the previous stage to get the resulting y value.

$$\begin{aligned} y &= 2x + 4 \\ y &= 2(-2) + 4 \\ y &= -4 + 4 \\ y &= 0 \end{aligned}$$

Perfect - now we know both the *x*- and *y*-coordinates of the intersect point. The exact (x, y) coordinate of their intersection point is $(-2,0)$.

I've included a picture of the entire situation below for reference. You can see the two lines and their *y*-intercepts. You can also observe their exact intersection point at (-2, 0).

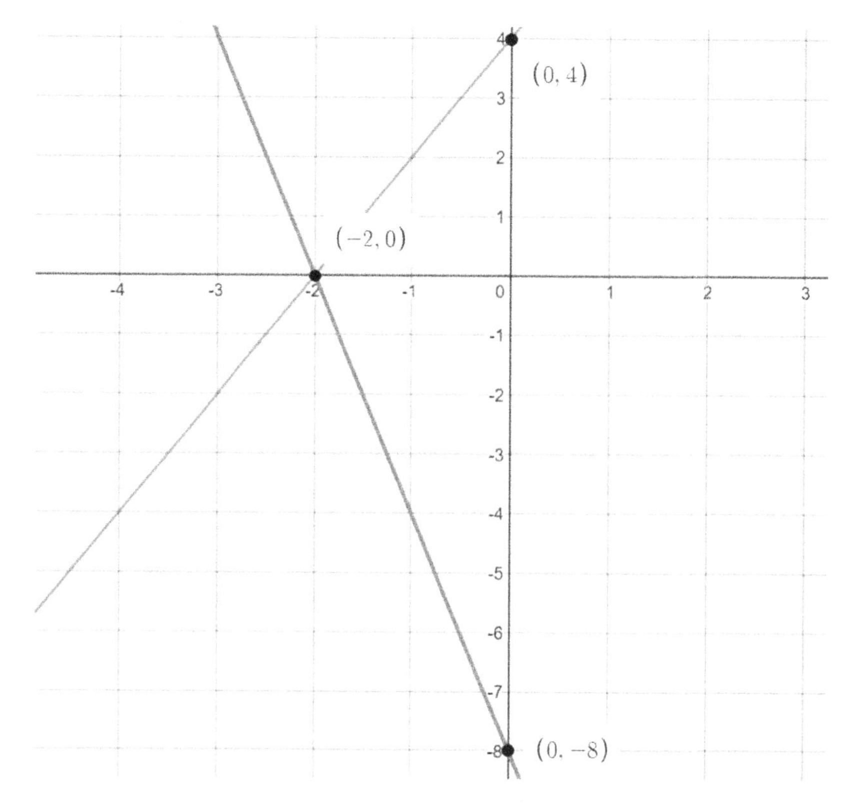

Remember - this is just a *picture* of the two algebra equations. The equations and the lines are the same things. The intersection point is the same as the Algebra equations being equal.

Lines and equations are just two different visual representations of the *same information*. Never forget this!

Pretest Question #1

Let's take a look at our first Pretest question on the topic of Linear Equations. It puts together everything we've learned so far about Linear Equations. Try it yourself before you look at my explanation.

Line A passes through the origin and the point $(-2, 2)$. Line B has a slope of 2 and passes through the point $(5, 4)$. If Lines A and B intersect at the point (t, n), what is the value of $2t - n$?

(A) -6

(B) -2

(C) 2

(D) 6

OK, so let's think backwards here. To *finish* the problem, we need to know the coordinates of the intersect point between Lines A and B. And, to calculate an intersect of two lines, the easiest way is usually to obtain both of their Linear Equations and set them equal to each other. Also, the question has given us enough information to create a Linear Equation for each line.

So, let's start with Line A. We already know the *y*-intercept will be $(0,0)$ since the line "passes through the origin" for its *y*-intercept. All we need is the slope to complete the Linear Equation.

Since we have two points on Line A, it's easy to calculate the slope:

$$\text{Slope}_A = \frac{y_2 - y_1}{x_2 - x_1}$$
$$\text{Slope}_A = \frac{2-0}{-2-0}$$
$$\text{Slope}_A = -\frac{2}{2}$$
$$\text{Slope}_A = -1$$

Now we know the slope of Line A is -1 and we can create the Linear Equation. We'll use the sub-label y_A to keep track of which line we're working with:

$$y_A = mx + b$$
$$y_A = -1x + 0$$
$$y_A = -x$$

The equation of Line A is $y_A = -x$.

OK, now on to Line B. We'll need to create a Linear Equation for this line, too. We already know the slope is 2 because the question tells us so, but we don't know its *y*-intercept. Here's how we'll handle it. First, create a $y = mx + b$ equation with as much information as we have:

$$y_B = mx + b$$
$$y_B = 2x + b$$

Now here's the trick. We know from the question that the (x, y) coordinate $(5,4)$ is on this line. Therefore, we can plug in 5 for x and 4 for y. Look what happens if we then apply some **Basic Algebra 1**:

$$y_B = 2x + b$$
$$4 = 2(5) + b$$
$$4 = 10 + b$$
$$-10 \quad -10$$
$$-6 = b$$

We've been able to find that the b-value (the *y*-intercept) is -6, simply by plugging in all the given information into the basic $y = mx + b$ equation and solving for b.

Now that we know the slope and *y*-intercept of Line B, we can finish setting up its Linear Equation:

$$y_B = 2x + (-6)$$
$$y_B = 2x - 6$$

Now we can move into the final phase of our plan, which we already had in mind from the beginning. We'll set the two linear equations equal to each other and solve for x, which is one of the best ways of calculating the point where they intersect:

$$-x = 2x - 6$$
$$-2x \quad -2x$$
$$-3x = -6$$
$$\frac{-3x}{-3} = \frac{-6}{-3}$$
$$x = 2$$

Now we know the *x*-coordinate of the intersect point is $x = 2$.

Almost done. The question asked for the value of $2t - n$, where t and n are the *x*- and *y*-coordinates of the intersection point.

We've just found the value of t, which is the *x*-coordinate of 2. But we still need the *y*-value of the intersect. That's easy - we can plug $x = 2$ back into either one of our Linear Equations. Either equation will work since we're calculating the intersection point, where both lines are equal to each other.

I'll just use Line A for this step, since it has a simpler equation:

$$y_A = -x$$
$$y_A = -(2)$$
$$y_A = -2$$

Now we know the *y*-value of the intersection point, which must be $(2, -2)$. Remember that this corresponds with (t, n), according to the question. Now, we can finally finish the question by calculating $2t - n$:

$$2t - n$$
$$= 2(2) - (-2)$$
$$= 4 + 2$$
$$= 6$$

Our final answer is 6, or **Choice D**.

I'll be honest: you have to be pretty satisfied with abstract mathematical rewards to stay focused on a question this long. If you're lucky (like me!) you'll find this sort of abstract math reasoning to be quite satisfying when you've successfully found a solution to the puzzle.

If you don't naturally feel that way, you can at least be proud of the achievement of solving this challenging, multi-step problem and taking yourself a big leap closer to a great SAT Math score.

Parallel & Perpendicular Lines

Linear Equation questions may make use of the terms "parallel" or "perpendicular." These are giveaway keywords to look out for, because they communicate valuable information about the slope of a line.

Parallel lines have the *same* slopes. Below is a picture of two parallel lines. Remember, two parallel lines will always have the *same* slope. These two lines both have a slope of 2.

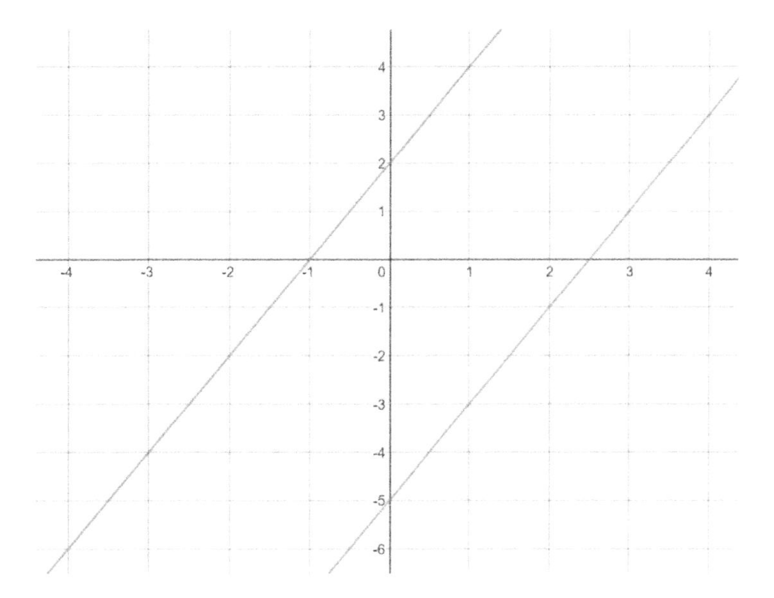

On the other hand, two perpendicular lines will intersect at a 90-degree right angle, and their slopes will be *negative reciprocals* of each other.

What's a "negative reciprocal?" Well, you know what "negative" means, and "reciprocal" means "upside-down". (For example, the reciprocal of $\frac{4}{7}$ is $\frac{7}{4}$). Take the first slope, make it negative, and turn it upside down.

One way to remember this is that Perpendicular lines are "as opposite as possible" to each other - and being "negative and upside down" versions of each other is about as opposite as two numbers can get.

Below is a picture of two perpendicular lines:

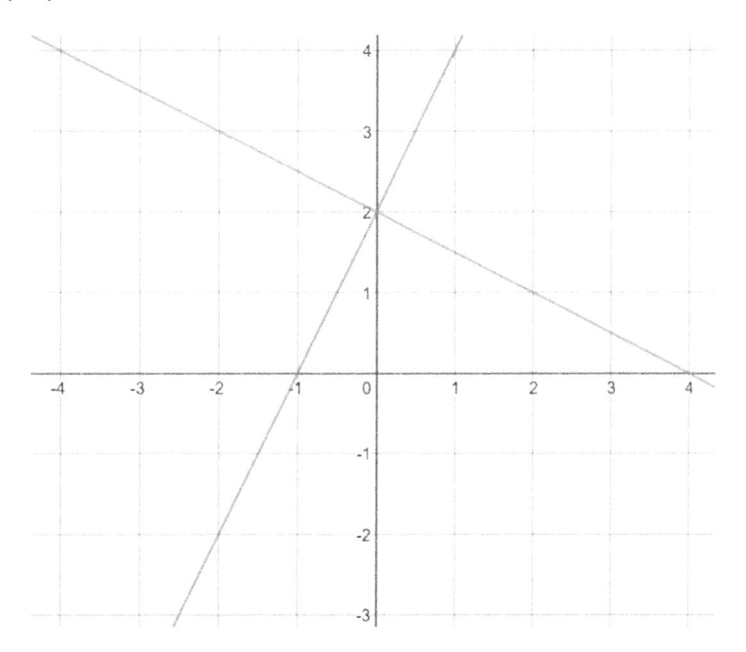

One line has a slope of 2, and the other has a slope of $-\frac{1}{2}$. For Perpendicular Lines, take one slope, make it negative and turn it upside-down, and you'll get the other slope. Try it for yourself and see!

Quadrants

Occasionally a Linear Equations question will refer to the "Quadrants" of the *xy*-plane. "Quad" means "four", and the "Quadrants" are the four sections of the *xy*-plane divided by the *x* and *y* axes.

The Quadrants are numbered 1-4, usually in Roman numerals (I, II, III, and IV). The numbering system starts with the top-right quadrant, where both *x* and *y* coordinates are positive, and moves counter-clockwise. Below I've included a picture for reference.

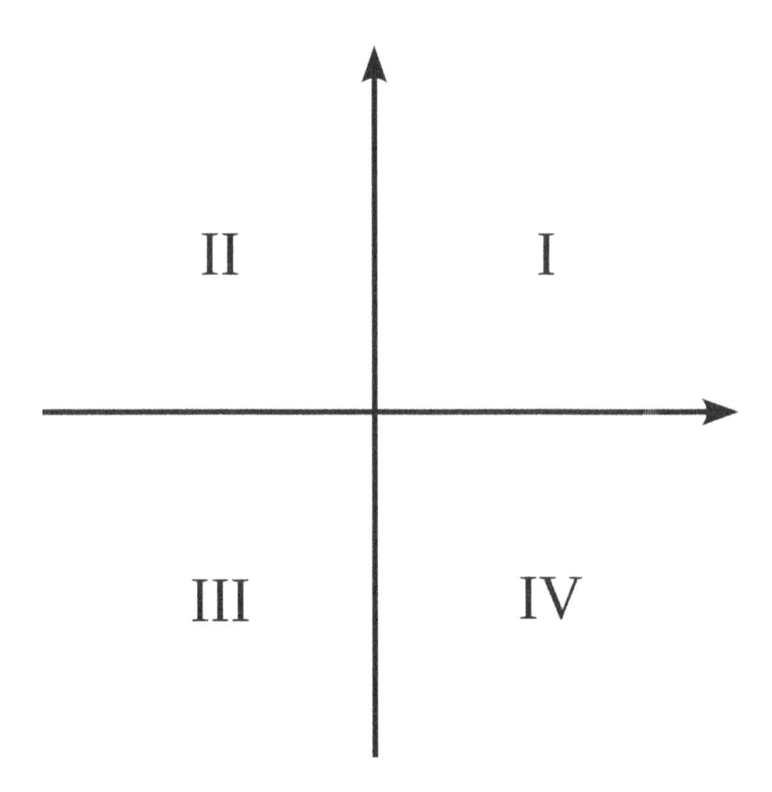

Quadrants are not a major feature of the SAT Math test, but they show up every now and then, so we at least need to be aware that they exist.

However, this topic isn't common enough to deserve an entire lesson to itself, and Linear Equations are one of the few topics that ever refer to Quadrants on the SAT Math test. It's very rare, trust me.

Inequality Graphing & Shading

Inequality symbols can also be (such as $>$ or $\leq$) can also be incorporated into Linear Equations. We first encountered these symbols in the lesson on **Basic Algebra 1**. Now, we'll also be using them to graph our Linear Equations.

The graphs of these inequalities will use shading above or below the line and a "solution" to an Inequality Graph is anywhere there is shading.

The best way to graph these Linear Inequalities is to keep using the $y = mx + b$ form to to find your slope and *y*-intercept and use them to sketch your basic line. The only difference is that, with inequalities, instead of just graphing a *line*, these Inequalities will shade an entire *region* either *above or below* the line.

You will draw a dotted line for the $<$ and $>$ symbols. You'll draw a solid line for the $\leq$ and $\geq$ symbols.

For example, if you need to graph the inequality $y > 2x + 1$, you would sketch a dotted line with slope of 2 and *y*-intercept of 1. Then shade everything *above* that dotted line, because "y is *greater* than but *not equal* to" the Linear Equation:

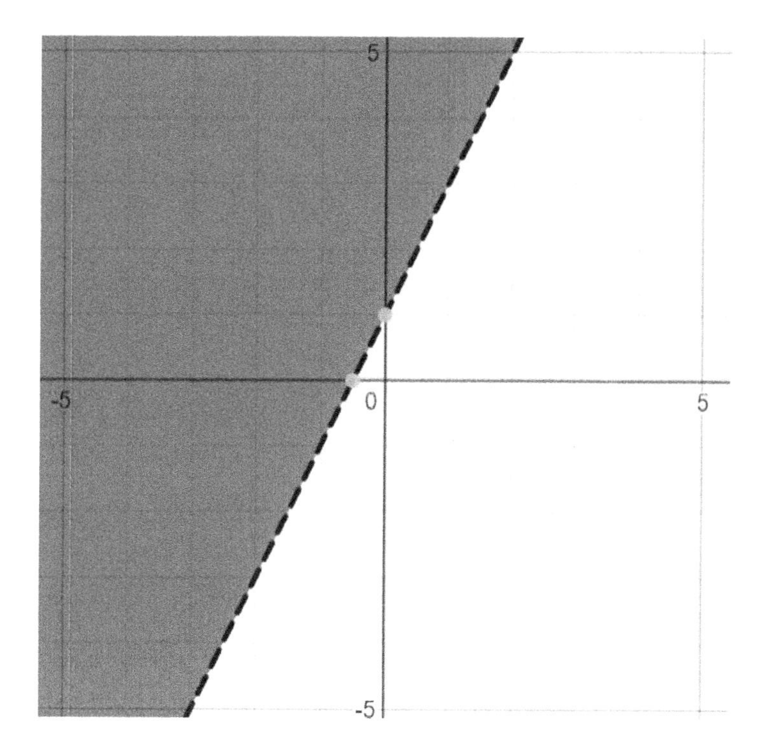

The symbols $<$ and $>$ do *not* include the line itself in their shaded solution region, which is why we graph them with a *dotted line*.

The inequality symbols $\leq$ and $\geq$ *do* include the line itself in the shaded region (remember, these symbols mean "greater/less than *or equal to*", so the line itself *is* included in the solution region).

Here's a graph of the inequality $y \leq \frac{1}{2}x + 2$. Note that we're using a *solid* line for this graph, because "y is less than *or equal to*" the Linear Equation:

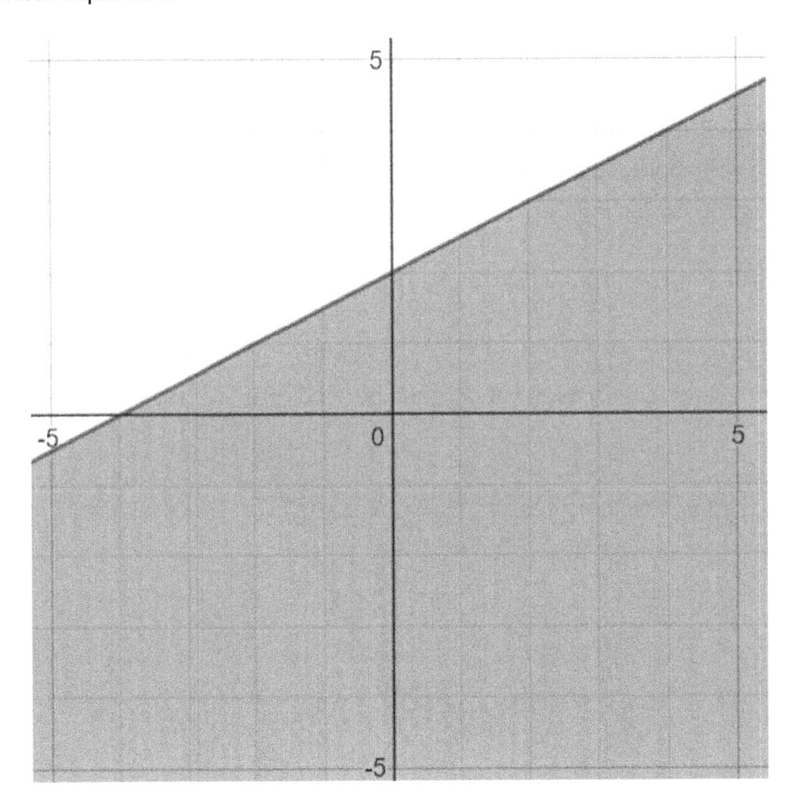

Graphing Multiple Inequalities

In questions where you graph more than one Inequality on the same *xy*-plane, the "solutions" are anywhere that the shaded regions overlap each other.

The graph below shows *both* of our previous Linear Inequalities on the same graph. Note that they overlap in the darkened area to the bottom-left. This overlapping region shows where our solutions to *both* inequalities are.

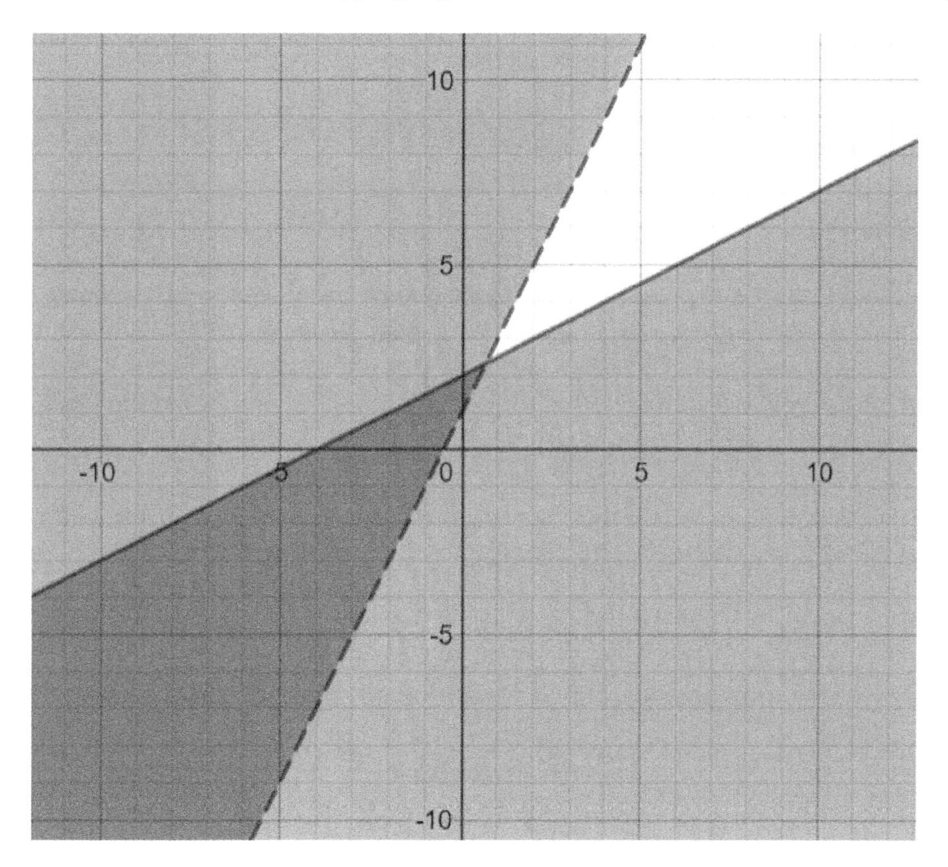

Make sure you're clear on the idea that the solution set includes the *solid* line on the top border of the overlapping dark region, but the solution set does *not* include the *dotted* line on the bottom border of the overlapping region.

Again, this is because the solid line came from a $\leq$ symbol, meaning "less than *or equal* to". The dotted line came from a $>$ symbol, meaning "greater than but *not equal* to."

To tie this back to Quadrants, I'll also point out that our System of Inequalities has solutions found in Quadrants I, II, and III - but has no solutions in Quadrant IV, where the two shaded regions never overlap.

Pretest Question #2

Let's take a look at another Pretest question. Try it yourself before you look at my explanation.

Line A passes through the point $(4,1)$ and is perpendicular to the line containing points $(-2,3)$ and $(6,2)$. Inequality B has the equation $y-2x \leq 6$. Which quadrants do not contain any solutions for Line A and Inequality B?

(A) I only

(B) II only

(C) II and III only

(D) I, III, and IV only

My plan of attack on this question is to gather enough information to be able to sketch a simple graph of the question. To do so, I'll need Linear Equations for both Line A and Inequality B.

Let's start with Line A. As always, to fill out our $y = mx + b$ equation, I'll need to find the slope and the *y*-intercept of the line. Unfortunately, it's going to take a bit of extra work this time.

To find the slope of Line A, I'll need to use the fact that Line A is "perpendicular to the line containing points $(-2,3)$ and $(6,2)$." So, first I'll calculate the slope between points $(-2,3)$ and $(6,2)$:

$$\frac{y_2 - y_1}{x_2 - x_1}$$
$$= \frac{2-3}{6-(-2)}$$
$$= -\frac{1}{8}$$

Now, remember that the slope of any *perpendicular* line will be the *negative reciprocal* of the first slope. What's the negative reciprocal of $-\frac{1}{8}$? Well, take the negative and flip it upside down. It becomes a slope of $+8$.

Now, we have the slope of Line A, and we also know one of the (x, y) points on Line A is $(4,1)$ from the question. Let's put that all together and see what we get:

$$y_A = 8x + b$$
$$1 = 8(4) + b$$
$$1 = 32 + b$$
$$-32 \quad -32$$
$$-31 = b$$

So we know the *y*-intercept of Line A is -31. Now we have enough info to create a complete Linear Equation for Line A:

$$y_A = mx + b$$
$$y_A = 8x - 31$$

Now we'll turn our attention to Inequality B, which also needs a clear Linear Equation so that we can sketch a graph of it. Luckily, it's already pretty close to being ready. The word problem gives the equation as:

$$y - 2x \leq 6$$

It doesn't take much work to put this into the $y = mx + b$ format - just a bit of **Basic Algebra 1**:

$$\begin{aligned} y - 2x &\leq 6 \\ +2x \quad &+2x \\ y_B &\leq 2x + 6 \end{aligned}$$

So, our two equations are:

$$y_A = 8x - 31$$

$$y_B \leq 2x + 6$$

Great - now we can sketch both of our equations using their *y*-intercepts and slopes. Remember to shade *under* the line for Inequality B, because it includes all *y*-values "less than or equal to" $2x + 6$.

I've taken the liberty of using a graphing calculator so you don't have to endure my poor artistic skills, but you could easily sketch a similar graph by hand. It doesn't have to be perfect, just "close enough."

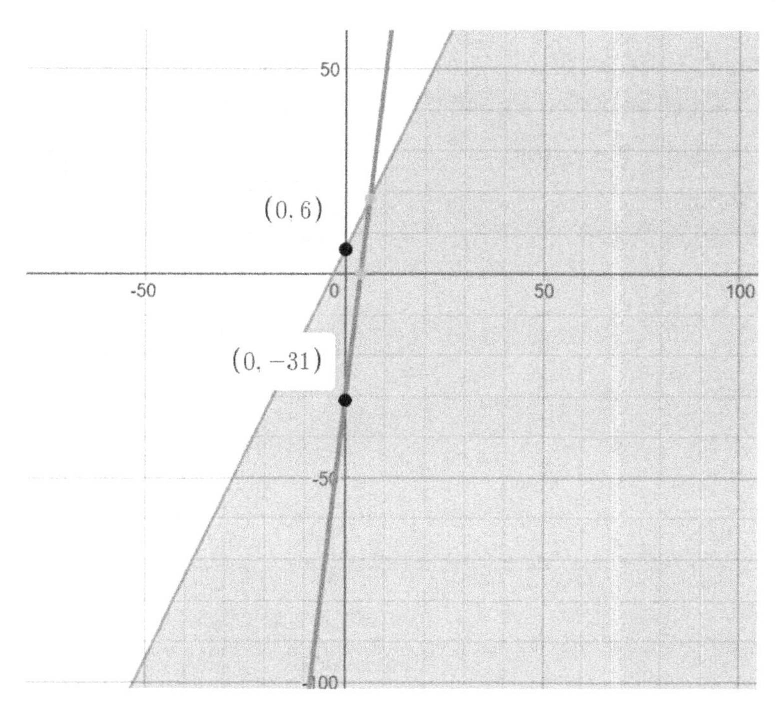

According to the question, we're looking for Quadrants *without* solutions to *both* Inequality B and Line A. A solution exists anywhere the line and the inequality region overlap each other.

Quadrant II definitely doesn't have any solutions, because Line A never even passes through this Quadrant. However, the other three Quadrants all have an overlap of both the inequality region and the Line A.

Therefore, our answer should *not* include Quadrants I, III, and IV. Only Quadrant II has *no* solutions - **Choice B**.

Review & Encouragement

As you can see, there's a lot to know about Linear Equations. After studying this chapter, I'm sure you can imagine why the SAT Math draws upon this concept so frequently.

There are many ways to test us on Linear Equations, from bare-bones-easy to extremely advanced and complex.

In just one lesson, we've covered everything on the list below:

- The $y = mx + b$ equation
- Slopes
- *y*-Intercepts
- Graphing Linear Equations
- Calculating Intersection Points of lines
- Parallel Lines
- Perpendicular Lines
- Quadrants
- Graphing Inequalities
- Solutions to Inequalities

If you take your time studying and absorbing this lesson, you'll be rewarded with more points - and more confidence - on the SAT Math test. Also, several concepts in this lesson will form the foundation for future lessons and more advanced math topics. Make sure you're clear on all of the topics listed above!

Now, work through the entire Practice Set of questions. You'll find a wide variety of question types, difficulty levels, skills and techniques - but all are based around the central concept of the $y = mx + b$ Linear Equation and the other topics we've studied in this lesson.

Linear Equations (Algebraic) Practice Questions

DO NOT USE A CALCULATOR ON ANY OF THE FOLLOWING QUESTIONS UNLESS INDICATED.

1. In the xy-plane, the graph of which of the following equations is a line with a slope of 3?

(A) 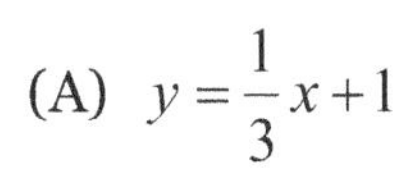 $y = \frac{1}{3}x + 1$

(B) $y = x + 3$

(C) $y = -3x + 3$

(D) $y = 3x + 1$

2. FREE RESPONSE: Line p is shown in the xy-plane below. What is the slope of line p?

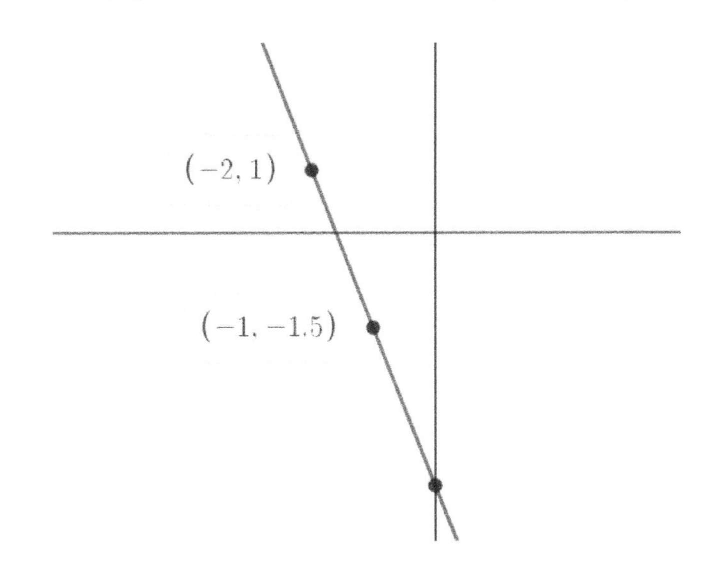

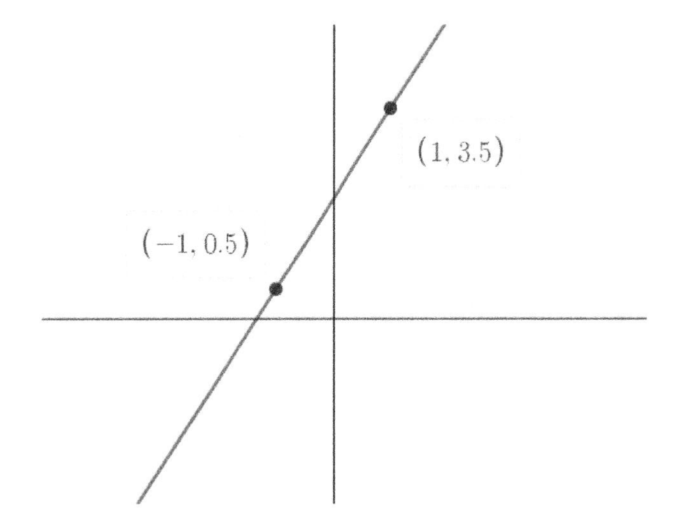

3. Which of the following is an equation of Line l in the xy-plane above?

(A) $y = 2x + \frac{3}{2}$

(B) $y = \frac{3}{2}x + 2$

(C) $y = \frac{3}{2}x - \frac{3}{2}$

(D) $y = 2x - \frac{3}{2}$

4. Which of the following equations relates y to x for the values in the table below?

x	-2	-1	0	1	2
y	-5	-2	1	4	7

(A) $y = -2x + 1$

(B) $y = -2x - 6$

(C) $y = 2x - 1$

(D) $y = 3x + 1$

5. Line w in the xy-plane contains the points $(-5,2)$ and $(0,7)$. Which of the following is an equation of Line w?

(A) $y - 7 = x$

(B) $y - 1 = 7x$

(C) $7y = x + 1$

(D) $y + 7 = x$

$$15x - 5y = -10$$

6. In the xy-plane, the graph of which of the following equation is perpendicular to the graph of the equation above?

(A) $y - 3x = 4$

(B) $-\frac{1}{3}x = 4 - y$

(C) $3y - 4 = -x$

(D) $y + 3x = 4$

7. FREE RESPONSE: Some values of the linear function f are shown in the table below. What is the value of $f(6.5)$?

x	-5	3	11
$f(x)$	-7	9	25

8. The line $y = nx + 2$, where n is a constant, is graphed in the xy-plane. If the line contains the point (a, b), where $a \neq 0$ and $b \neq 0$, what is the slope of the line in terms of a and b?

(A) $\frac{a-b}{2}$

(B) $\frac{a-2}{b}$

(C) $\frac{b-a}{2}$

(D) $\frac{b-2}{a}$

$$-4x - 3y = 12$$

9. In the xy-plane, the graph of which of the following equation is parallel to the graph of the equation above?

(A) $y = -\frac{4}{3}x - 6$

(B) $y = -\frac{3}{4}x - 2$

(C) $y = -\frac{3}{4}x + 8$

(D) $y = \frac{4}{3}x + 2$

10. A line in the xy-plane has a slope of $-\frac{9}{2}$ and passes through the origin. Which of the following points lies on the line?

(A) $(0,9)$

(B) $(2,9)$

(C) $(-2,-9)$

(D) $(-4,18)$

11. Line q in the xy-plane contains points from each of the Quadrants I, II, and III, but no points from Quadrant IV. Which of the following must be true?

(A) The slope of line q is zero.

(B) The slope of line q is positive.

(C) The slope of line q is negative.

(D) The slope of line q is undefined.

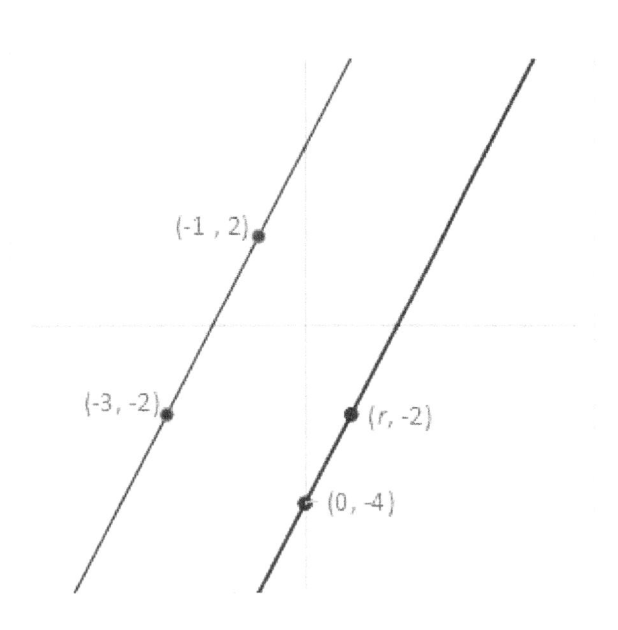

12. FREE RESPONSE: In the xy-plane above, line t is parallel to line g. What is the value of r?

$$y < 2x + 1$$
$$y > 3$$

13. When the system of inequalities above is graphed in the xy-plane, which quadrant contains solutions to the system?

(A) I

(B) II

(C) III

(D) IV

14. FREE RESPONSE: The graph of a line in the xy-plane crosses the x-axis at x-coordinate -2 and passes through the point $(5,3)$. The line crosses the y-axis at the point $(0, g)$. What is the value of g?

15. FREE RESPONSE: The line with the equation $\frac{2}{3}x - \frac{5}{4}y = 1$ is graphed in the xy-plane. What is the x-coordinate of the x-intercept of the line?

16. (CALCULATOR) In the *xy*-plane, the line containing points $(-4, k)$ and $(k, -36)$ passes through the origin. Which of the following could be the value of k?

(A) 4

(B) 6

(C) 12

(D) 18

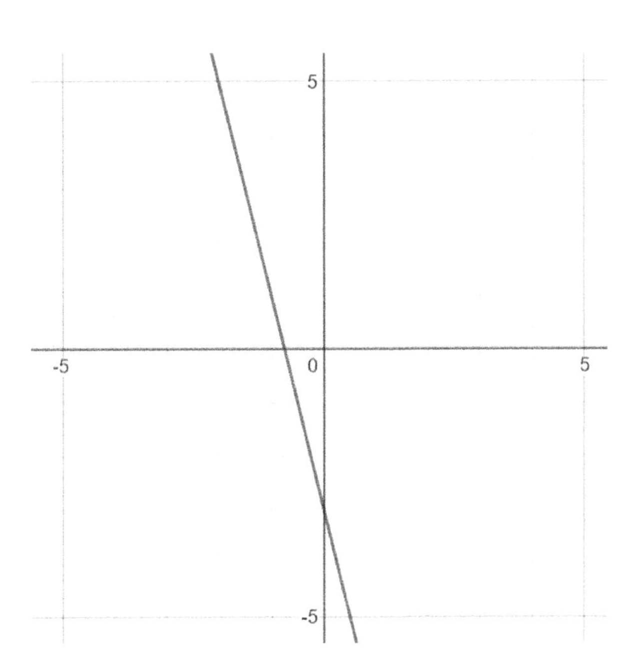

17. FREE RESPONSE: The graph of linear function f is shown in the *xy*-plane above. The graph of the linear function g (not shown) is perpendicular to the graph of f and passes through the point $(-2,1)$. What is the value of $g(-4)$?

$$y < -x$$
$$y > -.5x + 1$$

18. When the system of inequalities above is graphed in the *xy*-coordinate plane, which of the quadrants contains solutions to the system?

(A) I

(B) II

(C) III

(D) IV

19. (CALCULATOR) FREE RESPONSE: In the *xy*-plane, line t has an *x*-intercept of -40 and is perpendicular to the line with equation $y + 2 = \frac{3}{2}x$. If the point $(27.5, -n)$ is on line t, what is the value of n?

$$2x + y = 6$$
$$8 - 2y = ax$$

20. FREE RESPONSE: If there are no solutions to the system of equations above, what is the value of a?

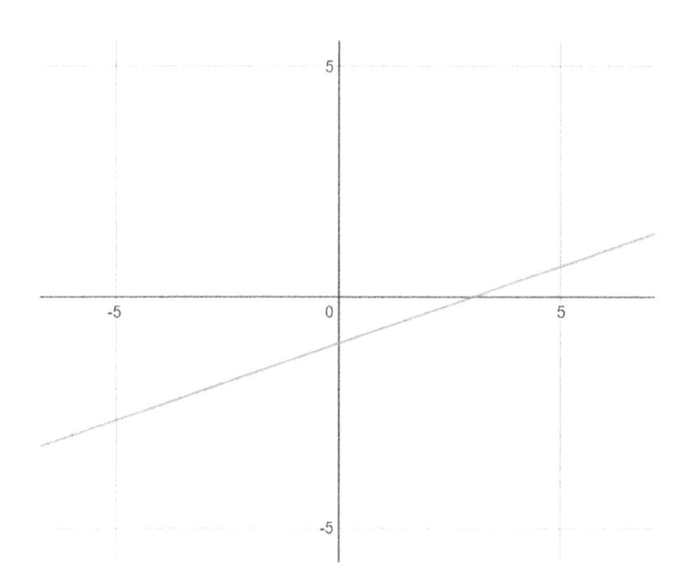

21. The graph of the linear function f is shown in the xy-plane above. The slope of the graph of the linear function g is 3 times the slope of the graph of f. If the graph of g passes through the point $(-3, 6)$, what is the value of $g(15)$?

$$tx + y = 5$$
$$-c = y - 5x$$

22. FREE RESPONSE: If the system of equations above has infinitely many solutions, what is the value of $t - 4c$?

Linear Equations (Algebraic) Answers

1. D
2. -2.5 or $-\frac{5}{2}$
3. B
4. D
5. A
6. C
7. 16
8. D
9. A
10. D
11. B
12. 1
13. A
14. $\frac{6}{7}$ or .857
15. $\frac{3}{2}$ or 1.5
16. C
17. $\frac{1}{2}$ or .5
18. B
19. 45
20. 4
21. 24
22. 15

Linear Equations (Algebraic) Explanations

1. **D.** This question is extremely basic. We know that in the Linear Equation $y = mx + b$, the value of m is the slope. To find a line with a slope of 3, we need to pick the only answer choice that includes the term "$3x$", which is Choice D.

2. **-2.5** or $-\frac{5}{2}$. We're just warming up, for now. To calculate slope between two points, we use the formula $\frac{y_2 - y_1}{x_2 - x_1}$.

Go ahead and plug in the coordinates from our two points. Remember, with this formula it doesn't matter which point you designate as "point 1" or "point 2," just remember to remain *consistent*. To be safe, it always helps to label or write down which coordinates you're calling "point 1" and "point 2" before you plug them into the equation.

Here's how I'll do it:

Point 1: $(-2,1)$ and Point 2: $(-1,-1.5)$

$$\frac{y_2 - y_1}{x_2 - x_1}$$
$$= \frac{(-1.5)-(1)}{(-1)-(-2)}$$
$$= \frac{-2.5}{-1+2}$$
$$= \frac{-2.5}{1}$$
$$= -2.5$$

So, the slope of this line is "-2.5" or, if you prefer, $-\frac{5}{2}$.

3. **B.** For this question, we need to provide a complete equation of the form $y = mx + b$. That means we'll need to know both the slope and the *y*-intercept.

Let's start with the slope, since we can easily calculate that from the two points that we've started with. Use the equation $\frac{y_2 - y_1}{x_2 - x_1}$. I'll use $(-1, .5)$ as Point 1 and $(1, 3.5)$ as point 2.

$$\frac{(3.5)-(.5)}{(1)-(-1)}$$
$$= \frac{3}{1+1}$$
$$= \frac{3}{2}$$

OK, good. Now we know that the slope of this line is 1.5 or $\frac{3}{2}$. But how to find the *y*-intercept?

The trick is to set up our $y = mx + b$ with the information we currently have, then plug in either one of the two (x, y) points given on the graph. I'll plug in the point $(1, 3.5)$:

$$y = mx + b$$
$$y = \frac{3}{2}x + b$$
$$3.5 = \frac{3}{2}(1) + b$$

Now solve for b, the *y*-intercept value:

$$3.5 = \frac{3}{2}(1) + b$$
$$3.5 = 1.5 + b$$
$$-1.5 \quad -1.5$$
$$2 = b$$

With our slope and *y*-intercept calculated, we can now finish our $y = mx + b$ Linear Equation:

$$y = mx + b$$
$$y = \frac{3}{2}x + 2$$

4. **D.** For this question, we have two options. First, we could calculate the slope from the given x and y values, then use the $y = mx + b$ equation to solve for our *b*-value.

But to be honest, it will be faster to test the answer choices by plugging in values from the table into the provided answer choices, and eliminating an answer choice whenever it doesn't work.

Let's start by plugging in $x = 0$. The table tells us that any resulting *y*-value must equal 1.

If you plug in to Choice A, you'll get $y = 1$ - which is what we wanted - so we'll keep this option. Plug 0 into Choice B and you get -6, which isn't what we want. Plug into Choice C and you get -1, which isn't what we want either. However, when we plug into Choice D, we also get $y = 1$, which is the correct value.

Now we know that only Choices A and D are possibly correct. Let's pick another pair of values from the table and try again. I'll use $x = 1$, and the table says we're supposed to get $y = 4$ as a result.

In Choice A, when I plug in 1, I get -1, which isn't what we want. When I plug into Choice D, I get 4, which is what I'm looking for. By process of elimination, Choice D must be correct.

In this type of question, it's faster and easier just to test the answer choices. The reason I used $x = 0$ and $x = 1$ is that these are very quick and simple values to plug into the equations given in the answer choices.

5. **A.** To give the equation of a line, we need to create a $y = mx + b$ equation. To do that, we need the slope and the *y*-intercept. In this case, it's easiest to find the slope first. We can calculate slope using $\frac{y_2 - y_1}{x_2 - x_1}$. Let's set that up:

$$\begin{aligned} &\frac{y_2 - y_1}{x_2 - x_1} \\ &= \frac{7-2}{0-(-5)} \\ &= \frac{5}{5} \\ &= 1 \end{aligned}$$

Now we know the slope of the line is 1. The next step is to set up everything we know about this line's equation, plug in one of the two points we're given, and calculate the *b*-value (the *y*-intercept) from that setup.

I'll use the point $(-5, 2)$ for x and y, but you can also use the point $(0, 7)$ if you want:

$$\begin{aligned} y &= 1x + b \\ 2 &= 1(-5) + b \\ 2 &= -5 + b \\ +5 & \quad +5 \\ 7 &= b \end{aligned}$$

Now we know our *b*-value, and can finish setting up our $y = mx + b$ equation:

$$\begin{aligned} y &= mx + b \\ y &= 1x + 7 \end{aligned}$$

Notice that the answer choices have been slightly rearranged, but it only takes a few steps of **Basic Algebra 1** to realize that Answer Choice A is the same equation as the one we're looking for.

6. **C.** This Linear Equation question focuses on a pair of "perpendicular" lines. Remember that the slope values will be *negative reciprocals* of each other. Therefore, if we know the slope of one line, we can easily calculate the slope of the other line.

The first thing to do is take the given equation and rearrange it into a more useful format. Let's do some **Basic Algebra 1** to convert the given $15x - 5y = -10$ equation into $y = mx + b$ format. This will tell us the slope of the first line:

$$\begin{aligned} 15x - 5y &= -10 \\ -15x \qquad & \quad -15x \\ -5y &= -15x - 10 \\ \frac{-5y}{-5} &= \frac{-15x - 10}{-5} \\ y &= 3x + 2 \end{aligned}$$

OK, wonderful. We can see that the slope of the original line is 3. Now, we can find the slope of the perpendicular line with the negative reciprocal - the negative, upside-down version of 3 - which is $-\frac{1}{3}$.

Now, STOP for a second. Do you see how Choice B gives $-\frac{1}{3}x$ as part of the answer? It's a *trap*. None of these Answer Choices are currently in $y = mx + b$ form, which means they can't be trusted until we rearrange the equations into the proper format.

So, now's the time to go down the list of answer choices, rearranging them each into $y = mx + b$ form, and looking for the one line with a slope of $-\frac{1}{3}$.

Here are my conclusions. Choice A rearranges to $y = 3x + 4$, which does *not* have a slope of $-\frac{1}{3}$. Choice B rearranges to $y = \frac{1}{3}x + 4$, which does *not* have a slope of $-\frac{1}{3}$. Choice C rearranges to $y = -\frac{1}{3}x + \frac{4}{3}$, which *does* have a slope of $-\frac{1}{3}$. Choice D rearranges to $y = -3x + 4$, which does *not* have a slope of $-\frac{1}{3}$.

Note that, throughout our reasoning, we did *not* put any focus on the *y*-intercept value. That's because *y*-intercepts are not related to perpendicular lines; perpendicular lines are only about slope.

7. **16.** OK, this question has a lot in common with Question 4. However, this time we aren't given any multiple choice answers, so we'll have to actually figure out the $y = mx + b$ Linear Equation for this table.

We can treat $f(x)$ values exactly like *y*-values (study the upcoming lesson on **Functions** for more on this topic). So, where the table says $f(x)$, just imagine it says "y" instead.

Let's start by calculating the slope, since we have several coordinates to draw from. Use the $\frac{y_2 - y_1}{x_2 - x_1}$ equation. I'll select $(3,9)$ as my Point 1, and $(-5,-7)$ as my Point 2.

$$\begin{aligned} &\frac{y_2 - y_1}{x_2 - x_1} \\ &= \frac{(-7)-(9)}{(-5)-(3)} \\ &= \frac{-16}{-8} \\ &= 2 \end{aligned}$$

Alright, now we've found that this linear function has a slope of 2.

Now use the $y = mx + b$, plug in the slope of 2, and then plug in any of the three (x, y) coordinates given in the table. I'll use $(3,9)$.

$$\begin{aligned} y &= mx + b \\ y &= 2x + b \\ 9 &= 2(3) + b \\ 9 &= 6 + b \\ -6 & \quad -6 \\ 3 &= b \end{aligned}$$

Great - now we know the *y*-intercept and can complete our $y = mx + b$ equation:

$$y = 2x + 3$$

The final step is to evaluate $f(6.5)$. This means to plug $x = 6.5$ into our Linear Equation. Again, be sure to study the lesson on **Functions** if this notation form is confusing or unclear to you.

$$\begin{aligned} y &= 2x + 3 \\ y &= 2(6.5) + 3 \\ y &= 13 + 3 \\ y &= 16 \end{aligned}$$

8. **D.** This question provides us with a Linear Equation in the form $y = nx + 2$, and asks us to give the slope. Instead of using numbers, the Answer Choices are mainly based on variables. How to make sense of it?

The first step may seem strange to you: we're going to plug in the point (a,b) for our *x*- and *y*-values:

$$\begin{aligned} y &= nx + 2 \\ b &= na + 2 \end{aligned}$$

I know it may seem strange to replace two letters with two more letters, but the point (a,b) is on our line, which makes it valid to plug into our Linear Equation for x and y.

We've been asked to solve for the slope of the line. Now, in the original equation $y = nx + 2$, where do we find the slope of the line? It's the coefficient of x, of course - in this case, that would be the letter n.

So, we'll take our equation and solve for n:

$$b = na + 2$$
$$-2 \qquad -2$$
$$b - 2 = na$$
$$\frac{b-2}{a} = \frac{na}{a}$$
$$\frac{b-2}{a} = n$$

And we're done. The slope of this line can be expressed as $\frac{b-2}{a} = n$, or Choice D.

By the way, the reason the question includes the warnings that $a \neq 0$ and $b \neq 0$ is simply to prevent any of our answer choices from dividing by zero, which is never allowed in math. It's basically info that we can ignore in this question.

9. **A.** This question tasks us with finding a "parallel" line to an existing line. Remember that *parallel* lines always have *equal* slopes. First, let's find the slope of the original line by putting it into $y = mx + b$ form:

$$-4x - 3y = 12$$
$$+4x \qquad +4x$$
$$-3y = 4x + 12$$
$$\frac{-3y}{-3} = \frac{4x+12}{-3}$$
$$y = -\frac{4}{3}x - 4$$

Now we can read the original line as having a slope of $-\frac{4}{3}$. So, our new line must also have the same slope. Luckily, it's easy to find: only Choice A has a slope of $-\frac{4}{3}$.

10. **D.** There are two options for solving this problem. The "low-tech" way is to basically count on your fingers and eliminate wrong answers. We start at the origin, point $(0,0)$, as given in the question. This eliminates Choice A, because if our line is already at $(0,0)$, it can't also be at $(0,9)$ at the same time. Then imagine tracking your line from left to right. Our line has a slope of $-\frac{9}{2}$, so it will go *down* 9 units for every 2 units we move to the right. That eliminates Choice B, which went *up* 9 units when we moved to the right by 2 units.

Now return to $(0,0)$ and start moving right to left. When we move left by 2 units, the line will go *up* by 9 units. That eliminates Choice C. But, if we move to the left by 4 units, the line will go up 18 units, which puts us right at point $(-4,18)$.

The other way to solve this question is using the $y = mx + b$ form. Plug in the values we know: slope is $-\frac{9}{2}$ and the line contains the point $(0,0)$ - which means the *y*-intercept is 0:

$$y = mx + b$$
$$y = -\frac{9}{2}x + 0$$
$$y = -\frac{9}{2}x$$

Now you could test each of the answer choices by plugging them into this formula. If the point does not return a true equality when plugged in, then eliminate it. You'll again find that only Choice D returns a true equality.

11. **B.** Probably the easiest way to deal with this question is to play around with sketching a few possible graphs. First draw a simple *xy*-axis. Then place a point in each of Quadrants I, II, and III - but not in Quadrant IV (review the section of the lesson on Quadrants if you need a reminder of which is which).

Now try to sketch a single straight line that passes relatively close to each of your three points, without dipping into Quadrant IV at all. It may be difficult or impossible to make your straight line pass exactly through all three points, but that's OK. You're just using them as a guide for your sketch.

Within a couple of tries you'll realize that only lines with *positive slopes* can pass through these three quadrants.

12. **1.** Two parallel lines will always have equal slopes. Therefore, let's calculate the slope of the upper line, which has provided us with exact coordinates for two points. I'll use $(-3,-2)$ as Point 1 and $(-1,2)$ as Point 2:

$$\frac{y_2 - y_1}{x_2 - x_1}$$
$$= \frac{(2)-(-2)}{(-1)-(-3)}$$
$$= \frac{2+2}{-1+3}$$
$$= \frac{4}{2}$$
$$= 2$$

The slope of the upper line must be 2. OK, now we can set up another slope equation for the lower line. We know the value of this slope must also equal 2. We will also use the unknown *x*-coordinate r in our slope equation, since we haven't figured out its value yet. Then solve for r using **Basic Algebra 1**:

$$\frac{y_2 - y_1}{x_2 - x_1} = 2$$
$$\frac{(-2)-(-4)}{(r)-(0)} = 2$$
$$\frac{-2+4}{r-0} = 2$$
$$\frac{2}{r} = 2$$
$$(r)\frac{2}{r} = 2(r)$$
$$2 = 2r$$
$$\frac{2}{2} = \frac{2r}{2}$$
$$1 = r$$

There we go - the value of r must be 1.

13. **A.** The safest way to solve this problem is probably to actually draw it out. Sketch an *xy*-coordinate plane with around 5 tick marks per "arm" of the axes. It's easy to graph $y > 3$. Simply draw a horizontal dotted line at $y = 3$, then lightly shade everything above it, like this:

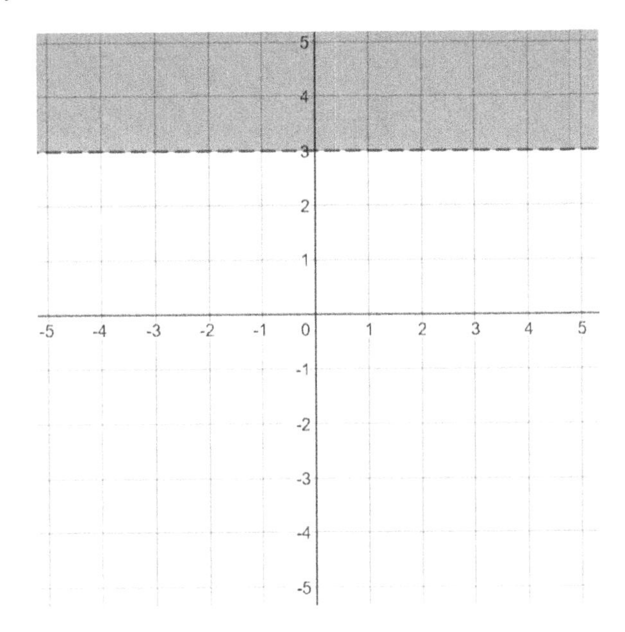

Next, we'll add the line for $y < 2x + 1$. This will have a *y*-intercept at 1, and a slope of 2. Sketch this line (it should be a dotted line), then shade *under* it. It looks something like this:

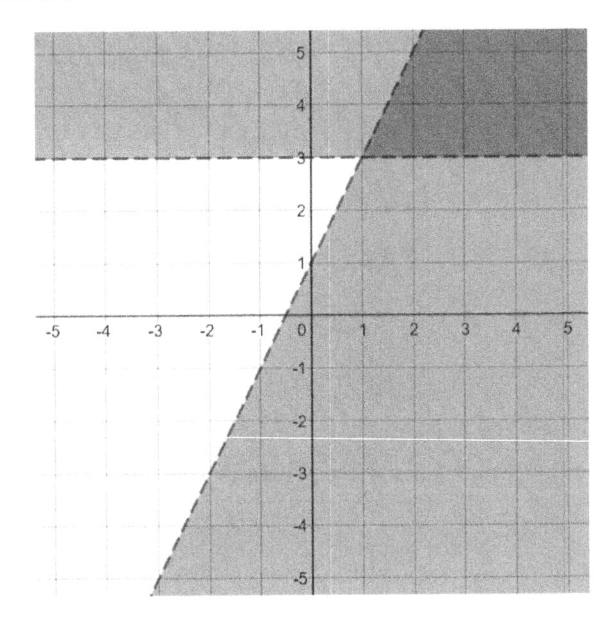

Almost done. Now that we've sketched the situation, we can find the region where *both* inequalities have shading. That's in the upper-right corner of our graph, in Quadrant I.

14. **.857** or $\frac{6}{7}$. The first step of this question - like most others in this lesson - is to set up the $y = mx + b$ equation. But before we can do that, we have to determine the slope of the line. It may seem impossible at first, because there only seems to be the point $(5,3)$ that gives us complete values to work with.

Luckily, if you think carefully, we also have another point at $(-2,0)$. This is where the line "crosses the *x*-axis," which means it must have a *y*-coordinate of 0.

Let's calculate the slope using the points $(5,3)$ as Point 1 and $(-2,0)$ as Point 2.

$$\frac{y_2 - y_1}{x_2 - x_1}$$
$$= \frac{(0)-(3)}{(-2)-(5)}$$
$$= \frac{-3}{-7}$$
$$= \frac{3}{7}$$

OK, now we know the slope is $\frac{3}{7}$:

$$y = \frac{3}{7}x + b$$

Now, remember that we're solving for g, which comes from the coordinates $(0, g)$. Notice that this point happens to be the *y*-intercept since the x-coordinate is 0, so we can plug g directly in for b:

$$y = \frac{3}{7}x + b$$
$$y = \frac{3}{7}x + g$$

Now, let's use our point $(5,3)$ to plug in for x and y, then solve for g. We'll also need our knowledge of **Fractions** to finish the subtraction in the last few steps.

$$y = \frac{3}{7}x + g$$
$$3 = \frac{3}{7}(5) + g$$
$$3 = \frac{15}{7} + g$$
$$-\frac{15}{7} - \frac{15}{7}$$
$$3 - \frac{15}{7} = g$$
$$\frac{21}{7} - \frac{15}{7} = g$$
$$\frac{6}{7} = g$$

15. **1.5** or $\frac{3}{2}$. Solving this question won't be as hard as it might look. We know that an *x*-intercept is where the line crosses the *x*-axis. Therefore, any *x*-intercept will always have a *y*-coordinate of 0. Let's represent this desired coordinate as $(X,0)$ where X represents the *x*-coordinate we're looking for. Now, let's try plugging this coordinate into the given equation for the line:

$$\frac{2}{3}x - \frac{5}{4}y = 1$$
$$\frac{2}{3}X - \frac{5}{4}(0) = 1$$
$$\frac{2}{3}X - 0 = 1$$
$$\frac{2}{3}X = 1$$
$$(\frac{3}{2})\frac{2}{3}X = 1(\frac{3}{2})$$
$$X = \frac{3}{2}$$

And we're done! This line crosses the *x*-axis at $(\frac{3}{2}, 0)$ and our final answer is $\frac{3}{2}$ or 1.5.

16. **C.** To solve the question, we have to have a crucial realization: although it seems like there are only two given points, there's actually a *third* point hiding within the words: this line passes through the origin, which means it also has the point $(0,0)$.

So, how can we use this? Instead of creating a $y = mx + b$, this time we'll focus only on the Slope equation $\frac{y_2 - y_1}{x_2 - x_1}$.

Since all three points are on the same line, the slope between all three points should be exactly the same. So, we can set up *two* Slope equations and set them equal to each other. Something like this:

$$\frac{y_2 - y_1}{x_2 - x_1} = \frac{y_3 - y_1}{x_3 - x_1}$$

Notice that I've adjusted the right side of the equation to use a *third* point. Let's call $(-4, k)$ Point 1, we'll make $(k, -36)$ Point 2, and $(0,0)$ will be Point 3. Plus them all into the equation above and solve for k, using techniques we learned in the first two lessons on **Algebra 1**.

$$\frac{(-36)-(k)}{(k)-(-4)} = \frac{(0)-(k)}{(0)-(-4)}$$
$$\frac{-36-k}{k+4} = \frac{0-k}{0+4}$$
$$\frac{-36-k}{k+4} = \frac{-k}{4}$$
$$4(-36-k) = -k(k+4)$$
$$-144 - 4k = -k^2 - 4k$$
$$+k^2 + 4k \quad +k^2 + 4k$$
$$k^2 - 144 = 0$$
$$+144 \quad +144$$
$$k^2 = 144$$
$$\sqrt{k^2} = \sqrt{144}$$
$$k = 12$$

17. $\frac{1}{2}$ or **.5.** As always, the word "perpendicular" in the context of a Linear Equation question means that we'll use one slope to calculate the negative reciprocal slope of the perpendicular line.

So, first we'll need the original slope of function f, shown in the graph. It may seem low-tech, but we're just going to measure rise over run on the gridlines.

Notice the line for f passes directly through coordinates $(-1,1)$ and $(0,-3)$. We can calculate the slope by using the $\frac{y_2 - y_1}{x_2 - x_1}$ equation. Let's do so now:

$$\frac{(-3)-(1)}{(0)-(-1)}$$
$$= \frac{-3-1}{0+1}$$
$$= \frac{-4}{1}$$
$$= -4$$

Now we know that the slope of our original line for function f is -4. To calculate the slope of function g, we'll take the negative reciprocal of -4, which is $\frac{1}{4}$.

We're making progress. Now we'll set up a $y = mx + b$ Linear Equation for function g and plug in the slope, along with the point $(-2,1)$, which the question tells us is a point on the line for g. This will allow us to solve for the *y*-intercept or *b*-value of function g:

$$y = mx + b$$
$$y = \frac{1}{4}x + b$$
$$1 = \frac{1}{4}(-2) + b$$
$$1 = -\frac{1}{2} + b$$
$$+\frac{1}{2} \quad +\frac{1}{2}$$
$$\frac{3}{2} = b$$

Now we know both the slope and the *y*-intercept of the Linear Equation for function g, so let's put that all together:

$$y = mx + b$$
$$y = \frac{1}{4}x + \frac{3}{2}$$

And now, we can finish the question, which asks for the value of $g(-4)$. That means to plug in -4 for x into the equation for function g that we've just created (the lesson on **Functions** contains additional practice on this topic). Then, finish evaluating to get the final value of $g(-4)$:

$$y = \frac{1}{4}x + \frac{3}{2}$$
$$y = \frac{1}{4}(-4) + \frac{3}{2}$$
$$y = -1 + \frac{3}{2}$$
$$y = \frac{1}{2}$$

It may have taken a lot of work, but now we know the value of $g(-4)$ is $\frac{1}{2}$ - or in other words, when you plug $x = -4$ into the equation, you get $y = \frac{1}{2}$.

18. **B.** Like Question 13, the easiest thing to do here is actually graph the two inequalities. By this point you should be comfortable graphing equations based on $y = mx + b$.

The first line has a *y*-intercept of 0 and a slope of -1. It will be a dotted line, with shading *underneath* it.

The second line has a *y*-intercept of 1, and a slope of $-\frac{1}{2}$. It will also be a dotted line, with shading *above* it.

The result looks like this:

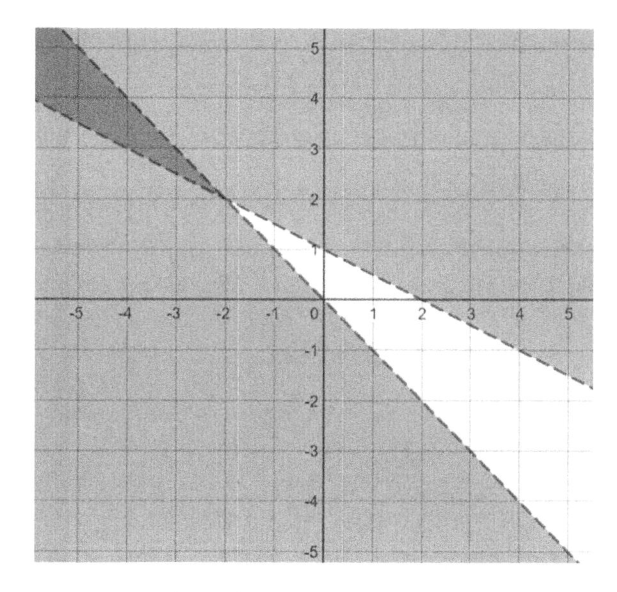

As you can see, the only solutions (where both shaded regions overlap) are in Quadrant II.

19. **45.** As with all Linear Equations questions involving a "perpendicular" line, we'll need to calculate the slope of the first line and then take the negative reciprocal to get the slope of the perpendicular line. Let's use the equation $y + 2 = \frac{3}{2}x$ and put it into $y = mx + b$ form:

$$y + 2 = \frac{3}{2}x$$
$$-2 \qquad -2$$
$$y = \frac{3}{2}x - 2$$

Now we can see that the slope of the first line is $\frac{3}{2}$. That means the perpendicular line t will have a slope of $-\frac{2}{3}$.

Next we'll set up a $y = mx + b$ equation for line t.

$$y = mx + b$$
$$y = -\frac{2}{3}x + b$$

We also know that line t passes through the x-intercept of -40, which has the coordinates $(-40, 0)$. We can plug these coordinates into the equation for line t, then solve for the *y*-intercept or *b*-value:

$$y = -\frac{2}{3}x + b$$
$$0 = -\frac{2}{3}(-40) + b$$
$$0 = \frac{80}{3} + b$$
$$-\frac{80}{3} \quad -\frac{80}{3}$$
$$-\frac{80}{3} = b$$

It's an ugly fraction, but now we know the *y*-intercept of line t is $-\frac{80}{3}$. Put this together with the slope to complete our Linear Equation for line t.

$$y = -\frac{2}{3}x - \frac{80}{3}$$

Last but not least, finish the question by plugging in the point $(27.5, -n)$:

$$-n = -\frac{2}{3}(27.5) - \frac{80}{3}$$
$$-n = -\frac{55}{3} - \frac{80}{3}$$
$$-n = -\frac{135}{3}$$
$$-n = -45$$
$$n = 45$$

20. **4.** This question is an interesting sort of challenge that you only really see on the SAT Math test. I had never seen a question like this in my high school classes, for example - and most of my students feel the same way. Now, it does also include a System of Equations, which means there's more than one way to solve it (you can study the upcoming lesson on **Systems of Equations** for more information).

But here's why this question relates to Linear Equations, and why I think it belongs in this chapter.

Now, follow me through on some logic. When you graph two Linear Equations, what does their solution look like on the graph?

It's the point where they intersect, right?

So, if two Linear Equations have *no* solution, what does that mean? It means that they *never* intersect.

And how do you get two lines that never intersect?

That's right, they must be *parallel* to each other.

And if two lines are parallel, what else do you know about them?

Yup - they have the exact same slopes.

So, what we'll do is put both of the given equations into $y = mx + b$. Then we'll take their slopes and set them equal to each other, and see what happens.

First, let's get them both into $y = mx + b$ form.

$$2x + y = 6$$
$$-2x \qquad -2x$$
$$y = -2x + 6$$

$$8 - 2y = ax$$
$$-8 \qquad -8$$
$$-2y = ax - 8$$
$$\frac{-2y}{-2} = \frac{ax - 8}{-2}$$
$$y = -\frac{a}{2}x + 4$$

Now, compare the slopes of these two equations. The first one has a slope of -2. The second one has a slope of $-\frac{a}{2}$. Set those two slopes equal to each other, since the lines are parallel, then solve for a:

$$-2 = -\frac{a}{2}$$
$$(-2) - 2 = -\frac{a}{2}(-2)$$
$$4 = a$$

There we go: the value of a is 4.

Remember the mental process and tricks we used for this question with "no solutions"! This type of question *is* tested on the SAT - and like I said, most students have never seen it before. But, it's quite simple when you realize how "no solutions" relates to Linear Equations, parallel lines, and equal slopes.

21. **24.** This is a multistage Linear Equation problem, but it uses all the same concepts we've been practicing throughout this lesson.

We're being asked for a value of $g(15)$, which is a function - so, this overlaps a bit with the upcoming lesson on **Functions**. But, we can easily handle it with Linear Equations if you understand that $g(15)$ just means "plug $x = 15$ into the equation for g."

Before we can do that, we'll need to actually *find* the equation for g. Since it's a linear equation, we'll have to find the slope and the *y*-intercept.

To find the slope for function g, we'll have to use "3 times the slope of the graph of f." But that's easy - we can just use the given graph to pick two points from the graph and calculate the slope of f.

Let's use the points $(0,-1)$ and $(3,0)$ since the line goes cleanly through those points:

$$\frac{(0)-(-1)}{(3)-(0)}$$
$$= \frac{1}{3}$$

The slope of line f is $\frac{1}{3}$. Since the slope of line g is "3 times that", we know that the slope of line g must be $3(\frac{1}{3})$, which equals a slope of 1.

Now, how to find the *y*-intercept for g? Let's set up the $y = mx + b$ for function g with as much information as we have:

$$y = mx + b$$
$$y = 1x + b$$

The question also gives us an (x, y) coordinate $(-3, 6)$ that g passes through. Plug that in for x and y:

$$y = 1x + b$$
$$6 = 1(-3) + b$$
$$6 = -3 + b$$
$$+3 \quad +3$$
$$9 = b$$

OK, now we know the *y*-intercept for function g is 9, and the slope is 1, so we can finish our $y = mx + b$ setup for g:

$$y = 1x + 9$$

Now finish the question off by plugging in 15 for x, since we were asked for $g(15)$:

$$y = 1x + 9$$
$$y = 1(15) + 9$$
$$y = 15 + 9$$
$$y = 24$$

Finished! The value of $g(15)$ is 24.

22. **15.** Compare this question to Question 20. It's almost exactly the same, with one difference: instead of having "*no* solutions," this system has "*infinitely many* solutions." Let's think this through one more time.

What does the solution to a pair of lines look like on a graph?

Right, it's the intersection point where both lines cross.

What does it mean if a pair of lines have *infinitely many* solutions?

It means they cross at infinite points!

How do two lines cross at infinite points? Well, only if they're the *exact same line*.

Therefore, these two equations must be the same line with the same $y = mx + b$ equations.

First, let's put them both into $y = mx + b$ form:

$$tx + y = 5 \qquad\qquad -c = y - 5x$$
$$-tx \quad\quad -tx \qquad\qquad +5x \quad\quad +5x$$
$$y = -tx + 5 \qquad\qquad 5x - c = y$$

So, these two equations are the same as each other. Set them equal:

$$-tx + 5 = 5x - c$$

Now compare them. They must have the same slopes. Therefore, $-t = 5$, so $t = -5$.

Since they're the same line, they must also have the same *y*-intercepts. Therefore, $5 = -c$, so $c = -5$.

The question asked for the value of $t - 4c$. If we plug in our values, we get:

$$t - 4c$$
$$= (-5) - 4(-5)$$
$$= -5 + 20$$
$$= 15$$

Finished! $t - 4c$ must equal 15.

Remember, if two Linear Equations have "*no* solutions", they must be parallel lines and have the same slopes. If two Linear Equations have "*infinitely-many* solutions", they must be the *exact same lines*, with the same slopes *and* the same *y*-intercepts.

Lesson 10: Linear Equations (Words & Tables)

Percentages

- 8.6% of Whole Test
- 8% of No-Calculator Section
- 8.9% of Calculator Section

Prerequisites

- Linear Equations (Algebraic)
- Algebra 1 Word Problems
- Basic Algebra 1

In the last lesson, we studied the essentials of Linear Equations, but we're not done with them yet. The SAT Math test includes an enormous amount of Linear Equation questions.

Just *look* at the massive percentage of questions that Linear Equation problems represent - almost 10% of the entire SAT Math test from this chapter, with another 5% from the last chapter!

In the previous chapter we explored the algebraic basics of Linear Equations. In this chapter, we'll be exploring Linear Equations based on word problems, charts, and tables.

All the essential groundwork and math concepts were already laid out in the last chapter. This time, we'll just be working through Linear Equation questions that convey their information through real-life situations, word problems, charts and tables.

Linear Equations (Words & Tables) Quick Reference

- The SAT Math frequently tests Linear Equations in word problems, charts & tables, based around "real-world" situations.
- Most of the math concepts in this lesson come directly from the previous chapter.
- These questions can be recognized by key phrases like "at a constant rate," "linear," "line of best fit," etc. You may also notice graphs of straight lines or answer choices in $y = mx + b$ equation formats.
- In most questions, you will either *create* a $y = mx + b$ equation from words, graphs, or tables, or *interpret* a Linear Equation and explain its relationship to a real-world situation.
- Sometimes, testing the answer choices is faster than creating an equation from scratch.

Word Problems and Linear Equations

In this lesson, we're studying Word Problems based on Linear Equations. This chapter also includes Linear Equations that involve Charts and Tables.

These are "information-rich" problems with detailed Word Problems that translate mathematically into Linear Equations (compared to the last chapter, which was based on abstract & algebraic Linear Equations with no connection to real-world situations or word problems).

Most of these Linear Equation Word Problems are easy-to-intermediate questions describing a real-world situation that can be modeled by a basic $y = mx + b$ Linear Equation. These problems describe situations that can be modeled with Linear Equations - things like the cost of a boat rental, savings accounts, or the heights of growing plants.

These situations typically start at a certain value - meaning a *y-intercept* (such as the initial cost of renting a boat, the initial deposit into a savings account, or the starting height of a plant). Then they continue increasing or decreasing at a constant rate, meaning *slope* (for example, the hourly cost of the boat rental, the recurring monthly deposits into the savings account, or the growth per week of the plant).

So, you'll see lots of questions based around the question "which of the following Linear Equations correctly relates Value A and Value B?" For example, distance and time; monthly deposits and total savings; initial cost plus hourly cost; plant height vs. plant mass.

You may also need to *interpret* or *explain* one or more components of a Linear Equation. For example, what does the slope *mean* in relation to the word problem? What does the *y*-intercept *represent* in the real-world?

Setting Up Word Problems Into Linear Equations

A key skill for these questions is the ability to quickly and accurately translate words into Linear Equations.

These questions all revolve around the $y = mx + b$ equation. Luckily, this equation only has three components, which we studied in the last lesson: a slope, a *y*-intercept, and an (x, y) coordinate to plug into the equation.

Oftentimes, the *y*-intercept is the easiest value to identify first. Think of the *y*-intercept as representing a "starting point" or "initial value." For example, if a boat costs $30 to begin renting, before any hourly cost kicks in, then $30 would be the "initial value" and would plug in as the *y*-intercept.

The slope usually isn't hard to find either. You're looking for something representing a "constant rate of change." For example, the slope could be the hourly cost of renting the boat. If the hourly cost of rental is $5, then the slope of this Linear Equation would be 5.

Or, we may be required to calculate the slope from two given points by using the $\frac{y_2 - y_1}{x_2 - x_1}$.

As for the (x, y) coordinate to plug into the $y = mx + b$ equation, this depends on the situation. In some of these questions it's not even important to do so, while in other questions it's a vital step.

In many of these word questions, the letters x and y are changed to other letters that fit the word problem, such as h for "hours" and C for "cost." In any event, just remember that a Linear Equation allows you to plug in an input value for x (or a similar variable) and get out a result for y (or another similar variable).

Pretest Question #1

Let's take a look at our first Pretest question on this topic. Try it yourself if you got it wrong the first time.

(CALCULATOR) Henry is training for a reality show involving multiple physical competitions. In one of these competitions, Henry will be required to swing across a minimum of 300 monkey bars without stopping. Henry's training plan calls for him to increase the number of monkey bars he can swing across by a constant amount each week. If Henry predicts he will be able to swing across 168 monkey bars by the end of week 3 and be able to complete the competition at the end of week 7, what was the original number of monkey bars that Henry could swing over without stopping?

(A) 68

(B) 69

(C) 70

(D) 71

The giveaways that this is a Linear Equation with Word Problem are the phrases "increase... by a constant amount each week" and "original number." These are the key words that remind us to look for a *slope* and a *y-intercept*. Remember that an "initial value" makes a great *y*-intercept, and a "constant increase" (or decrease) is another word for slope.

So, where to start? Probably the best place in this question is to calculate slope using the $\frac{y_2 - y_1}{x_2 - x_1}$ equation.

After all, we have two "coordinate points" to use: (Week 3, 168 monkey bars) and (Week 7, 300 monkey bars). Let's treat these like (x, y) coordinates where x is weeks and y is monkey bars, and set up a slope equation from these coordinate points:

$$\frac{y_2 - y_1}{x_2 - x_1}$$
$$= \frac{300 - 168}{7 - 3}$$
$$= \frac{132}{4}$$
$$= 33$$

This gives a slope of $+33$ monkey bars per week, meaning that Henry can cross an additional 33 monkey bars after each week of training.

The question is asking us to solve for the "original value" or *y*-intercept of monkey bars that Henry could cross before he started his training. To do so, we'll need to set up a $y = mx + b$ equation. We can already plug in the slope for m:

$$y = mx + b$$
$$y = 33x + b$$

Next, we'll use either of our two points $(3,168)$ or $(7,300)$ and plug them in for x weeks and y monkey bars. I'll use $(3,168)$ but either point works equally well:

$$\begin{aligned} y &= 33x + b \\ 168 &= 33(3) + b \\ 168 &= 99 + b \\ -99 & \quad -99 \\ 69 &= b \end{aligned}$$

We've found that the Linear Equation has a *y*-intercept or "starting value" of 69 monkey bars at the start of Week 0. So, we have shown that Henry was able to swing over 69 monkey bars before his training began - **Choice B**.

Charts, Tables, and Linear Equations

Another common type of question provides data in the form of a bar chart, table, or scatterplot instead of a word problem. In fact, there may be a word problem *and* a chart or table - just to add to the "fun."

There are a variety of questions that the test can ask that require us to relate the $y = mx + b$ equation to the data.

There really aren't any additional tricks or secrets to these questions; it's just a matter of whether the slope, *y*-intercept, and (x, y) coordinates are delivered by words, table, or some combination.

Lines of Best Fit

In a scatterplot of "real-world" data, there is typically some variation or difference between the *actual* data compared to the "ideal" or "perfect" line modeled by $y = mx + b$.

You may have encountered this personally in your science classes in school: perhaps you were recording data on the growth of a plant in Biology, or the distance a model rocket traveled in Physics, and your pesky real-world data just wouldn't quite fit perfectly into the equation it was supposed to follow.

This is common in all science experiments or indeed, in any collection of real-world data. Errors in measurement and other random variations commonly cause real data to disagree with the "ideal" equation.

A "Line of Best Fit" is simply a Linear Equation that provides the closest possible "ideal" equation to fit the imperfect, real-world data. A Line of Best Fit travels as close as possible to all the real data points, but does not pass perfectly through all the data points. You can think of it as a kind of compromise that tries to fit a "perfect" equation to the "real-world" data as closely as possible.

Below is an example of a Line of Best Fit for six data points. The six data points represent "real-world" measurements while the Line of Best Fit, modeled as the Linear Equation $y = x + 1.5$, provides an "ideal" line that doesn't pass *exactly* through the actual data points, but is a fairly accurate approximation that can "model" or represent the data effectively. You can even *see* that this Line of Best Fit is "pretty close" to the actual data.

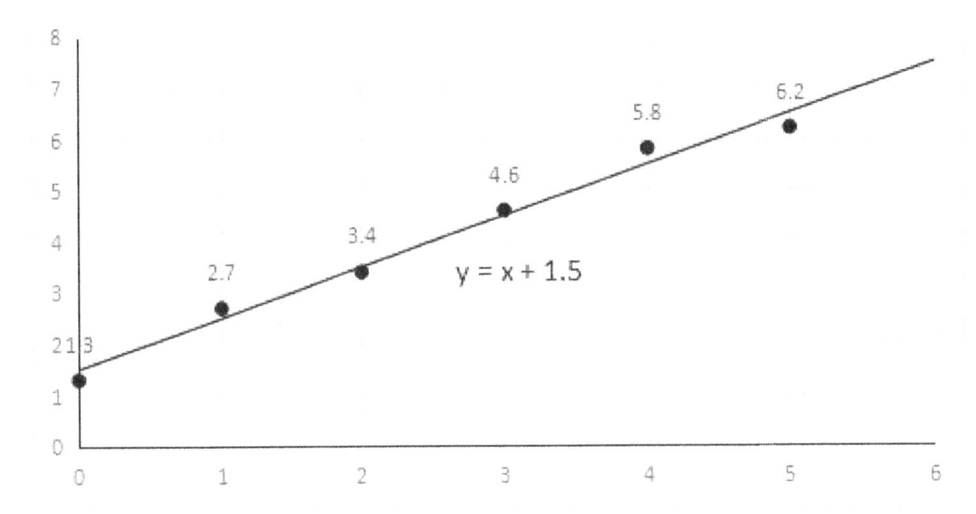

This type of setup can lead to some interesting SAT Math questions. For example, we may be asked the difference between an *actual* data point and the *ideal* data predicted by the Line of Best Fit.

In such a case, we would use the $y = mx + b$ equation of the Line of Best Fit to calculate an "ideal" result, then we would read the "actual" data point from the scatterplot, and then use subtraction to find the difference between the "actual" and "ideal" results.

For example, consider the scatterplot above and try to answer the following question:

For the scatterplot above, what is the difference between the measured value for $x = 4$ and the value predicted by the line of best fit?

First, let's see what the actual data says for the point at $x = 4$. We can simply read off the data point at $x = 4$: the data point has a *y*-value of 5.8.

Now, let's use the Line of Best Fit equation to calculate the "predicted" or "ideal" value. We'll just plug $x = 4$ into the Line of Best fit equation, $y = x + 1.5$

$$y_{\text{ideal}} = x + 1.5$$
$$y_{\text{ideal}} = (4) + 1.5$$
$$y_{\text{ideal}} = 5.5$$

So what's the difference between the "predicted" and the "actual" data? It will be the **Absolute Value** of subtracting the actual data from the predicted result.

$$| y_{\text{ideal}} - y_{\text{actual}} |$$
$$= | 5.5 - 5.8 |$$
$$= | -.3 |$$
$$= .3$$

The difference between the actual data point and the predicted result for $x = 4$ is **.3**. This is a perfect example of the difference between actual, real-world data and the Line of Best Fit that "models" or approximates that data with a Linear Equation.

Pretest Question #2

Let's take a look at another Pretest question. Try it yourself before you look at my explanation below the question.

t	4	7	10
v	95	131	167

(CALCULATOR) A sports car is in a straight-line race that begins with the car already moving at a certain speed. Once the race begins, the car accelerates at a constant rate per second. The table above shows a set of values for time t since the race began, in seconds, and the car's resulting velocity v in miles per hour. Which of the following functions best models the relationship of t and v?

(A) $v = 95 + 12t$

(B) $v = 67 + 7t$

(C) $v = 57 + 11t$

(D) $v = 47 + 12t$

This type of question provides a data table and a word problem that gives a backstory for the data. But, you can disregard most of that word problem as "fluff" - the most important keys are that the car accelerates at a "constant rate" and "begins... at a certain speed." These are key phrases that represent a *slope* and a *y*-intercept. Furthermore, the four answer choices are all clearly variations upon our familiar $y = mx + b$ equation.

All the signs point directly to a Linear Equation. Regardless of word problem stories about cars, boats, savings accounts, or anything else, it's all just decoration for the underlying math concepts that we know so well.

There are a few ways we can go about solving a problem like this. One smart move would be to calculate the slope right away from any two coordinates in the table. Let's use $(10,167)$ and $(7,131)$. Now put them into our familiar $\frac{y_2 - y_1}{x_2 - x_1}$ equation:

$$\frac{y_2 - y_1}{x_2 - x_1}$$
$$= \frac{(167) - (131)}{(10) - (7)}$$
$$= \frac{36}{3}$$
$$= 12$$

Now we know the slope is 12. This is extremely valuable, because we can eliminate two of the Answer Choices right off the bat: both Choice B and Choice C do *not* have slopes of 12, so they're out. (Remember from the previous lesson, we can read slope as the coefficient of x, or in this case, of t).

My next move - and this may surprise you - will be to *test* my remaining two answer choices by plugging in values of t. If either equation fails to return the proper v value, we'll know that we can eliminate it.

Let's use the data point $(4,95)$ and test Answer Choices A and D. First, Choice A:

$$v = 95 + 12t$$
$$95 = 95 + 12(4)$$
$$95 = 95 + 48$$
$$95 \neq 143$$

Immediately, we can see that Choice A doesn't fit the data. Let's eliminate it, and move onto Choice D. We better hope it works, because we're out of other options! Let's test it to be sure:

$$v = 47 + 12t$$
$$95 = 47 + 12(4)$$
$$95 = 47 + 48$$
$$95 = 95$$

Nice! The correct equation to model the data in the table must be **Choice D**. First we calculated the slope, and quickly eliminated two choices. Then, we tested our remaining two equations by plugging in a data point, which eliminated a third answer, and showed that the fourth answer works.

Two other ways of solving this question: First, you could just start testing the Answer Choices from the very beginning, without even calculating the slope first. However, this would be fairly time-consuming because we'd have to test 4 answer choices instead of just 2. Remember that testing answer choices is a strategy that can't prove answers *correct*; it only allows you to eliminate *wrong* answers.

The other option is to calculate the $y = mx + b$ equation by figuring out the *y*-intercept. It's not even hard - we've done it many times before in the previous lesson. We already know the slope is 12, so let's start by plugging that in for m. We'll also adjust to the correct variables t and v instead of x and y:

$$v = 12t + b$$

Now plug in a data point for t and v. I'll use $(7,131)$ but anything from the table would work:

$$v = 12t + b$$
$$131 = 12(7) + b$$

And now solve for the value of b, the *y*-intercept:

$$131 = 12(7) + b$$
$$131 = 84 + b$$
$$-84 \quad -84$$
$$47 = b$$

Now we know the initial speed, the *y*-intercept, is 47. Put this together with the slope and you'll get the correct equation to model the data in the table:

$$v = 12t + 47$$

Notice it's the same as the equation for Choice D, so we definitely got it right the first time!

Review & Encouragement

As you may have noticed, this lesson doesn't contain any major new concepts - it merely presents the Linear Equations of last lesson in a different context.

We may need to create or interpret $y = mx + b$ equations or relate them to real-world situations. Tables or charts may supply essential information, or everything can just be given in word problem form instead. We've also learned about Lines of Best Fit and how "ideal" equations can differ from the "actual" data that it models.

It's good news for you that these Linear Equation questions are such a major percentage of the SAT Math test. They're easy and plentiful. Remember, almost 15% of your entire SAT Math score will come down to your understanding of Linear Equations from this chapter and the previous one.

Now, work through the following practice question set to make sure you can apply the essential concepts of Linear Equations to Word Problems, Charts, and Tables.

Linear Equations (Words & Tables) Practice Questions

DO NOT USE A CALCULATOR ON ANY OF THE FOLLOWING QUESTIONS UNLESS INDICATED.

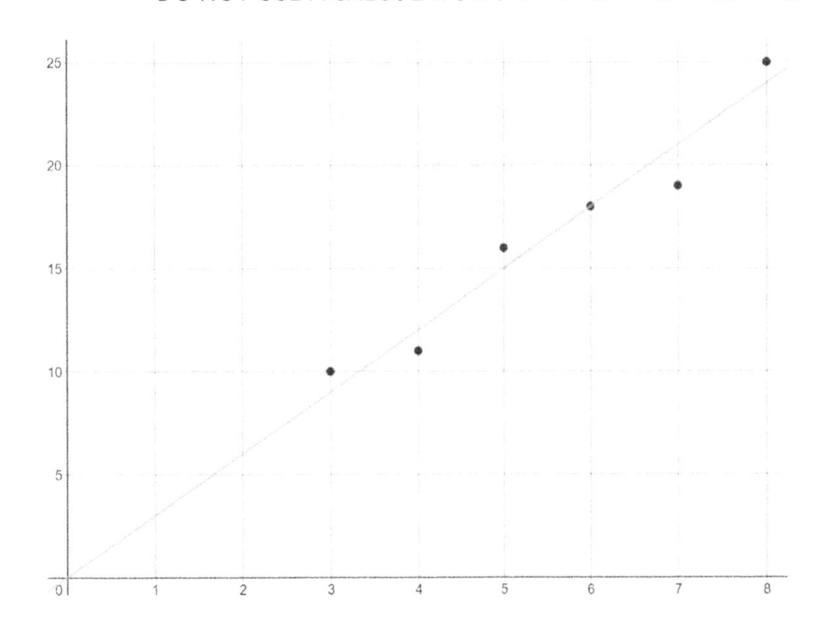

1. The scatterplot to the left shows the number of white dogs, x, and total dogs of all colors, y, for six dog shelters in a large city. A line of best fit for the data is also shown. Which of the following could be the equation of the line of best fit?

(A) $y = 3x$

(B) $y = \frac{1}{3}x$

(C) $y = -3x$

(D) $y = -\frac{1}{3}x$

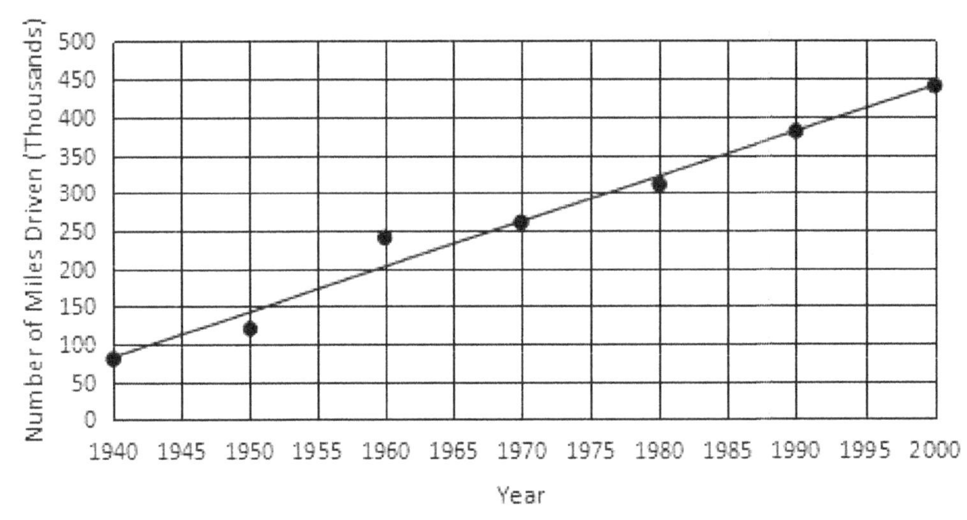

2. According to the line of best fit in the scatterplot above, which of the following best approximates the year in which the number of miles driven by car in Country X was estimated to be 300 thousand?

(A) 1975

(B) 1976

(C) 1980

(D) 1985

3. A building contractor estimates the price of building a shed with the equation $2400 + 22nh$, where n is the number of workers building the shed and h is the number of hours it takes them to finish the project. Which of the following is the best interpretation of the number 22 in the expression?

(A) Each worker is paid $22 per hour.

(B) The total cost of the project increases by $22 per hour, regardless of how many workers are building the shed.

(C) The total cost of the project increases by $22 per worker.

(D) Each worker works on the project for a total of 22 hours.

4. Jesse is purchasing tickets to a concert from an online vendor. The vendor charges a one-time processing fee for the purchase of the tickets. The equation $T = 25n + 10$ represents the total cost T, in dollars, that Jesse will pay for n tickets. What does 10 represent in the equation?

(A) The price of a single ticket, in dollars

(B) The cost of the processing fee, in dollars

(C) The total amount, in dollars, that Jesse will pay for a single ticket

(D) The total amount, in dollars, that Jesse will pay for n tickets

$$l = 3 + 2n$$

5. The equation above represents the length l of a rubber band that is stretched with a force of n newtons. What is n when l is 11?

(A) 4

(B) 5

(C) 8

(D) 25

6. (CALCULATOR) A certain candy store charges 50 cents for an empty bag and 25 cents per piece of candy. Jeremy's aunt spent $6.75 on a bag of candy for Jeremy from this store. How many pieces of candy did Jeremy's aunt buy?

(A) 13

(B) 25

(C) 26

(D) 50

$$L = 3.37H + 14.05$$

7. The formula above can be used to approximate the length L, in feet, of a jet plane constructed by a certain company based on the height off the ground H, in feet, of the plane's wingtips. What is the meaning of 3.37 in this context?

   (A) The approximate height off the ground, in feet, of the plane's wingtips.

   (B) The approximate increase of the height off the ground of the plane's wingtips, in feet, for each increase of 14.05 feet in the length of the plane.

   (C) The approximate increase of the plane's length, in feet, for each one-foot increase in the height off the ground of the plane's wingtips.

   (D) The approximate length of the plane, in feet, for a plane with wingtips at a height of 14.05 feet off the ground.

Year	Widget Production
1940	5,000
1960	6,200

8. The table above shows the widget production of a certain factory for the years 1940 and 1960. If the relationship between widget production and year is linear, which of the following functions W models the widget production of this factory in year t?

   (A) $W(t) = 5000 - 60t$

   (B) $W(t) = 5000 + 60t$

   (C) $W(t) = 5000 + 60(t - 1940)$

   (D) $W(t) = 5000 + 60(t + 1940)$

9. FREE RESPONSE: $I = 500t + 2000$ Ian made an initial deposit to his bank account. Each month after that, he makes an additional deposit for the same amount each month. The equation above models the amount, I, in dollars, that Ian has deposited after t months. According to the model, how many dollars was Ian's initial deposit?

$$c = 21.99 + .99m$$

10. The equation above models the total cost c in dollars that a company charges a customer to rent a dirt bike for one day and ride for m miles. The total cost consists of a flat rental fee plus a charge per mile driven. When the equation is graphed in the *mc*-plane, what does the *c*-intercept of the graph represent in terms of the model?

   (A) A charge per mile of $0.99

   (B) A charge per mile of $21.99

   (C) A total charge of $22.98

   (D) A flat rental fee of $21.99

11. The average annual cost for water utilities in a certain office building is \$3,833. The office manager plans to spend \$12,500 to install a rainwater collector and purification system. The office manager estimates that the new average annual cost for water utilities will be \$3,353. Which of the following inequalities to be solved to find t, the number of years after installation at which the total savings on water utilities will pay for itself?

(A) $12{,}500 \leq \dfrac{3{,}833}{480}$

(B) $12{,}500 \leq (3{,}833 - 3{,}353)t$

(C) $12{,}500 - 3{,}353 \geq 3{,}833t$

(D) $12{,}500 \leq (3{,}833 - 480)t$

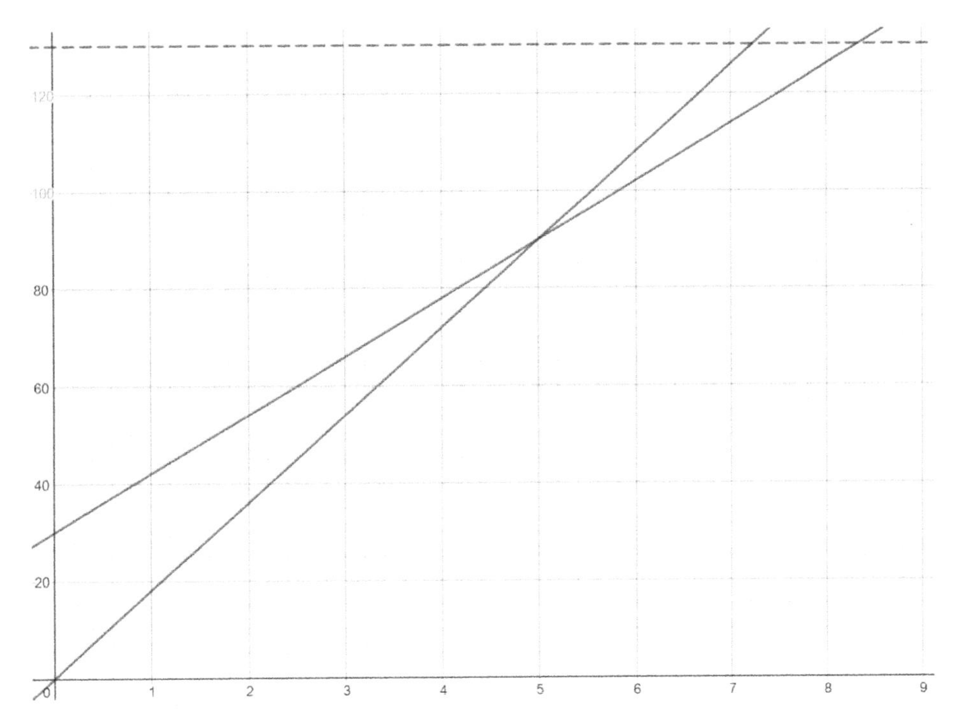

12. The graph above shows the positions of Ian and Christian during a scooter race. The y-axis gives distance from the starting line in feet, and the x-axis shows the time, in seconds, since the race began. Ian and Christian both ride their scooters at a constant rate, and Christian was given a head start to shorten the distance he needed to ride. According to the graph, Christian was given a head start of how many feet?

(A) 5

(B) 25

(C) 30

(D) 130

13. The velocity V of a certain spacecraft in meters per second is a linear function of the length of time in seconds s the thrusters have been firing since launch, and is given by $V(s) = 397.4 + 58s$. Which of the following statements is the best interpretation of the number 397.4 in this context?

(A) The velocity of the spacecraft, in meters per second, at launch.

(B) The velocity of the spacecraft, in meters per second, after 58 seconds.

(C) The increase in the spacecraft's velocity, in meters per second, that corresponds to an increase in one second of thrusters being fired.

(D) The increase in the spacecraft's velocity, in meters per second, that corresponds to an increase of 58 seconds of thrusters being fired.

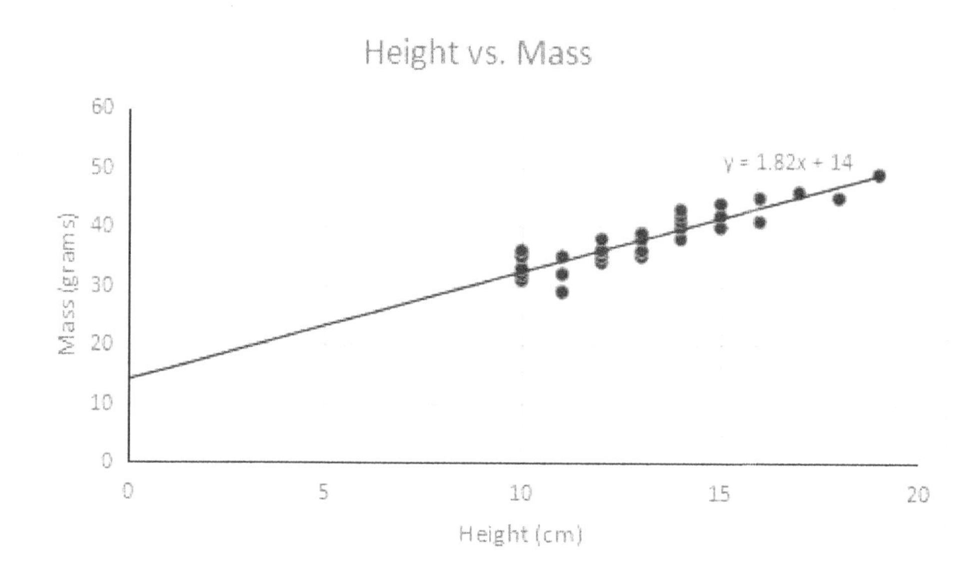

14. (CALCULATOR) The scatterplot above shows data collected on the height in centimeters and mass in grams of a certain group of small plants. A line of best fit for the data is also shown. Based on the line of best fit, if the height of one of these plants is 5 centimeters, what is the predicted weight in grams of that plant?

(A) 22

(B) 22.7

(C) 23.1

(D) 23.8

15. FREE RESPONSE: Christian is practicing to play speed metal guitar. His goal is to reach 230 notes per minute. Currently, he can play 160 notes per minute. If Christian believes he can improve his speed by 7 notes per minute each week, how many weeks does Christian estimate it will take him to reach his goal?

QUESTIONS 16 AND 17 REFER TO THE FOLLOWING GRAPH.

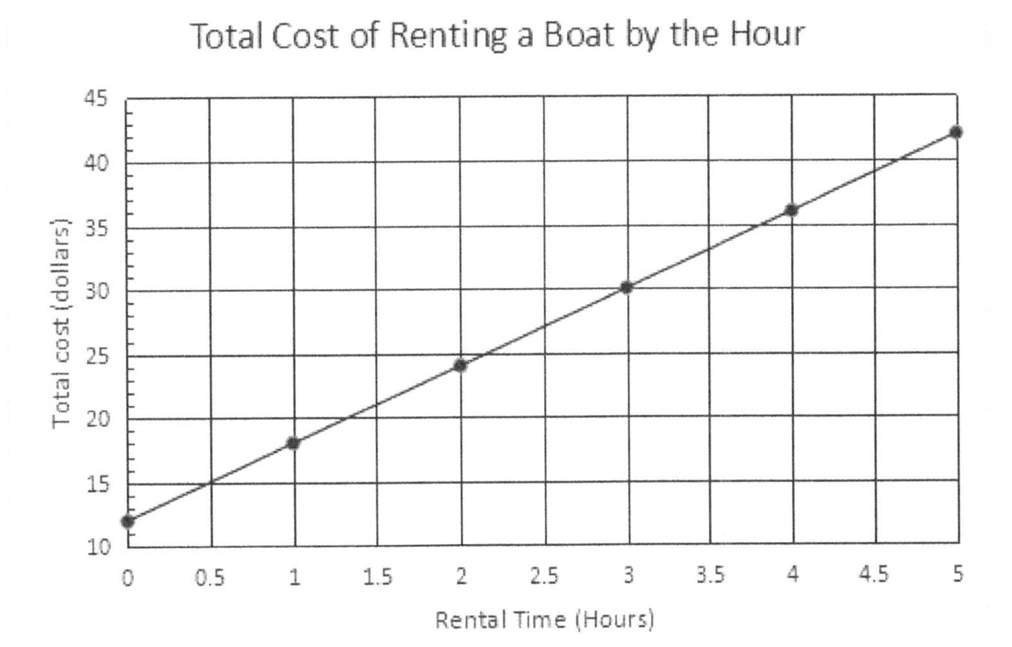

The graph above displays the total cost C, in dollars, of renting a sailboat for h hours.

16. What does the C-intercept represent in the graph?

(A) The total cost of renting the sailboat.

(B) The initial cost of renting the sailboat.

(C) The cost per hour of renting the sailboat.

(D) The total number of hours the sailboat is rented.

17. Which of the following equations represents the relationship between h and C?

(A) $C = 10h$

(B) $C = 12 + 6h$

(C) $C = 6 + 12h$

(D) $C = 12 + 3h$

18. (CALCULATOR) Coal mining production in a certain region dropped from 3.7 million tons in 1990 to 1.3 million tons in 2002. Assuming that the coal mining production decreased at a constant rate, which of the following functions f best models the production, in millions of tons, t years after the year 1990?

(A) $f(t)=\frac{1}{5}t-3.7$

(B) $f(t)=\frac{13}{120}t+3.7$

(C) $f(t)=-\frac{1}{5}t+3.7$

(D) $f(t)=-\frac{13}{120}t-3.7$

$$m=4.5t+65.4$$

19. A botanist uses the model above to estimate the mass m of a raspberry bush, in grams, in terms of the bush's age t, in months, between 6 months and 12 months. Based on the model, what is the estimate increase, in grams, of the raspberry bush each month?

(A) 4.5

(B) 27

(C) 60.9

(D) 65.4

20. FREE RESPONSE: The distance L a wind-up toy car travels in inches can be modeled by the equation $L=\frac{t+8}{3}$, where t is the time in seconds since the car was released from rest for the first 13 seconds of travel. According to the model, for every one second of time between zero and thirteen seconds of travel, by how many inches did the distance increase?

21. (CALCULATOR) While practicing for a breath-holding competition, Jerry created a training schedule in which the length of time he could hold his breath increased each week by a constant amount. If Jerry's training schedule requires that his longest duration of held breath is 80 seconds in week 2 and 110 seconds in week 10, which of the following best describes how the length of time Jerry can hold his breath changes between week 2 and week 10 of his training schedule?

(A) Jerry increases the length of time he can hold his breath by $\frac{1}{3}$ of a second each week.

(B) Jerry increases the length of time he can hold his breath by 3 seconds each week.

(C) Jerry increases the length of time he can hold his breath by 3.75 seconds every 4 weeks.

(D) Jerry increases the length of time he can hold his breath by 15 seconds every 4 weeks.

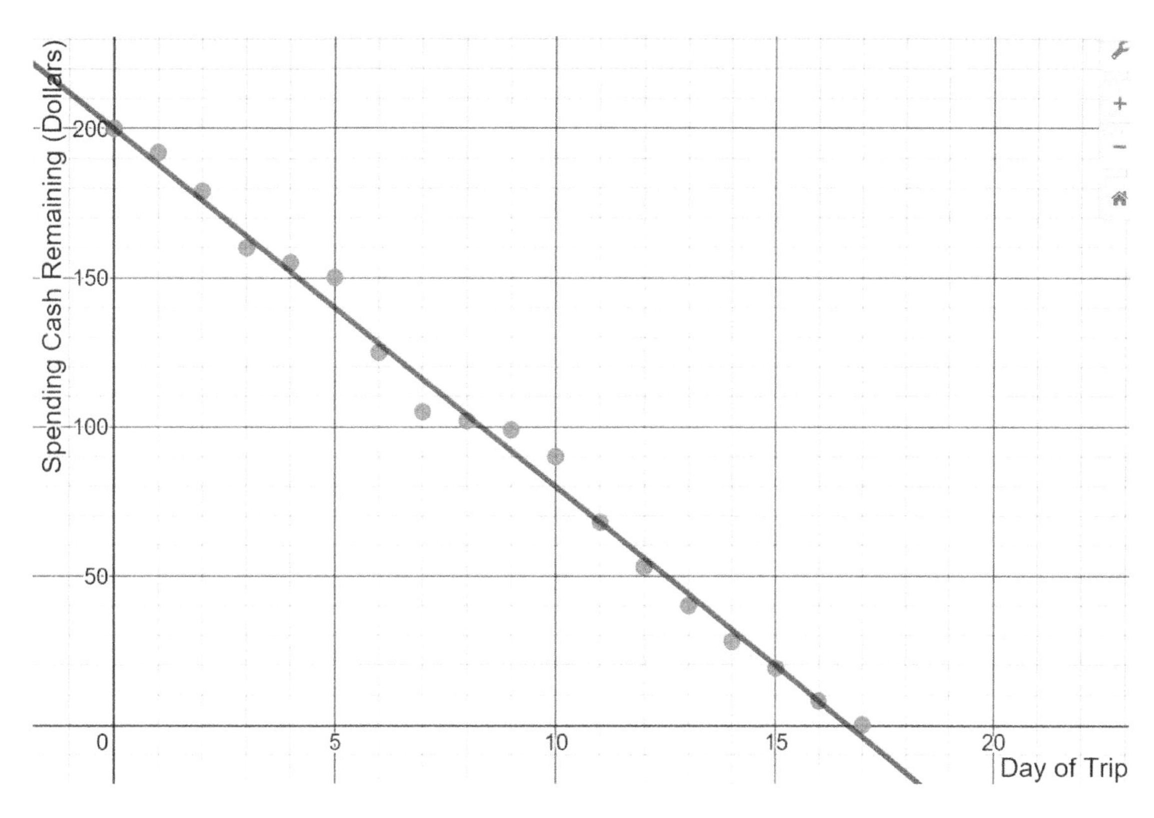

22. Christian took a vacation out of the country. The scatterplot above shows his amount of spending cash remaining after each day of travel. The line of best fit for the data is also shown. On day 10 of his trip, Christian's spending cash remaining was about how many dollars less than the amount predicted by the line of best fit?

(A) 0

(B) 5

(C) 10

(D) 20

23. FREE RESPONSE: A tutoring company opened with 5 employees. The company's owner assumes that 1 new employee will be hired each quarter (every 3 months) for the first 4 years. If an equation is written in the form $y = ax + b$ to represent the number of employees, y, employed by the company x quarters after the company opened, what is the value of b?

24. (CALCULATOR) A recording studio that supplies practice space for musicians purchases music equipment for $44,800. The equipment depreciates in value at a constant rate for 20 years, after which it is considered to have no financial value. How much is the music equipment worth 8 years after it is purchased?

(A) $5,600

(B) $17,920

(C) $26,880

(D) $39,200

25. The manager of a coffee shop in a certain city plans to increase the number of musicians that perform in the coffee shop each month by a total of p performers per month. There were x performers in the coffee shop at the beginning of this month. Which function best models the total number of performers, t, the manager plans to have as performers m months from now?

(A) $t = pm + x$

(B) $t = px + m$

(C) $t = x(p)^m$

(D) $t = p(x)^m$

Apartment Name	Purchase Price (Dollars)	Monthly Rental Income (Dollars)
Oakview Apartments	1,600,000	12,000
The Hacienda	2,100,000	15,000
Peaceful Cove	1,200,000	9,600
Sunlawn	3,300,000	22,200
The Vistas	2,800,000	19,200

26. (CALCULATOR) The table above shows five apartment complexes purchased by a real-estate investment company. The table shows the amount, in dollars, paid for each apartment complex and the corresponding monthly rental income, in dollars, the company receives from each of the five properties. The relationship between the monthly rental price, in dollars, and the apartment complex's purchase price, in thousands of dollars, can be represented by which of the following linear functions, where p is purchase price in thousands of dollars and $r(p)$ is rental income?

(A) $r(p) = .006p + 2400$

(B) $r(p) = 8p - 800$

(C) $r(p) = 7p - 400$

(D) $r(p) = 6p + 2400$

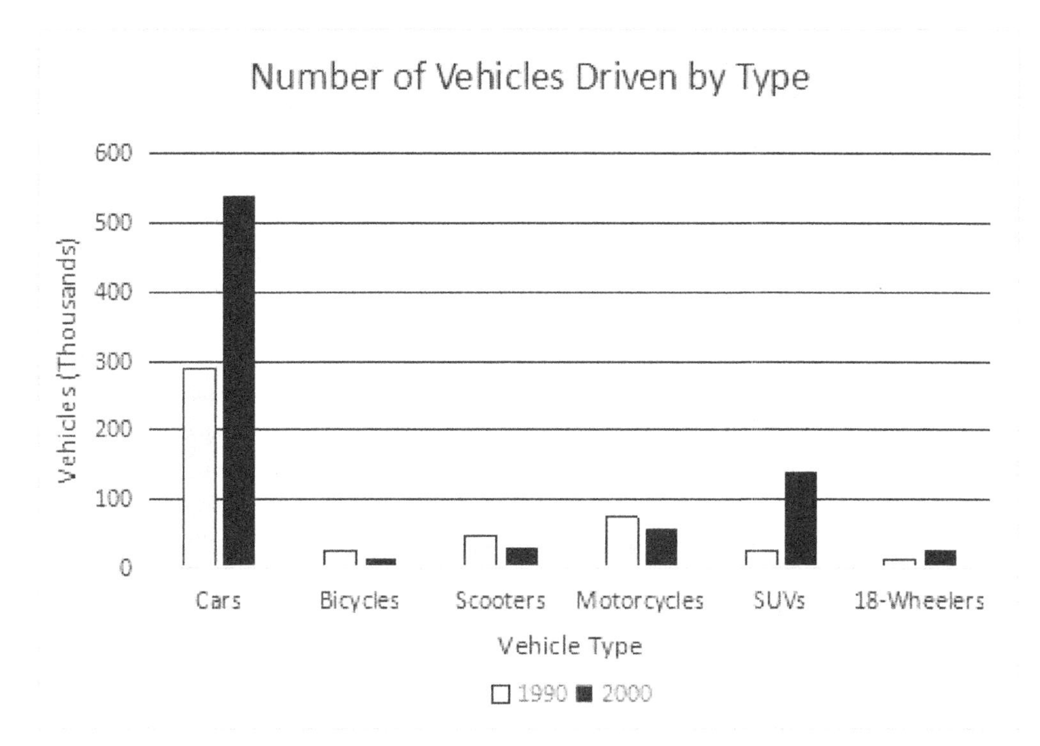

27. The bar chart above shows the types of vehicles driven on the road in a certain region (in thousands) in the years 1990 and 2000. In a scatterplot of this data, where vehicles driven in the year 1990 is plotted along the x-axis and vehicles driven in the year 2000 is plotted along the y-axis, how many data points would be above the line $y = x$?

(A) 2

(B) 3

(C) 4

(D) 5

$$4nx + 27 = 6(3 + x) - 2(x - 1)$$

28. FREE RESPONSE: In the equation above, n is a constant. If no value of x satisfies the equation, what is the value of n?

29. (CALCULATOR) An RV (recreational vehicle) has a gas tank that is initially filled with 400 quarts of gasoline. The driver sets out on the freeway and drives at a constant speed, consuming a steady volume of gasoline per mile. After 110 miles, it has used 22% of the gasoline. Which of the following equations models the volume of gasoline v, in quarts, remaining in the tank m miles after the RV started driving?

(A) $v = 400 - 110m$

(B) $v = 400 - .8m$

(C) $v = 400(.22)^{\frac{m}{110}}$

(D) $v = 400(.78)^{\frac{m}{110}}$

30. (CALCULATOR) FREE RESPONSE: The mesopelagic zone is a depth of the ocean that ranges from approximately 200 to 1000 meters below the surface of the ocean. A scientist finds that the temperature of the water in the mesopelagic zone is 67 degrees Fahrenheit at 200 meters and the temperature of the mesopelagic zone at a depth of 1000 meters is 39 degrees Fahrenheit. For every additional 100 meters below the ocean's surface, the temperature in the mesopelagic zone decreases by k degrees Fahrenheit, where k is a constant. What is the value of k?

Linear Equations (Words & Tables) Answers

1. A
2. B
3. A
4. B
5. A
6. B
7. C
8. C
9. 2000
10. D
11. B
12. C
13. A
14. C
15. 10
16. B
17. B
18. C
19. A
20. $\frac{1}{3}$ or .333
21. D
22. C
23. 5
24. C
25. A
26. D
27. B
28. 1
29. B
30. 3.5

Linear Equations (Words & Tables) Explanations

1. **A.** The line of best fit is represented by a linear equation; therefore, the form $y = mx + b$ is what we should work with. We know from the previous chapter that the two most important elements of a $y = mx + b$ equation are the Slope and the *y*-intercept. This line goes cleanly through the origin - point $(0,0)$ - so we know the *y*-intercept or *b*-value is 0.

As for the slope, we should look for a second point that the line goes cleanly through; in other words, a point that we can easily read from the gridlines of the graph. The coordinates $(5,15)$ will do nicely. Now use the $\frac{y_2 - y_1}{x_2 - x_1}$ equation to calculate Slope (again, something we learned to do in the previous chapter). I'll use $(0,0)$ as Point 1 and $(5,15)$ as Point 2:

$$\frac{15-0}{5-0}$$
$$=\frac{15}{5}$$
$$=3$$

Now we know the Slope is 3, the *y*-intercept is 0, and we can put this all together into a finished $y = mx + b$ equation:

$$y = mx + b$$
$$y = 3x + 0$$
$$y = 3x$$

Done! We know the equation of the line of best fit is $y = 3x$.

2. **B.** In this question, we're simply tasked with finding a specific coordinate on the line of best fit and reading it off the scatterplot. Specifically, we're looking for where "number of miles" is closest to 300 (thousand).

Use your pencil - not your eyes - to mark out a horizontal line starting from "300" on the *y*-axis until it reaches the line of best fit. When your horizontal line reaches the line of best fit, stop and mark that point.

Now drop a vertical line with your pencil down from your point to the *x*-axis. Notice that we don't land perfectly on a "tick mark" for the year, so we have to make our best estimate. It looks like we're just a bit beyond the year 1975, but definitely not to the year 1980, so our best available choice would be the year 1976.

Note: we should always use our pencil, rather than just looking with our eyes or tracing with our finger, not only because it's much more accurate, but also because it breaks the work into stages and allows us to review our work if we want to check over the problem later.

3. **A.** This word problem asks us to interpret the variables in an equation that is based on the $y = mx + b$ format. We're asked what the meaning of the value "22" is and how it connects the equation into the real-world situation.

Notice that 22 is a *coefficient* of variables n and h, which represent the number of workers and the number of hours they work. Each time either n or h goes up by "+1," the final cost will increase by "+22." This must mean that each worker gets paid $22 per hour.

FYI, we can also understand the "2400" as the base price of building the shed - the initial cost - even without any worker-hours. This isn't necessary to know for this problem, but it's good to understand that 2400 is the *y*-intercept.

4. **B.** This question connects a word problem to a linear equation in $y = mx + b$ form. The *y*-intercept or "starting point" is $10, and the slope or "constant rate of increase" is $25 per ticket. If the cost of Jesse's tickets has a starting point or *y*-intercept of $10, that must mean he has to pay $10 no matter how many tickets he buys. And a slope of 25 per ticket means each ticket costs $25. Therefore, the number 10 represents the initial cost: the one-time processing fee for buying his tickets online.

5. **A.** In this question, we're provided a $y = mx + b$ equation (although it uses the letters l and n instead of y and x, it's still the exact same type of equation), and the words define what each of the variables represents. Then we're required to plug in "11" for the value of l and find n. This is relatively simple. Let's set it up by plugging in 11 for l:

$$l = 3 + 2n$$
$$11 = 3 + 2n$$

Now finish by solving the **Basic Algebra 1**.

$$11 = 3 + 2n$$
$$-3 \quad -3$$
$$8 = 2n$$
$$\frac{8}{2} = \frac{2n}{2}$$
$$4 = n$$

And we're done. All we needed to do was find the value of n when $l = 11$. Notice that the word problem itself didn't contribute anything to our work, except for helping us set up our algebra at the very beginning.

6. **B.** It might not be immediately obvious, but this question is based on a Linear Equation. How can you tell? Well, there's a *fixed* cost and a *steadily-increasing* cost. The bag itself costs 50 cents, and then each piece of candy costs an additional 25 cents. The fixed cost of the bag is the y-intercept and the price-per-candy is the slope.

Set up a $y = mx + b$ with the given values. Be sure to check your units! It's OK to work with either dollars or cents - but your entire equation must be consistent, regardless of which unit you choose. I'll stick with dollars. Note that x represents the pieces of candy that Jeremy's aunt is purchasing:

$$y = mx + b$$
$$6.75 = .5 + .25x$$

Now use **Basic Algebra 1** to solve for x.

$$6.75 = .5 + .25x$$
$$-.5 \quad -.5$$
$$6.25 = .25x$$
$$\frac{6.25}{.25} = \frac{.25x}{.25}$$
$$25 = x$$

Now we know that 25 pieces of candy were purchased, and we're done!

7. **C.** This question relates a $y = mx + b$ Linear Equation to a real-life word problem situation. As with other questions we've done in this lesson, it's important to establish the fixed starting value (or y-intercept) and what it means. It's also important to determine what the slope is, and what it means.

Comparing to the $y = mx + b$ equation, we can see that "3.37" must be the slope value. Every time H increases by 1, the value of L will increase by 3.37. Tying this to the word problem, we can say that every time the height H of the wingtips increases by 1 foot, the length L of the plane will increase by 3.37 feet.

Reading the answer choices carefully, that matches with Choice C: "the increase of the plane's length for each 1-foot increase in height."

The problem isn't difficult mathematically, but there are a lot of words. If you got the answer wrong, make sure you understand what the slope of this Linear Equation represents, but most of all, be sure you were reading the question and answer choices carefully. A Misreading **Careless Mistake** is extremely likely on a detailed word question like this one.

8. **C.** In this question, we're given a table and word problem, and asked to find a $y = mx + b$ Linear Equation that accurately represents the data.

There are two ways to handle this, and both have their pros and cons: you can either test the answer choices by plugging in the values from the table, or you can attempt to create your own $y = mx + b$ equation from scratch.

To be honest, this is a situation where I might just test the answer choices. It won't take very long and has a low risk of careless mistakes. Remember one *crucial* thing, though: testing the answer choices on the SAT never provides "positive proof" that an answer is actually *correct*; this strategy can only identify and eliminate *wrong* answers. In other words, don't just pick the first equation that works when you plug in: you need to test all four answer choices to be sure you can eliminate three of them.

Let's get started. I'll use the year 1940 for t, along with its 5,000 widgets for $W(t)$.

Testing Choice A shows that it's definitely wrong:

$$W(t) = 5000 - 60t$$
$$5{,}000 = 5000 - 60(1940)$$
$$5{,}000 = 5000 - 116{,}400$$
$$5{,}000 \neq -111{,}400$$

Testing Choice B shows that it also doesn't work:

$$\begin{aligned} W(t) &= 5000 + 60t \\ 5{,}000 &= 5000 + 60(1940) \\ 5{,}000 &= 5000 + 116{,}400 \\ 5{,}000 &\neq 121{,}400 \end{aligned}$$

Now testing Choice C shows that this equation might work:

$$\begin{aligned} W(t) &= 5000 + 60(t - 1940) \\ 5000 &= 5000 + 60(1940 - 1940) \\ 5000 &= 5000 + 60(0) \\ 5000 &= 5000 + 0 \\ 5000 &= 5000 \end{aligned}$$

We still need to test Choice D to see what happens:

$$\begin{aligned} W(t) &= 5000 + 60(t + 1940) \\ 5000 &= 5000 + 60(1940 + 1940) \\ 5000 &= 5000 + 60(3880) \\ 5000 &= 5000 + 232{,}800 \\ 5000 &\neq 237{,}800 \end{aligned}$$

OK, of our four answer choices, only Choice C provided a true equation. It must be the right answer.

If, by chance, two equations had worked on our first set of tests, we would then move to year 1960 and 6,200 widgets, plug those into whatever equations had passed the first set of tests, and again eliminate any equations that didn't return a true equality for the second round of testing. Luckily, we didn't have to deal with that this time.

9. **2000.** Like several other questions in this lesson, we've been given a Linear Equation based on $y = mx + b$. Also notice the signs that this is a Linear Equations question: there was an "initial deposit" (which will be the *y*-intercept) and an "additional deposit for the same amount each month," (which will be the slope).

Once you realize this, it's easy to finish. The *y*-intercept of this equation must be the original amount of Ian's initial deposit. The *y*-intercept is 2000, so we're done.

10. **D.** Yet another question that relates a linear $y = mx + b$ equation to a real-world situation. As usual, we should focus on the *slope* (the constant increase) and the *y-intercept* (the starting value).

In this question, the *y*-intercept is 21.99 - although they've re-titled the axes "c" and "m" instead of "y" and "x". That doesn't matter; these are just different labels for the same ideas we've grown used to in the past two lessons.

The slope of .99 means that for each additional mile m traveled, the cost c increases by 99 cents.

But if the *c*-intercept is 21.99, that must represent the fixed cost of renting the bike, *before* traveling any miles on the bike. Therefore, the *c*-intercept represents the flat rental fee of $21.99 before the bike actually goes anywhere.

11. **B.** This question combines a word problem, a linear equation, and an inequality. The inequality is important because we are looking for the elapsed time before we "break even" on our investment. As soon as our savings equal or *exceed* the cost of installation, we will break even.

On the left, we'll put the installation cost of $12,500. On the right, we need the amount of money we've *saved* with the new system - and we need it to be equal to or *greater than* the cost of installation, so that we save more total money than we spent.

$$\text{Cost of Installation} \leq \text{Total Savings}$$

How much money will be saved per year? That's the original annual cost minus the new annual cost:

$$(3833 - 3353) = \text{annual savings on water utilities}$$

These annual savings will accrue year after year - a steady slope. Multiply this slope of increasing yearly savings by the number of years t:

$$(3{,}833 - 3{,}353)t$$

Put everything together and you get this, our final expression:

$$\begin{aligned} \text{Cost of Installation} &\leq \text{Total Savings} \\ 12{,}500 &\leq (3{,}833 - 3{,}353)t \end{aligned}$$

The same equation is giving in **Choice B**.

12. **C.** This word problem, combined with the graph, describes a situation of two racers: Christian, who gets a head start but goes slower, and another racer, Ian, who starts behind, goes faster, catches up, and passes Christian.

Although the lines on the graph aren't labeled, it's clear that Christian's line - since he had a head start - must be the one that starts higher on the *y*-axis, which gives distance from the starting line.

The *y*-intercept of Christian's line would give his starting position or initial distance from the starting line. Reading the tick marks carefully, and using our pencil to mark so that we don't make a careless mistake, we can see that Christian's line has a *y*-intercept of 30 feet, while Ian's line has a *y*-intercept of 0 feet, and therefore Christian must have had a 30-foot head start.

13. **A.** Again, we're being asked to relate a Linear Equation to the real-life word problem that it describes. As always, the most important elements of the Linear Equation are the *slope* and the *y-intercept*.

The slope of this line must be "58". It's the coefficient of the s variable, so for every time s increases by +1, the velocity V will increase by +58.

On the other hand, the 397.4 must be the *y*-intercept. This is the initial value or starting velocity before any time has elapsed. It does not change based on elapsed time. In other words, even when $s=0$, the spacecraft will already have a velocity of 397.4 meters per second.

That's why why the correct answer is Choice A, the velocity of the spacecraft at launch (before any time has elapsed).

14. **C.** This question provides a scatterplot with a line of best fit, along with the equation for that line.

Initially, this question may send a bit of a false signal: some students will simply try to use the graph to visually read the expected mass. But there's a big problem with that method: the grid on the scatterplot is very wide, and although we can easily find the point on the line corresponding to "5 centimeters in height," it's impossible to confidently read the exact value of the mass in grams without finer grid lines. Furthermore, the answer choices are all extremely close. That makes visually estimating the value very risky.

However, we have a trump card: the graph itself gives us the exact Linear Equation for the line of best fit: $y=1.82x+14$.

We can use that equation as a better alternative to a visual estimate. Simply plug in 5 for the *x*-value, and evaluate for the corresponding *y*-value:

$$y=1.82x+14$$
$$y=1.82(5)+14$$
$$y=9.1+14$$
$$y=23.1$$

Using the given equation for the line of best fit allows us to calculate an *exact* value for the mass of the plant, rather than estimating between extremely close answer choices. Now we know that the predicted weight of the plant should be 23.1 grams.

15. **10.** This word problem provides the perfect scenario for a Linear Equation: we've got a starting value (160 notes), a slope (+7 notes per week), and a target value (230 notes). That means we can make a $y=mx+b$ equation with 160 as our y-intercept and a slope of 7, then set it equal to the final value of 230. We can use a new variable x to represent the number of weeks Christian has been practicing for:

$$y=mx+b$$
$$230=7x+160$$

Now solve for x:

$$230=7x+160$$
$$-160 \qquad -160$$
$$70=7x$$
$$\frac{70}{7}=\frac{7x}{7}$$
$$10=x$$

Great! After 10 weeks of adding +7 notes each week, building upon his initial speed of 160 notes per minute, Christian will hit his target of 230 notes per minute.

16. **B.** Remember, with a Linear Equation problem, the *y*-intercept is always the initial or starting value. True, this question has renamed our normal axes from "x" and "y" to "h" and "C", but that's just a cosmetic difference - the underlying concept is unchanged. Our "*C*-intercept" is just a *y*-intercept.

This question asks the *meaning* of the *C*-intercept - how it fits into the real-life situation. Well, the intercept must be the "initial value" of this boat rental, before any time has elapsed.

Carefully reading this value off the tick marks on the graph, we can see that this initial value is $12. This must mean that we have to pay $12 even before any rental time has elapsed on the rental. In other words, we have to pay $12 simply to take the rental boat out. It must be the initial cost of renting the sailboat, Choice B.

17. **B.** This question refers to the previous graph showing the cost of renting a boat by the hour. The graph clearly shows a linear relationship, and our four answer choices are all Linear Equations based on the essential $y = mx + b$ format.

What's that mean? Of course - we need to find our *y*-intercept and our slope to finish our equation.

First, let's find the *y*-intercept. It's not difficult: the line crosses the *y*-axis at a value of 12. (Remember, the question calls this the "*C*-intercept", but that's just an alternative name for a familiar concept.)

Now we need to find the slope. In this case, it's probably best to find two clearly-marked points and then calculate the slope from those two points.

Can you find two clear, definitive points to read off the graph? I'll use $(0,12)$ and $(3,30)$. These fall clearly on the grid lines and there's no worry about misreading the exact coordinates. Now let's use the slope equation:

$$\frac{y_2 - y_1}{x_2 - x_1}$$
$$= \frac{(30)-(12)}{(3)-(0)}$$
$$= \frac{18}{3}$$
$$= 6$$

Now we know the slope is 6, or $6/hr as the cost of the boat rental per hour.

Put together the slope and the *y*-intercept and you get:

$$y = mx + b$$
$$y = 6x + 12$$

For consistency, let's change the letters of the variables to the same letters used in the question:

$$y = 6x + 12$$
$$C = 6h + 12$$

We can see that our equation is the same as Choice B.

18. **C.** On the surface, this question is just another Linear Equation based on a word problem. And basically, that's all it is. Let's get started with the basics: we'll need to figure out the initial *y*-intercept value, and the slope as well.

The *y*-intercept is just the starting value, which is given in the word problem as "3.7 million tons."

The slope must be negative, because the production is *decreasing* over time. To calculate the exact slope, we'll need to use the $\frac{y_2 - y_1}{x_2 - x_1}$ equation. I'll use the *later* year $(2002, 1.3)$ as Point 2, and the earlier year $(1990, 3.7)$ as Point 1.

$$\frac{y_2 - y_1}{x_2 - x_1}$$
$$= \frac{(1.3)-(3.7)}{(2002)-(1990)}$$
$$= \frac{-2.4}{12}$$
$$= .2$$
$$= -\frac{1}{5}$$

Just to be totally clear, I used my calculator to calculate the value of $\frac{-2.4}{12}$, which came out as "$-.2$". Then I put this back into fractional form as $-\frac{1}{5}$ to fit the format of the answer choices, which use fractions for the slope instead of decimals.

In any event, we're basically done with the question. The slope is $-\frac{1}{5}$ and the *y*-intercept is 3.7.

Put it together and we get:

$$y = mx + b$$
$$y = -\frac{1}{5}x + 3.7$$

The only remaining change is to update our variables from "x" and "y" to "t" and "$f(t)$", according to the question:

$$y = -\frac{1}{5}x + 3.7$$
$$f(t) = -\frac{1}{5}t + 3.7$$

Note: if the notation "$f(t)$" is confusing to you, be sure to study the upcoming lesson on **Functions**.

19. **A.** There are two ways to deal with this question. The easier way is to understand that this is a Linear Equation based on the equation $m = 4.5t + 65.4$, which fits the Linear Equation form $y = mx + b$.

The word problem is also asking for the "increase in grams each month." Whenever you've got a $y = mx + b$ equation with a word question asking about a consistent increase or decrease, you can be sure it's a Linear Equation asking for a Slope value.

That makes this easy: we can just read the *m*-value of 4.5 and know that it's the slope. That gives us the final answer of +4.5 grams per month.

20. $\frac{1}{3}$ or **.333.** This question is very similar to the previous one. Again, we have a Linear Equation supported by a real-life word problem, and again we're being asked for the value of the slope.

On the other hand, something looks a little "off" about this original equation. It would help to adjust the format of the original equation so that it better fits the $y = mx + b$ format that we've become so comfortable with:

$$L = \frac{t+8}{3}$$
$$L = \frac{1}{3}t + \frac{8}{3}$$

Now we can read the slope of this line directly: it's $\frac{1}{3}$. In other words, the car travels an additional $+\frac{1}{3}$ inches for each +1 second of travel time. And that's our final answer: for each second of time, the distance increases by $\frac{1}{3}$ of an inch.

You could also enter your answer as .333, since the SAT Free Response questions allow up to four characters for entry, and fractions and decimals are both acceptable. However, on test day you must remember to fill all four Free Response spaces, since $\frac{1}{3}$ is an infinitely repeating decimal value.

21. **D.** This word problem perfectly describes a linear equation of the form $y = mx + b$. There is an "initial value", which we know by now can be called a *y*-intercept or *b*-value. That initial value is 80 seconds, the length of time Jerry can hold his breath starting in Week 2.

There's also a consistent increase; the question itself tells us so - "the length of time... increased each week by a constant amount." That consistent weekly increase would be represented by the *slope* of our Linear Equation, or *m*-value.

Furthermore, the word problem asks us to describe how the length of time *changes* from Week 2 to Week 10. The *change* in a Linear Equation is always due to its slope. In other words, this word problem is asking us for details regarding the slope of the equation.

To proceed any further, we should probably calculate the precise slope using the $\frac{y_2 - y_1}{x_2 - x_1}$ equation. We should use the Week 10 value as Point 2, and the Week 2 value as Point 1. Also note that I'll put the seconds on top, and the weeks on bottom, to yield final units of "seconds per week," which fits the format of the answer choices better than the alternative which would be "weeks per second":

$$\begin{aligned}&\frac{y_2 - y_1}{x_2 - x_1}\\&= \frac{(110 \text{ seconds}) - (80 \text{ seconds})}{(\text{Week } 10) - (\text{Week } 2)}\\&= \frac{30 \text{ seconds}}{8 \text{ weeks}}\\&= 3.75 \text{ seconds per week}\end{aligned}$$

Now that we've calculated the slope of this situation, we know that Jerry will increase his breath-hold time by 3.75 seconds per week.

Unfortunately, none of the answers perfectly fit our work so far. But we can eliminate several choices. For example, Choice A is obviously incorrect. So is Choice B.

However, don't misread Choice C out of haste. The number "3.75" is correct, but this choice says he increases by 3.75 seconds every *four* weeks, although we just calculated an increase of 3.75 seconds for *each* week of practice. Therefore, Choice C is also wrong. It's a trap!

Choice D may fly under the radar at first, but if we check the math we can see it is correct. Try multiplying 3.75 seconds per week by 4 weeks:

$$\begin{aligned}&(3.75 \text{ seconds per week})(4 \text{ weeks})\\&= 15 \text{ seconds per 4 weeks}\end{aligned}$$

The math checks out. For every four weeks of training, Jerry will increase his breath-holding time by 15 seconds.

22. **C.** This question gives us a scatterplot with "real" data (represented by the individual points on the graph) and an idealized "line of best fit". The question asks us for the *difference* between one of the *actual* data points and the *predicted* value from the line of best fit.

Specifically, we're focused on Day 10. This question is actually quite easy. We just need to find the value of the actual data point on Day 10, the predicted value of the line of best fit at Day 10, and use subtraction to find the difference between them.

The data point for Day 10 lands exactly at $90 remaining. Be sure to carefully check your grid and not misread it. The line of best fit for Day 10 predicts exactly $80 remaining. The difference of the actual data and the predicted line of best fit is $90 - 80 = 10$ or $10, our final answer.

23. **5.** This is another variation on the common "relationship of $y = mx + b$ equation with a real-world situation" question that we've seen throughout this lesson. In this case, we're asked to turn the word problem into a $y = mx + b$ while making sure to interpret how each part fits together.

Remember that the *b*-value is the "*y*-intercept," which can also be thought of as the "initial" or "starting" value. That actually makes this question *really* easy - the company starts with 5 employees, so that's the initial value or *y*-intercept already. We don't even need to calculate the slope. The final answer is 5.

24. **C.** There's a giveaway in the words that tells us this question needs a Linear Equation setup. That giveaway is the phrase "depreciates... at a *constant rate*." The value of the music equipment starts an initial value of $44,800 (this is perfect for a *y*-intercept). Then it decreases at a constant rate for 20 years, at which point it has a value of $0.

That's the perfect information to calculate slope using $\frac{y_2 - y_1}{x_2 - x_1}$. We'll use the original value at zero years $(0, 44800)$ as Point 1 and the final value at 20 years $(20, 0)$ as Point 2, then plug in and calculate the slope:

$$\begin{aligned}&\frac{y_2 - y_1}{x_2 - x_1}\\&= \frac{(0) - (44800)}{(20) - (0)}\\&= \frac{-44800}{20}\\&= -2240\end{aligned}$$

Now we know the slope is -2240, meaning that each year, the equipment decreases in value by a consistent $2240 until it goes to a final value of zero dollars.

This allows us to set up a complete $y = mx + b$ equation based on time, with $44,800 as the starting value or *y*-intercept:

$$\begin{aligned}y &= mx + b\\y &= -2{,}240x + 44{,}800\end{aligned}$$

Note that in this equation, x represents the number of years since the equipment was purchased, and y indicates the current value of the equipment.

Now plug in "8 years" for x to give the question what it wants, "the value of the equipment 8 years after it is purchased":

$$\begin{aligned}y &= -2{,}240x + 44{,}800\\y &= -2{,}240(8) + 44{,}800\\y &= -17{,}920 + 44{,}800\\y &= 26{,}880\end{aligned}$$

Finished! By using the information given in the word problem, we were able to calculate the slope, identify the starting value, and set up a $y = mx + b$ Linear Equation. From there, we plugged in the desired number of 8 years for x and found the resulting value of $26,880 for y.

25. **A.** This question contains a giveaway in the wording that reveals it's a linear equation: "increase the number... by p performers per month." As you probably know by now, that means that p must be the slope. This will multiply times m, the number of months that have passed since the manager began increasing the performers per month.

Furthermore, the coffee shop already begins with x performers before the manager ever starts increasing the number of performers. That means x will represent the initial value or *y*-intercept.

We can easily set up a $y = mx + b$ equation for this situation. We don't know any actual numbers, but it's no problem to use variables in their place. You should be very comfortable with all the components of this equation by now, so I'll just make the substitutions below:

$$\begin{aligned}y &= mx + b\\t &= pm + x\end{aligned}$$

I've plugged in t for y, because t is the "total" number of performers in any given month after the new performers are added to the initial number of performers.

The only thing that could possibly be confusing is the way we're using *only* unknown variables instead of any numbers. But that's OK - a variable like p can represent the slope just as easily as a specific number could.

Note that Choices C and D represent **Exponential Growth**, which will be covered in later chapters, and contrasts with Linear Equations.

26. **D.** This question asks us to identify the correct $y = mx + b$ for the data given in chart form. The word problem is only there to explain the table and guide us as we set up that equation.

To tell the truth, the easiest thing to do here is *not* to try and create a $y = mx + b$ equation yourself - although you could. It will be faster and simpler to test the four answer choice by plugging in values from the table into each answer choice and eliminating any that don't work.

Remember - *whenever* we use a "testing answer choices" strategy, we *cannot* prove answers true - we can only prove them *false*. This is an elimination-based strategy.

Take any line of the table you want - it shouldn't matter. All of them should work. Let's just use the top one, for Oakview Apartments.

Note well: we are required to enter the purchase price in thousands of dollars, as the question states. That means instead of entering 1,600,000 for p, we need to divide by 1000 and use 1,600 instead.

Now let's start testing. First, Choice A:

$$r(p) = .006p + 2400$$
$$12{,}000 = .006(1600) + 2400$$
$$12{,}000 = 9.6 + 2400$$
$$12{,}000 \neq 2409.6$$

We see that Choice A doesn't provide a true equation, and we can eliminate it from consideration. Onto Choice B:

$$r(p) = 8p - 800$$
$$12{,}000 = 8(1600) - 800$$
$$12{,}000 = 12{,}800 - 800$$
$$12{,}000 = 12{,}000$$

This equation returns a true value (for now) so we'll keep it. On to Choice C:

$$r(p) = 7p - 400$$
$$12{,}000 = 7(1600) - 400$$
$$12{,}000 = 11{,}200 - 400$$
$$12{,}000 \neq 10{,}800$$

OK, good news. Choice C does not return a true equation and can be eliminated. Let's move to Choice D:

$$r(p) = 6p + 2400$$
$$12{,}000 = 6(1600) + 2400$$
$$12{,}000 = 9600 + 2400$$
$$12{,}000 = 12{,}000$$

Alright, unfortunately Choice D also works. That means both Choice B and D are both still possibilities.

What to do? Let's move down the table to the next row and use The Hacienda. Remember again to divide the purchase price of 2,100,000 by 1000 to get 2,100.

Test Choice B again:

$$r(p) = 8p - 800$$
$$15{,}000 = 8(2100) - 800$$
$$15{,}000 = 16{,}800 - 800$$
$$15{,}000 \neq 16{,}000$$

Well, it seems that Choice B no longer works. Perhaps we should test Choice D to be sure:

$$r(p) = 6p + 2400$$
$$15{,}000 = 6(2100) + 2400$$
$$15{,}000 = 12{,}600 + 2400$$
$$15{,}000 = 15{,}000$$

We're in luck! Choice D has worked twice, and the other three choices were all eliminated. It might seem like a lot of work, but it's simple and repetitive work - just plugging in values and testing the results. This is a very effective strategy, as long as you work efficiently and accurately - and yes, it's exactly how I'd handle this question on my own SAT test (even with the timer running).

27. **B.** I've seen a question like this on the real SAT, and it's caused many of my students some grief. The main difficulty with this question lies in its *weirdness* and confusion over what exactly it's asking for.

The idea is that we take the bar chart and convert it into a scatterplot. There will be one point for each of the six categories of vehicles. The *x*- and *y*-coordinates of each point will be based on the year: the *x*-coordinate is the number of vehicles in 1990, and the *y*-coordinate is the number of vehicles in 2000, like so: (# in 1990, # in 2000).

So, for example, the coordinate for cars would be at approximately $(290, 540)$. The coordinate for motorcycles might be at approximately $(80, 70)$.

OK, make sure you understand how we find the coordinates for each of the vehicles. Now, the next part of the question asks "how many data points would be *above* the line $y = x$?"

Imagine that line. It has a *y*-intercept of 0, so it goes straight through the origin. It has a slope of 1, so it goes up +1 for each unit that it goes to the right by +1.

Here's the connection we need to make: any of our scatterplot points that has a *y*-value (or "number in 2000") *higher* than its *x*-value (or "number in 1990") would be *above* that line.

So, any vehicle category with more vehicles in 2000 than it had in 1990 will count as a point above the line $y = x$. Make sure you understand this. Reread it and sketch out a graph, if that's what it takes.

But, once you understand this, it's easy to finish: just count out all the vehicles that had higher numbers in 2000 than they had in 1990. That includes cars, SUVs, and 18-wheelers - a total of three vehicle categories.

Therefore, our answer is 3. There are three types of vehicles that had higher numbers in 2000 than in 1990, and are therefore above the line $y = x$.

28. **1.** This question calls back to several questions we saw in the previous lesson. The central issue of this question is that "no value of x satisfies the equation," which seems pretty weird by most students' standards.

But, if you saw these in the previous lesson, you'll remember what to do. Think of this equation as a pair of Linear Equations - one on the left side, and another on the right side. The "solution" to a pair of Linear Equations is the point where the two lines intersect. But if there is *no* solution to the pair of equations, it means they *never* intersect. And the only way two lines *never* intersect is if they are *parallel* to each other - which means they must both have *the same slope*.

So, our plan should be to split this into two $y = mx + b$ equations, identify their two slopes, and then set those two slopes equal to each other.

The left side of the equation is ready to go: $4nx + 27$.

On the right side, we'll need to clean up a bit first:

$$6(3+x) - 2(x-1)$$
$$= 18 + 6x - 2x + 2$$
$$= 4x + 20$$

So, our pair of Linear Equations can be written as:

$$y_1 = 4nx + 27$$
$$y_2 = 4x + 20$$

Now, compare the two slopes of these equations. Line 1 has a slope of $4n$, and Line 2 has a slope of 4. Set the slopes equal to each other to make the lines parallel:

$$4n = 4$$

Now just solve for n:

$$4n = 4$$
$$\frac{4n}{4} = \frac{4}{4}$$
$$n = 1$$

Done : $n = 1$.

I want you to make sure you completely understand the thought process behind this type of question, which is why I tested it again after the previous lesson.

The concept itself - understanding that an equation without a solution is the same as two lines that never intersect, and therefore must be *parallel* - is more challenging than any of the actual algebra.

29. **B.** This has to be one of the toughest word problems in this chapter. But that's not saying much - most of the Linear Equations we've worked with so far have not tested our abilities much beyond what we developed in the previous lesson.

It's still based on the $y = mx + b$ equation. How do we know for sure? The word "consuming a steady volume of gasoline per mile" is a dead giveaway that there's a steady linear decrease that can be modeled with a Linear Equation.

That phrase also means that it's essential for us to calculate the *slope* of this Linear Equation before moving on. But there's a bit of a trick, because this question makes use of a percentage. That percent is a trap to make advanced students think the question is about **Exponential Growth** (which we cover in a future lesson). Furthermore, we have to use the basics of **Percents**, which will also explore more in a later chapter.

For now, if you're shaky on Percents, just trust me when I do the following maneuver. I'm going to figure out just how much gas is equivalent to "22%" of the original full tank.

$$400(.22)$$
$$= 88$$

I've found that 22% of the full tank of 400 quarts is equivalent to 88 quarts of gasoline.

Also, we know that these 88 quarts were consumed during the first 110 miles of driving. We can use this to determine the number of quarts per mile:

$$\frac{\text{quarts}}{\text{miles}}$$
$$= \frac{88 \text{ quarts}}{110 \text{ miles}}$$
$$= .8 \text{ quarts per mile}$$

The fuel consumption per mile is .8 quarts. This would be perfect for a slope. Remember that this must be a *negative* slope, since the volume of gas is *decreasing*.

Now that we have the slope, we need the *y*-intercept or initial value. The tank started with 400 quarts of gas, so that's our initial value.

Now we know the *y*-intercept and the slope, and we can set up our complete $y = mx + b$ equation. Let's make sure to fit the question and use v instead of y, and m instead of x.

$$y = mx + b$$
$$v = -.8m + 400$$

And, we're done. We didn't fall for the "exponential percent" trap and instead used the percent to calculate how much gas the RV used in the first 110 miles. Then we calculated the fuel consumption per mile, which became our slope of $-.8$. The question gave us our *y*-intercept as initial amount of gas (400 quarts).

Put it all together in the $y = mx + b$ equation and adjust the letters we're using as variables to fit the ones used in the word problem. Our final equation matches Choice B.

30. **3.5.** OK - luckily we're ready for this question. We can spot the telltale giveaway of a Linear Equation: "for each additional 100 meters... the temperature... decreases by k ...".

This is a dead giveaway that there's a Linear relationship between depth and temperature. But our work is just beginning. To calculate the constant decrease in the question, we'll need a slope.

We have enough information to set up a $\frac{y_2 - y_1}{x_2 - x_1}$ equation. For Point 1, we'll use the starting depth of 200 meters and starting temperature of 67 degrees. Since we're calculating temperature based on depth, the temperature will be our *y*-value and the depth will be our *x*-value. So, our starting Point 1 can be represented as $(200,67)$. Then, Point 2 can be written as the other depth and temperature given in the question, $(1000,39)$.

Now put them together and calculate slope:

$$\frac{y_2 - y_1}{x_2 - x_1}$$
$$= \frac{(39)-(67)}{(1000)-(200)}$$
$$= \frac{-28}{800}$$
$$= -0.035$$

But, there's a problem here. This number can't possibly be entered into the 4-character space available for SAT Math Free-Response questions. It's too long. Why is that?

Well, the wording of the question specified that we're supposed to be calculating the decrease k for every *100* meters, and we just calculated it for every *1* meter.

It's simple enough to fix the problem: just multiply our previous temperature-change-per-meter by 100:

$$(-0.035)(100)$$
$$= -3.5$$

Now we know that the temperature decreases by 3.5 degrees for every additional 100 meters in depth.

One last thing: note that the word problem states that the temperature *decreases* by k degrees. Although the temperature is decreasing, our k value should actually be a positive number.

Therefore, we have to be alert enough not to give -3.5 as our final answer. Enter it as 3.5 instead.

Lesson 11: Probability

Percentages

- 1.2% of Whole Test
- 0% of No-Calculator Section
- 1.8% of Calculator Section

Prerequisites

- Basic Algebra 1
- Algebra 1 Word Problems
- Careless Mistakes

Probability is a way of calculating the likelihood of a certain event happening or a certain choice being made from a variety of possibilities. We calculate Probability when there are a variety of possible outcomes, but we want to know the chance of a certain *specific* outcome happening from all those possibilities.

On the SAT Math test, Probability is a very simple topic calculated via a fraction I call "Desired over Total," which will explore thoroughly in this lesson.

Probability only seems to be encountered in the Calculator portion of the SAT Math Test. I've never seen this topic in the No-Calculator section. At most, you will probably only see a single Probability question in any SAT Math test. It may not even appear on every test. However, it's important to be ready in case it does.

Probability Quick Reference

- Probability is calculated with a "Desired over Total" setup.
- Probabilities may be expressed as fractions, decimals, or percentages.
- Probability Questions on the SAT always involve a Word Problem and almost always include a table or chart as well.
- Set up a "Word Fraction" to get your ideas clear before inputting any numerical values.
- Watch out for deceptive or confusing word problems. Don't make any hasty assumptions without reading carefully.

Intro to Probability

Probability questions on the SAT Math always begin with a word problem, and usually include a chart or table of some kind as well.

Probability is calculated with the fraction "Desired over Total." "Total" means the number of *all* possible choices. "Desired" means the group you *want* to pick, which is always a smaller subset of the total number.

Use a "Word Fraction" first (like we did in **Ratios & Proportions**). Turn the SAT's word problem into your own Word Fraction first, *then* turn your Word Fraction into numbers. It's safer than translating from the SAT's words directly into numbers.

Final probability answers are expressed either as a fraction, decimal, or percentage.

Try out this super-basic Probability question to see what I mean.

> There are 10 students in a class. The favorite color of three of these students is red. If a student is picked at random from the class, what is the probability that their favorite color is red?

This is very simple. First, set up a Word Fraction for "Desired over Total". The "total" is all the students in the class. The "Desired" is the students whose favorite color is red:

$$\frac{\text{desired}}{\text{total}}$$
$$=\frac{\text{red}}{\text{all students}}$$

Now plug in the numerical values. There are 10 total students in the class, and 3 of them prefer red:

$$\frac{\text{red}}{\text{all students}}$$
$$=\frac{3}{10}$$

The probability that a random student's favorite color will be red is $\frac{3}{10}$. It can be expressed as a fraction, or given as the decimal "$.3$" or the percentage "30%".

Of course, this was a very simple example. Watch out for deceptive wordings in harder questions. In longer or more complicated word problems, you must be extra-careful to double-check the value of the "Total."

Some Probability problems on the SAT will be purposefully confusing or deceptive about what to use for the "total". Some questions only use a *subset* of the data, rather than the entire set. So, just because a chart or table has a row marked "Total" does *not* always mean that's the same "total" you'll use for your Probability fraction.

Read carefully and match your "Desired" and "Total" values to the specific wording of each question. We'll explore this more within this lesson, so don't worry - you'll soon see what I mean. It will make more sense when we get to some specific examples.

Pretest Question #1

Let's take a look at our first Pretest question on this topic. Try it yourself if you got it wrong the first time. Be sure to read the Word Problem carefully!

Peaches	5
Cherries	2
Bananas	6
Apples	9
Grapes	8
Total	30

(CALCULATOR) The table above shows the favorite fruits for each student in a class of 30. If a student is picked at random from all students whose favorite fruit is not peaches or apples, what is the probability that their favorite fruit is either cherries or bananas?

(A) $\frac{4}{15}$

(B) $\frac{1}{3}$

(C) $\frac{1}{2}$

(D) $\frac{4}{7}$

Let's start by using the "Desired over Total" setup and translating the words of the question into our own "Word Fraction" to avoid any misreadings or careless mistakes.

$$\frac{\text{Desired}}{\text{Total}}$$

$$= \frac{\text{Cherries or Bananas}}{\text{All Except Peaches and Apples}}$$

We are asked to *exclude* any students whose favorite fruits are either peaches or apples. Any students who like peaches or apples will be completely excluded from our calculations.

Notice the "Total" on the bottom of my Word Fraction excludes "peaches" and "apples" from the "total," and thus avoids the mistake of using *all* 30 students in my setup.

Now plug in values from the table:

$$\frac{\text{Cherries or Bananas}}{\text{All Except Peaches and Apples}}$$
$$= \frac{2+6}{30-5-9}$$

Check my numbers. On top I've included all students who prefer Cherries (2 students) or Bananas (6 students). On bottom I've started with all 30 students, but *removed* any students who prefer Peaches (5 students) or Apples (9 students). Now simplify and reduce the fraction to finish the question:

$$= \frac{2+6}{30-5-9}$$
$$= \frac{8}{16}$$
$$= \frac{1}{2}$$

We can see that the probability for this question will be $\frac{1}{2}$, or **Choice C**.

Probabilities Involving Algebra 1

It's also possible for the SAT to *give* you a final probability value, then ask you to calculate a missing or unknown value from the data set - basically, to work backwards from our previous example.

In this type of problem, you'll set up our "Desired over Total" fraction first. As always, it's smart to use a "Word Fraction" to get your ideas down on paper clearly first.

Then, you'll fill in your Word Fraction with any values you can - but, you will also need to use a variable, like x, in place of the missing value you're trying to solve for.

Next, set your fraction equal to the probability given in the question. Now you'll have an Algebra setup that you can solve using **Basic Algebra 1** to find your missing value.

Let's take a look at a simple example to make this concept more clear.

Red	7
Blue	5
Yellow	x
Green	6

A bag is filled with marbles of four different colors. The number of marbles is shown in the table above. The probability of drawing a green marble is $\frac{3}{10}$. What is the value of x?

To solve this problem, we'll follow the steps explained above.

First, let's set up a "Word Fraction" for the Desired over Total:

$$\frac{\text{Desired}}{\text{Total}}$$
$$= \frac{\text{Green Marbles}}{\text{Total Marbles}}$$

Now, plug in the values we know from the table, including the variable x for yellow marbles:

$$= \frac{\text{Green Marbles}}{\text{Total Marbles}}$$
$$= \frac{6}{7+5+x+6}$$
$$= \frac{6}{18+x}$$

Now, set this equal to the probability of $\frac{3}{10}$, which was given by the word problem:

$$\frac{6}{18+x} = \frac{3}{10}$$

This is our complete Algebra setup that will enable us to solve this Probability question.

Now use Cross-Multiplication to solve for the value of x.

$$\frac{6}{18+x} = \frac{3}{10}$$
$$(6)(10) = (18+x)(3)$$
$$60 = 54 + 3x$$
$$-54 \quad -54$$
$$6 = 3x$$
$$\frac{6}{3} = \frac{3x}{3}$$
$$2 = x$$

Now we know the value of x is 2, so there must have been **2 yellow marbles**.

Pretest Question #2

Let's take a look at another Pretest question. Try it yourself before you look at my explanation below the question.

Dream Topic	Number of Students
Going to School	10
Surfing	3
Robots	2
Animals	x
Seeing Friends	8
Couldn't Recall	12

(CALCULATOR) FREE RESPONSE: A class of students took a survey on the topic of their dreams from the previous night. The table above gives the topics of their dreams with the number of students who dreamed about each topic. Among the students who could remember their dreams, the probability of selecting a student at random who dreamt about either animals or surfing was $\frac{1}{5}$. How many students dreamt about animals?

As with any Probability question, our first step is to create a "Word Fraction" with a "Desired over Total" setup:

$$\frac{\text{Desired}}{\text{Total}}$$
$$= \frac{\text{Animals or Surfing}}{\text{All Students Who Remembered Dreams}}$$

Note well: the question asked us to only consider students who could *remember* their dreams. When we fill in our Word Fraction with values, we must remember *not* to include anyone who "couldn't recall" in our total.

Now let's fill in the Word Fraction with values and variables:

$$\frac{\text{Animals or Surfing}}{\text{All Students Who Remembered Dreams}}$$
$$= \frac{x+3}{10+3+2+x+8}$$

Notice that I've had to use x on top *and* bottom of my fraction. On top I've included the number of students who dreamt about "Animals" or "Surfing." On bottom, I've combined *all* the students who could remember the topic of their dreams. We've completely excluded any students in the "Couldn't Recall" category.

Now clean up the fraction by Combining Like Terms:

$$\frac{x+3}{10+3+2+x+8}$$
$$=\frac{x+3}{23+x}$$

We know from the question that the value of this probability is $\frac{1}{5}$, so let's set our work equal to $\frac{1}{5}$.

$$\frac{x+3}{23+x}=\frac{1}{5}$$

Now we can solve for the value of x using **Algebra 1**, starting with Cross-Multiplication:

$$\frac{x+3}{23+x}=\frac{1}{5}$$
$$(x+3)(5)=(23+x)(1)$$
$$5x+15=23+x$$
$$-15 \quad -15$$
$$5x=8+x$$
$$-x \qquad -x$$
$$4x=8$$
$$\frac{4x}{4}=\frac{8}{4}$$
$$x=2$$

There we go! The value of x, and therefore the number of students who dreamt about animals, is **2.**

Review & Encouragement

Probability questions are an uncommon but simple topic on the SAT Math test. We should be glad every time we see one, because they are basically "free points" (as long as we read carefully and avoid **Careless Mistakes**).

The most important thing to remember for Probability is the "Desired over Total" setup. These magic words can untangle every Probability question on the SAT test.

Just be sure to read the word problem *very* carefully and use Word Fractions to clearly set up your fractions with words before proceeding to input any numbers.

The two most common issues my students have on these questions are either not remembering to use "Desired over Total" for their setup, or making hasty assumptions and rushing to plug in numbers before they get clear on exactly what the "Desired" and the "Total" are supposed to be. Again, adding a "Word Fraction" step into your setup process will cut down on these mistakes substantially.

We may also set up a **Basic Algebra 1** equation to solve for missing values in a Probability table. As we've seen, we can calculate *backwards* to find missing data values if we are already given the final probability.

Now practice with the following set of Probability questions!

Probability Practice Questions

YOU MAY USE A CALCULATOR ON ALL PROBLEMS IN THIS PRACTICE SET.

1. A certain hospital currently contains 319 patients, 25 nurses, 8 doctors, and 48 visiting family members. If a person is picked at random from every person currently in the hospital, which of the following choices is closest to the probability that they are a nurse?

(A) .063

(B) .066

(C) .25

(D) 16

Type of Snack	Number of Children
Popcorn	7
Licorice	1
Ice Cream	12
Chips	18
Sandwich	3
Carrots	6
No Response	3

2. The table above shows the results of a survey of a group of 50 children who were asked their favorite type of snack. If a child is picked at random from this group, which of the following is closest to the probability that their favorite snack is either popcorn or carrots?

(A) .12

(B) .14

(C) .26

(D) .35

Genre	Number of Respondents
Science Fiction	264
Fantasy	423
Mystery	637
Non-Fiction	168
Sports	142
Unsure	205

3. The table above shows the results of a survey of 1,875 readers and their favorite genre of books. If one of these readers is selected at random, which of the following is closest to the probability that their favorite genre is NOT mystery or sports?

(A) .34

(B) .42

(C) .47

(D) .58

	Drama	Comedy	Total
Animated	2	45	47
Live-Action	28	15	43
Total	30	60	90

4. A group of moviegoers were asked for their preferences in movies. The table above shows those moviegoers' preferences. Based on the data in the table, what is the probability, rounded to the closest percentage, that a member of this subset who prefers animated movies over live-action movies does NOT prefer drama to comedy?

   (A) 2%

   (B) 33%

   (C) 52%

   (D) 95%

Hours of Sleep	Night 1	Night 2	Night 3	Total
4-5 hours	1	2	1	4
5-6 hours	3	4	2	9
6-7 hours	4	1	3	8
7-8 hours	1	2	2	5
8+ hours	1	1	2	4
Total	10	10	10	30

5. FREE RESPONSE: 10 people took part in a sleep study for three consecutive nights. The table above shows the number of hours of sleep each participant got on each of the three nights. If a participant is selected at random, what is the probability that the selected participant slept for more than 8 hours on either Night 1 or Night 3, given that the participant slept for more than 8 hours on at least one night of the study?

	Yes	No	N/A	Total
Issue A	15	5	3	23
Issue B	9	13	1	23
Issue C	12	4	7	23
Issue D	13	8	2	23
Total	49	30	13	92

6. The table above shows the results of 23 voters who each voted "yes," "no," or "no answer" to four different issues. If one of these voters is chosen at random, what is the probability that the voter did NOT vote "yes" on Issue B, given that they did not vote "no" on Issue B?

(A) 0

(B) $\frac{1}{10}$

(C) $\frac{9}{23}$

(D) $\frac{13}{23}$

Pet	Number of Owners
Dogs	5
Cats	n
Birds	7

7. FREE RESPONSE: The table above shows the results for a survey of a small group of pet owners who each owned one pet per person. If a person is selected at random from the surveyed group, there is a 20% probability that they own a cat. What is the value of n?

	Male	**Female**
Freshman	15	17
Sophomore	12	x
Junior	16	18
Senior	11	14

8. FREE RESPONSE: The table above gives data about a group of high school students who participated in a community service project. If a sophomore or freshman from this community service project is chosen at random, the probability that they are a female freshman is $\frac{1}{4}$. What is the value of x?

	Years of Experience				
Career	0-2	2-5	5-10	10-20	20+
Lawyer	3	4	3	1	2
Business Owner	1	2	3	4	1
Architect	2	2	2	1	1
Engineer	4	5	6	x	2
Artist / Musician	3	2	1	2	2
Medical	4	3	5	6	4

9. The table above shows a survey of people in a range of careers, along with the years of experience they have working in that field. Each person is only represented once within the table. If the probability is $\frac{1}{3}$ that a person with more than 10 years of experience selected at random from the table is an engineer or architect, what is the value of x?

(A) 4

(B) 5

(C) 7

(D) 9

	Extreme Sport			
	Hang-Gliding	Motorcycle Racing	Skateboarding	Snowboarding
Visited Hospital	2	4	5	1
Not Visited Hospital	?	2	?	6

10. The table above shows the results of a survey of extreme sports participants. Each survey participant is only represented once in the table. The survey results were smudged in two categories and are illegible. If the probability is 50% that a survey participant has NOT been to the hospital and is NOT involved in motorcycle racing, what is the total number of hang-gliding and skateboarding participants?

(A) 8

(B) 15

(C) 19

(D) 23

Probability Answers

1. A
2. C
3. D
4. D
5. .75 or $\frac{3}{4}$
6. B
7. 3
8. 24
9. C
10. B

Probability Explanations

1. **A.** In all Probability questions on the SAT Math, we'll be using our trusty "Desired over Total" fraction. Set it up first as a Word Fraction:

$$\frac{\text{Desired}}{\text{Total}} = \frac{\text{Nurses}}{\text{All People}}$$

Now plug in the values we are given in the question and simplify:

$$= \frac{\text{Nurses}}{\text{All People}}$$
$$= \frac{25}{319 + 25 + 8 + 48}$$
$$= \frac{25}{400}$$
$$= .0625$$

This final value of .0625 rounds to the nearest answer .063, Choice A.

2. **C.** We'll be using our "Desired over Total" approach to Probabilities and setting it up as a Word Fraction first:

$$\frac{\text{Desired}}{\text{Total}} = \frac{\text{Popcorn or Carrots}}{\text{All Children}}$$

Now plug in values from the table (the word problem also tells us there are 50 total children in the group) and simplify:

$$= \frac{\text{Popcorn or Carrots}}{\text{All Children}}$$
$$= \frac{7 + 6}{50}$$
$$= \frac{13}{50}$$
$$= .26$$

From this short and sweet work, we can easily calculate that our desired probability is .26, or Choice C.

3. **D.** We'll use our "Desired over Total" approach with a Word Fraction setup:

$$\frac{\text{Desired}}{\text{Total}} = \frac{\text{NOT Mystery or Sports}}{\text{All Readers}}$$

Note that our "Desired" value should include all readers *except* those whose favorites are Mystery or Sports. Now input values from the table:

$$= \frac{\text{NOT Mystery or Sports}}{\text{All Readers}}$$
$$= \frac{1{,}875 - (637 + 142)}{1{,}875}$$

Notice how I've saved some time on top. Instead of totaling up everyone from the table *except* Mystery or Sports, I've simply subtracted Mystery (637) and Sports (142) readers from the total readers (1,875) on top of my fraction. You could also add everything up if you prefer - my way just saves a couple of seconds.

Now finish off the calculations and the problem will be done:

$$= \frac{1{,}875 - (637 + 142)}{1{,}875}$$
$$= \frac{1{,}875 - (779)}{1{,}875}$$
$$= \frac{1{,}096}{1{,}875}$$
$$= .5845333...$$

It's a bit of an ugly decimal, so we should be glad to round to the closest value in the answer choices, which is .58, or Choice D.

4. **D.** We'll create a setup for the word problem with the "Desired over Total" and a Word Fraction. This step is particularly important for this question because of the deliberately confusing wording of the question.

In this probability, we're only including the people "who prefer Animated movies over Live-Action movies." Anyone who prefers Live-Action movies will be completely excluded from our calculation.

$$\frac{\text{Desired}}{\text{Total}}$$
$$= \frac{\text{Animated but NOT Drama}}{\text{All Animated Movies}}$$

Now input values from the table. Notice that "Animated but NOT Drama" leaves only one value from the table, which is "Animated Comedy."

$$= \frac{\text{Animated but NOT Drama}}{\text{All Animated Movies}}$$
$$= \frac{45}{47}$$
$$= .957...$$

The decimal value, expressed as a percent, rounds closest to Answer D, or 95%.

5. **.75** or $\frac{3}{4}$. As with all SAT Math problems based on Probability, we'll start by setting up a "Desired over Total" fraction, using a Word Fraction to get our ideas arranged clearly before we start inputting values:

$$\frac{\text{Desired}}{\text{Total}}$$
$$= \frac{\text{More than 8 Hours on Night 1 or Night 3}}{\text{More than 8 Hours on Any Night}}$$

Be sure you understand the Word Fraction before proceeding to input values in the next step:

$$= \frac{\text{More than 8 Hours on Night 1 or Night 3}}{\text{More than 8 Hours on Any Night}}$$
$$= \frac{1+2}{1+1+2}$$

Can you follow where my numbers are coming from? They are the values in the table that correspond directly to my Word Fraction setup.

Now simplify and finish the calculation:

$$= \frac{1+2}{1+1+2}$$
$$= \frac{3}{4}$$
$$= .75$$

You can enter your final answer as either $\frac{3}{4}$ or .75.

6. **B.** As with other questions in this lesson, we'll keep using the combination of "Desired over Total" setup plus an initial Word Fraction to arrange our ideas before inputting values from the table. This is a deliberately-confusing word problem, which makes setting up a clear Word Fraction more important than ever.

Notice that this probability is *only* selecting from voters who did *not* vote "No" on Issue B. Therefore, our pool of possible people for this probability is absolutely limited to voters who did NOT vote "No" on Issue B.

In other words, any voters who voted "No" on Issue B cannot be a part of *either* our "Desired' or our "Total" values - they must be completely ignored for this calculation.

$$\frac{\text{Desired}}{\text{Total}}$$
$$= \frac{\text{Did NOT Vote Yes (or No) on Issue B}}{\text{All Who Did NOT Vote No on Issue B}}$$
$$= \frac{\text{Voted N/A on Issue B}}{\text{All Who Voted Yes or N/A on Issue B}}$$

This can be a bit confusing, so take a moment to get clear. Remember that it's possible to vote "N/A" on Issue B, which allows someone to vote *neither* "Yes" nor "No."

Now plug in values:

$$= \frac{\text{Voted N/A on Issue B}}{\text{All Who Voted Yes or N/A on Issue B}}$$
$$= \frac{1}{9+1}$$
$$= \frac{1}{10}$$

The hardest part of this question is simply getting your Word Fraction correct for the "Desired Over Total" values. It would be very easy to accidentally include the "No" voters, but they must be completely excluded from this calculation - it's "given that they did not vote 'no' on Issue B."

7. **3.** Like all these Probability questions, we'll be using a "Desired over Total" combined with a Word Fraction setup. However, one of our values is currently unknown. That's OK - we'll be fine.

Starting with the Word Fraction setup:

$$\frac{\text{Desired}}{\text{Total}}$$
$$=\frac{\text{Cat}}{\text{All Animals}}$$

Now we'll input our values; the only difference is we'll have to use n to represent our value for "Cats":

$$=\frac{\text{Cat}}{\text{All Animals}}$$
$$=\frac{n}{5+n+7}$$
$$=\frac{n}{12+n}$$

Good - our "Desired over Total" setup is complete. Now, the question *gives* us that the final probability for this is "20%" (a decimal value of .20). Set our fraction above equal to .2:

$$\frac{n}{12+n}=.2$$

Now use **Basic Algebra 1** to solve for the value of n.

$$\frac{n}{12+n}=.2$$
$$n=(.2)(12+n)$$
$$n=2.4+.2n$$
$$-.2n \qquad -.2n$$
$$.8n=2.4$$
$$\frac{.8n}{.8}=\frac{2.4}{.8}$$
$$n=3$$

There we go! The value of n must be 3.

8. **24.** We'll start by setting up a "Desired over Total" with a Word Fraction. We're asked to select a female freshman from all sophomores and freshman. Notice that our group or "Total" is limited to *only* sophomores and freshman:

$$\frac{\text{Desired}}{\text{Total}}$$
$$=\frac{\text{Female Freshman}}{\text{All Sophomores \& Freshmen}}$$

Now plug in the values we have from the table. We'll need to use the variable x as a placeholder for Female Sophomores:

$$=\frac{\text{Female Freshman}}{\text{All Sophomores \& Freshmen}}$$
$$=\frac{17}{12+x+15+17}$$
$$=\frac{17}{44+x}$$

Our "Desired over Total" setup is complete. Now set equal to the given probability of $\frac{1}{4}$ and solve for x:

$$\frac{17}{44+x}=\frac{1}{4}$$
$$(17)(4)=(44+x)(1)$$
$$68=44+x$$
$$-44 \quad -44$$
$$24=x$$

The value of x, or Female Sophomores, is 24, and we're done.

9. **C.** One of the biggest issues with this question is simply the "information overload" of a big table with a big word problem. Don't be intimidated.

Start with your trusty "Desired over Total" setup into a Word Fraction before you try to input any specific values.

$$\frac{\text{Desired}}{\text{Total}}$$
$$=\frac{\text{Engineers \& Architects with more than 10 Years}}{\text{All People with more than 10 Years}}$$

Notice that we are completely restricted to *only* people with more than 10 years of experience. That includes both the top and bottom of our setup - *only* people with more than 10 years of experience are under consideration.

If you're clear on the Word Fraction setup, it's time to input all the values we can. There are a lot of values in this setup; be sure not to miss any:

$$\frac{\text{Engineers \& Architects with more than 10 Years}}{\text{All People with more than 10 Years}}$$
$$=\frac{x+2+1+1}{1+2+4+1+1+1+x+2+2+2+6+4}$$

Now simplify the fraction down:

$$\frac{4+x}{26+x}$$

Now, we can set our fraction equal to the given probability of $\frac{1}{3}$:

$$\frac{4+x}{26+x}=\frac{1}{3}$$

Now solve for x using **Basic Algebra 1**:

$$\frac{4+x}{26+x}=\frac{1}{3}$$
$$(3)(4+x)=(1)(26+x)$$
$$12+3x=26+x$$
$$-12 \qquad -12$$
$$3x=14+x$$
$$-x \qquad -x$$
$$2x=14$$
$$\frac{2x}{2}=\frac{14}{2}$$
$$x=7$$

After a clear setup and some basic algebra, we have our final answer - x must equal 7.

10. **B.** As always with Probability questions, our first step is to set up a "Desired over Total" fraction with a Word Fraction setup:

$$\frac{\text{Desired}}{\text{Total}}=\frac{\text{Not Been to Hospital \& Not a Motorcycle Racer}}{\text{All Survey Participants}}$$

Before we can plug in values from the table, we should probably create some variables for the two question marks missing from our table. I'll use *a* for Hang-Gliders who have not visited the hospital, and *b* for Skateboarders who have not visited the hospital. Now let's plug in:

$$\frac{\text{Not Been to Hospital \& Not a Motorcycle Racer}}{\text{All Survey Participants}}$$
$$=\frac{a+b+6}{2+4+5+1+a+2+b+6}$$

Be very sure that you understand our setup so far before you proceed.

Now simplify and clean up:

$$=\frac{a+b+6}{2+4+5+1+a+2+b+6}$$
$$=\frac{a+b+6}{20+a+b}$$

This represents our final "Desired over Total" setup.

Now set it equal to the given probability of 50%, or a decimal value of .5.

$$\frac{a+b+6}{20+a+b}=.5$$

And solve for the value of $a+b$. Notice that we can't solve for the individual values of a or b - we can only solve for both of them together.

$$\frac{a+b+6}{20+a+b}=.5$$
$$a+b+6=.5(20+a+b)$$
$$a+b+6=10+.5a+.5b$$
$$-6 \quad -6$$
$$a+b=4+.5a+.5b$$
$$-.5a \qquad -.5a$$
$$.5a+b=4+.5b$$
$$-.5b \qquad -.5b$$
$$.5a+.5b=4$$
$$(2)(.5a+.5b)=4(2)$$
$$a+b=8$$

OK, now we know that the combined value of a and b is 8. We're now in a position to solve the final question, which was "the total number of hang-gliding and skateboarding participants."

This wording includes both the groups that have *and* have not been to the hospital. Also keep in mind that the *combined* total of the two "question marks" (which we've labeled "a" and "b") in the table is 8, as we've just determined. Below, I'm using the abbreviations "Hos" for hospital visitors, "HG" for hang-gliders and "SB" for skateboarders, just to save space.

$$\text{HosHGs} + \text{HosSBs} + (\text{total NON - Hos HG's \& SBs})$$
$$=2+5+(8)$$
$$=15$$

Our final total is 15, or Answer Choice B.

Lesson 12: Charts & Tables

Percentages

- 3.8% of Whole Test
- 0% of No-Calculator Section
- 5.8% of Calculator Section

Prerequisites

- Basic Algebra 1
- Algebra 1 Word Problems
- Ratios & Proportions
- Careless Mistakes

In this lesson we will explore simple Charts and Tables questions that we frequently see on the SAT Math test.

Many of these questions are quite basic. Some are just about reading the Chart and interpreting the Word Problem. Many Charts and Tables questions barely have any "steps" of math - they're just about reading correctly.

On the other hand, some of these questions can overlap with other topics such as **Ratios & Proportions**, **Probability**, **Percentages**, **Basic Algebra 1**, and **Algebra 1 Word Problems**.

Basically, these are a grab-bag variety of topics that start with a Chart or Table combined with a Word Problem. The focus is on using the Word Problem to find the correct information from the Chart or Table and use any supporting math skills from other chapters.

This topic seem to show up much more on Calculator than No-Calculator section. It's common to see a large quantity of Charts and Tables within the Calculator Math section of the SAT test.

Charts & Tables Quick Reference

- Charts and Tables questions are based on careful reading. Most of these questions rely on "Information Overload" to confuse, intimidate, or distract you. Keep a clear head, read carefully, and stay patient.
- These questions are often based just on reading a chart or table. More difficult questions may incorporate Percentages, Ratios & Proportions, Basic Algebra 1, and other math topics.
- It's often wise to first set up a "Word Fraction" before entering specific numbers, like we did in the lessons on **Ratios and Proportions** and **Probability.**
- Never rush these questions. Misreading these dense, high-information problems is your biggest enemy.

Intro to Charts & Tables

Charts and Tables are simply a way of organizing large quantities of data into a visual format. Tables are just grids of data arranged in columns and rows. Charts can be more visual (for example, Pie Charts or Bar Charts).

Most Table questions on the SAT Math are simple word problems that require carefully pulling basic data from the correct "cells" of the table. Frequently, the only challenge is locating the correct cells to correspond with the word problem. You may set up a simple fraction or ratio from the word problem, but there are typically very few actual *calculations* involved.

Chart questions can be a bit trickier - mainly because there are a variety of small tricks the SAT can play when arranging data in a chart. These tricks are primarily intended to cause common misreading errors, especially if you rush through the problem or make hasty assumptions.

Nevertheless, the vast majority of Charts and Tables questions are quite simple, if you just take your time and avoid careless Misreading errors. Never rush to finish the question, and never assume you've read the word problem and data correctly until you double-check your work.

I also recommend using "Word Fractions" to translate the combined Table or Chart and the word question into clear written ideas before you start reaching for the numbers. We studied the "Word Fraction" technique in the lessons on both **Probability** and **Ratios and Proportions**, so review them if necessary.

Pretest Question #1

Let's take a look at our first Pretest question on this topic. Try it yourself if you got it wrong the first time.

		Coloration Style					
		Solid	Mirrored	Swirl	Matte	Two-Tone	Total
Primary Color	Red	11	1	2	3	4	21
	Green	4	5	3	2	6	20
	Purple	2	4	1	4	5	16
	Orange	6	8	5	2	2	23
	Yellow	3	2	6	1	1	13
	Blue	7	5	9	3	3	27
	Total	33	25	26	15	21	120

(CALCULATOR) A boy has a collection of marbles in a variety of primary colors and coloration schemes. His collection of marbles is summarized in the table above. What fraction of his green marbles have a swirled coloration pattern?

(A) $\frac{1}{40}$

(B) $\frac{3}{26}$

(C) $\frac{2}{15}$

(D) $\frac{3}{20}$

$\frac{3}{20}$

In this question, all we're being asked for is the fraction of "green" marbles that are "swirled." Let's set up a Word Fraction first:

$$\frac{\text{Swirled Green Marbles}}{\text{All Green Marbles}}$$

Notice that we are *only* focusing on green marbles. The "swirled green marbles" are a subset of the total green marbles.

Let's just read these values off our chart, plug them into our fraction, and finish the question. Don't rush and misread!

$$\frac{\text{Swirled Green Marbles}}{\text{All Green Marbles}} = \frac{3}{20}$$

Our final fraction is $\frac{3}{20}$, or **Choice D**. It's as easy as that. A bit anticlimactic, perhaps - given how much information was contained in the table. But, think of all that extra information that acts like a dense forest, trying to prevent you from finding the only two cells you need to complete the question.

There really isn't much more to say about Charts and Tables questions. Just keep reading carefully, setting up your Word Fractions, and taking your time instead of rushing.

Pretest Question #2

Let's take a look at another Pretest question. Try it yourself before you look at my explanation below the question.

Time (minutes)	Concentration (milligrams per liter)
0	12.0
3	17.4
6	25.2
9	36.6
12	53.0
15	76.9

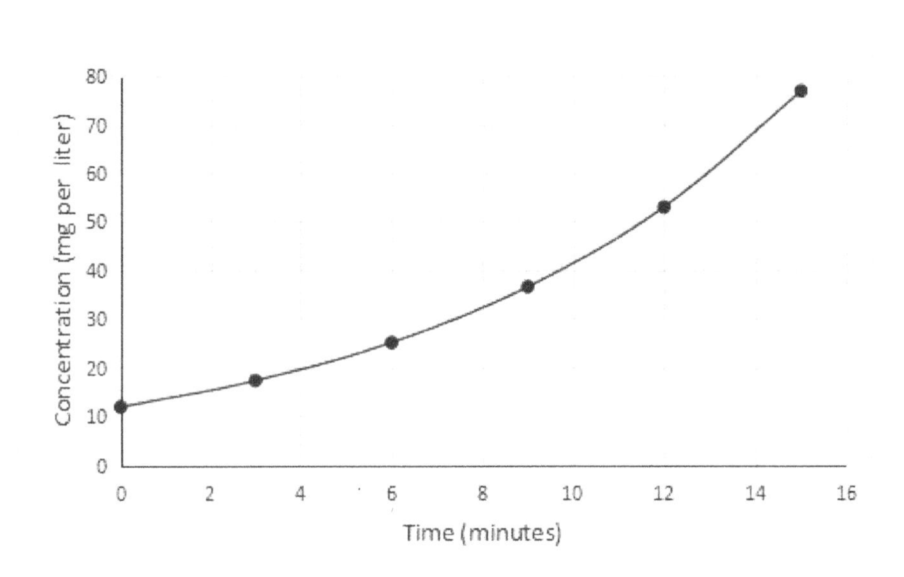

(CALCULATOR) FREE RESPONSE: The table and graph above show the concentration over time of a certain painkiller in the bloodstream of a hospital patient who is in surgery. According to the data, how many more milligrams of painkiller are present in 2 liters of the patient's blood after 12 minutes than in 1.5 liters of the patient's blood after 9 minutes? (Round your answer to the nearest tenth)

This question tries to overwhelm us with information from the very start. There's a table, a graph, *and* a word problem.

The most important thing in this situation is to determine exactly what you're trying to *find* amidst this sea of information.

We're asked for the *difference* of two amounts of painkiller. The first amount is present "after 12 minutes" and the second amount is present "after 9 minutes."

The table makes it easy to locate this information. At 12 minutes, the concentration is 53 mg/liter, and at 9 minutes the concentration is 36.6 mg/liter.

However, we're not done yet. The word problem specifies two *different volumes* of the patient's blood.

At 12 minutes, we're using "2 liters" of blood. This requires a bit of simple math that also calls back slightly to **Unit Conversions**.

We'll need to multiply the concentration of 53 mg/liter by the volume of 2 liters to get the total milligrams of painkiller, like so:

$$(53\frac{\text{mg}}{\text{liter}})(2\text{ liters})$$
$$=106\text{ mg}$$

Now let's calculate the total milligrams for the concentration at 9 minutes and "1.5 liters" of blood.

$$(36.6\frac{\text{mg}}{\text{liter}})(1.5\text{ liters})$$
$$=54.9\text{ mg}$$

Finally, we take these two values and use subtraction to find the difference between them:

$$106\text{ mg}-54.9\text{ mg}=51.1\text{ mg}$$

The question asks us to round to the nearest tenth, but it's not necessary in this case. Our final answer is **51.1**.

Review & Encouragement

Even in Pretest Question 2, which seems awfully confusing when you first see it, take note that there was very little *actual math* involved. The hardest part is reading and understanding the question - beyond that, there are only a few basic calculations.

These Charts and Tables questions are almost entirely easy to medium difficulty. When I see a student get this type of question wrong, it is almost always due to a Misreading mistake, *not* because they lack the necessary math skills. Typically they're either rushing the question or they're overconfident and make assumptions about what the question is asking for (or what the chart or table is saying).

Be sure to read carefully and check your work before giving your final answer, or you run the risk of throwing away free points and feeling foolish later.

Charts & Tables Practice Questions

YOU MAY USE A CALCULATOR ON ALL OF THE FOLLOWING QUESTIONS.

	Seeds	Pellets	Total
Finches	12	3	15
Parakeets	9	14	23
Total	21	17	38

1. The table above shows the food preferences of two types of bird in an aviary. What fraction of the parakeets prefer to eat seeds rather than pellets?

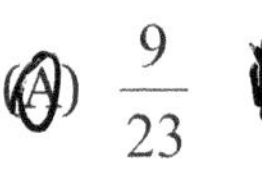

(A) $\frac{9}{23}$

(B) $\frac{3}{7}$

(C) $\frac{9}{14}$ ✗

(D) $\frac{4}{5}$

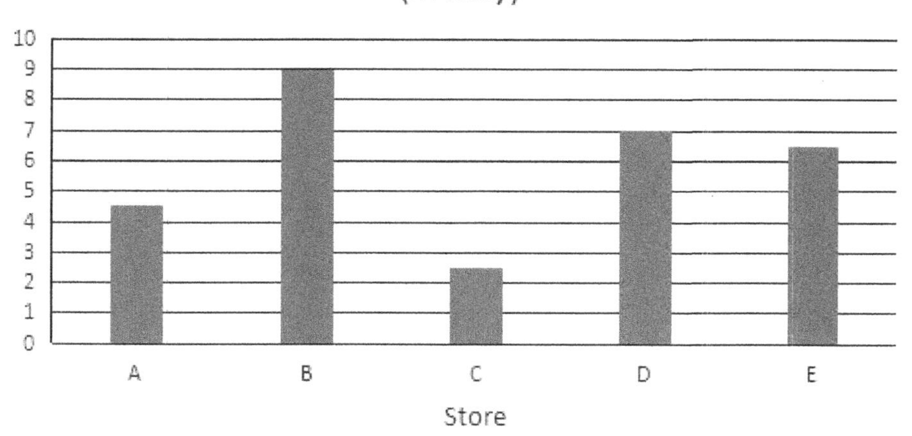

2. The total sales volume of potato chips in five different convenience stores for the month of May is shown in the bar chart above. If the total volume of sales in dollars is $29,500 for the month of May, what is an appropriate label for the vertical axis of the chart?

(A) Potato chip sales (in tens of dollars)

(B) Potato chip sales (in hundreds of dollars)

(C) Potato chip sales (in thousands of dollars)

(D) Potato chip sales (in tens of thousands of dollars)

		Preferred Winter Clothing				
		Sweater	Jacket	Sweatshirt	Poncho	Total
Gender	Male	6	11	9	4	30
	Female	15	7	8	1	31
	Total	21	18	17	5	61

3. A group of high school students were surveyed about their favorite winter clothing. The survey data was broken down as shown in the table above. Which of the following categories represents approximately 25 percent of all the survey respondents?

(A) Males who prefer sweatshirts

(B) Males who prefer ponchos

(C) Females who prefer jackets

(D) Females who prefer sweaters

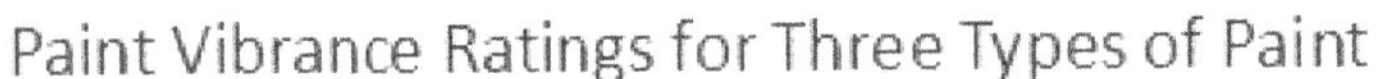
QUESTIONS 4 AND 5 RELATE TO THE FOLLOWING BAR CHART:

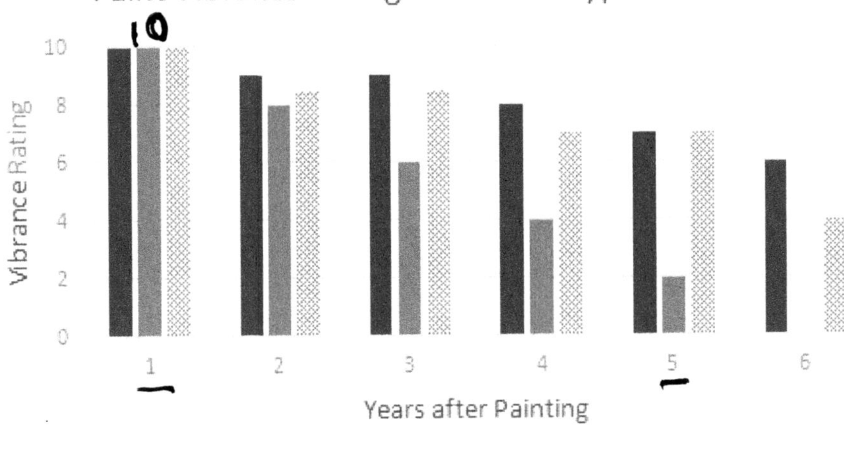

4. Three houses were each painted with a different type of newly-developed paint designed to resist rain and sun without fading. For six years after the initial painting, the vibrance of the paint was measured on a 1-10 scale, with 10 being the most vibrant. The bar chart above shows the vibrance ratings for each of the three types of paint from 1 to 6 years after the paint was applied. Which of the following paint types showed a decrease in vibrance every year after the initial paint was applied?

I. Paint A
II. Paint B
III. Paint C

(A) II only

(B) III only

(C) II and III only

(D) I, II, and III

5. Of the following, which is the closest to the ratio of the total vibrance rating of all three paints in the fifth year after painting to the total vibrance rating in the first year after painting?

(A) 1 to 3

(B) 8 to 15

(C) 3 to 5

(D) 4 to 5

	Adults	Children	Total
Soccer	29	57	86
Baseball	46	18	64
Total	75	75	150

6. Two researchers were studying the preferred sport of children and adults in America as part of a research paper on the changing preferences of Americans over generations. The table above shows the results of a survey that recorded the preferred sport of a group of adults and children between soccer and baseball. What proportion of children reported that baseball was their preferred sport?

(A) $\frac{6}{25}$

(B) $\frac{9}{32}$

(C) $\frac{18}{57}$

(D) $\frac{57}{75}$

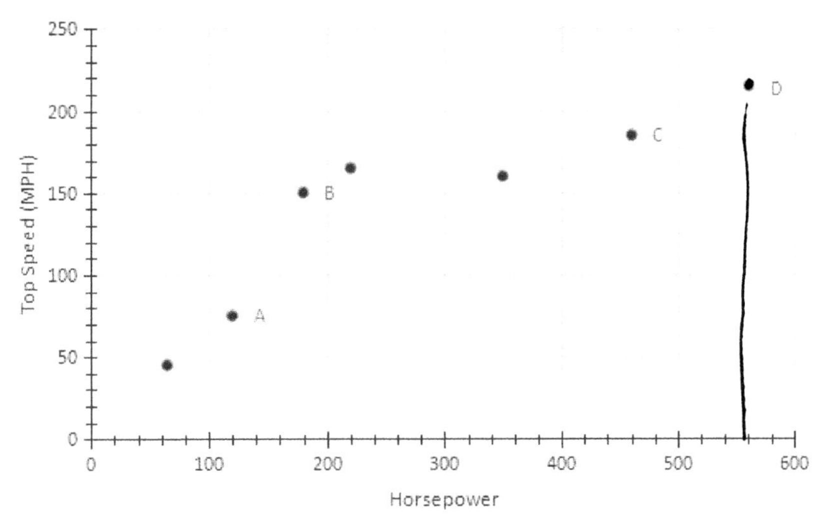

7. The scatterplot above charts the relationship of horsepower to top speed for seven different vehicles. What is the horsepower of the vehicle with the highest top speed?

(A) 60

(B) 220

(C) 560

(D) 600

Energy per Gram of Macronutrient

Macronutrient	Calories
Protein	4
Fat	9
Carbohydrate	4

8. The table above shows the typical number of calories per gram of three major macronutrients present in food. If the 1200 calories in a cheeseburger come entirely from p grams of protein, f grams of fat, and c grams of carbohydrate, which of the following expresses c in terms of p and f?

(A) $c = 300 - p - \frac{4}{9}f$

(B) $c = 300 - p - \frac{9}{4}f$

(C) $c = 1200 - p - \frac{9}{4}f$ X

(D) $c = 1200 - \frac{9}{4}(p - f)$ X

Seconds after drop	Height above ground
0	10000
15	6000
30	3600
45	2160
60	1296

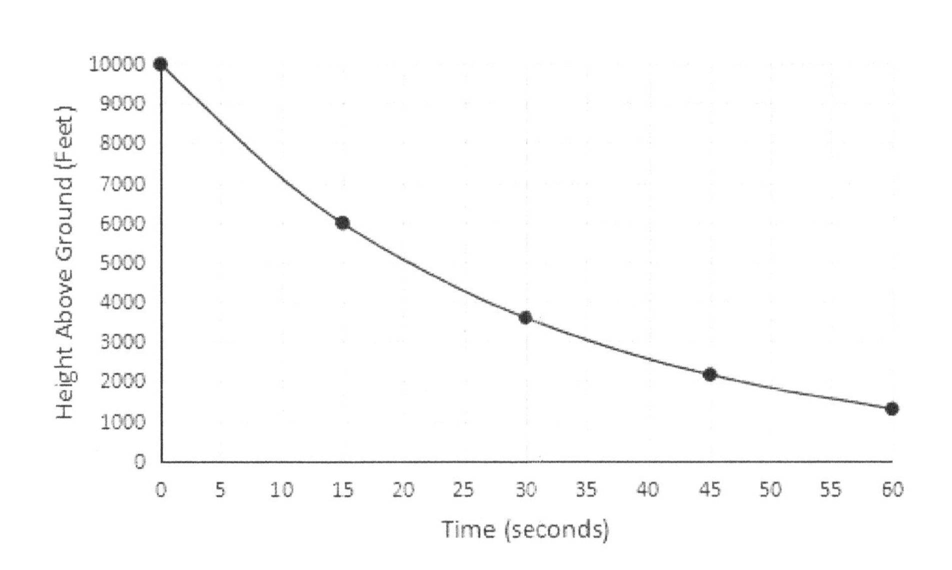

9. FREE RESPONSE: A certain experimental type of parachute is tested from a height of 10,000 feet above ground. The table and graph above show the height above ground for a fall lasting 60 seconds before the parachute is recovered mid-air. Assuming the rate of descent in feet per second is the same if the parachute is dropped from a height of 12,000 feet, what is the difference in height above ground (in feet) between the parachute dropped at 12,000 feet after 15 seconds compared to the height above ground of the parachute dropped at 10,000 feet after 30 seconds?

2000 ?

Time (minutes)	Salt concentration (grams per liter)
0	50.0
10	60.0
20	72.0
30	86.4
40	103.7
50	124.4
60	149.3

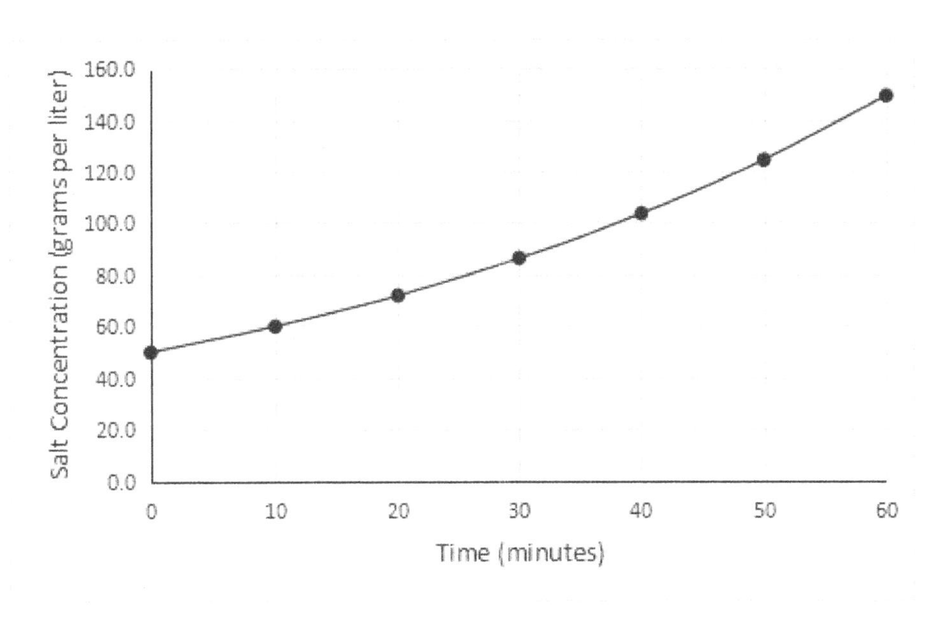

10. FREE RESPONSE: A student is performing an experiment on the salt concentration over time in three liters of saline solution. She adds salt to the solution continuously over a 60 minute period using a mechanical dispenser that operates according to an exponential formula. The table and graph above show the concentration of salt in the solution at 10-minute intervals over time. According to the data, how many more grams of salt are present in .7 liters of the solution after 10 minutes of her experiment than are present in .3 liters of the solution after 30 minutes? (Round your answer to the nearest whole number).

16 grams.

Charts & Tables Answers

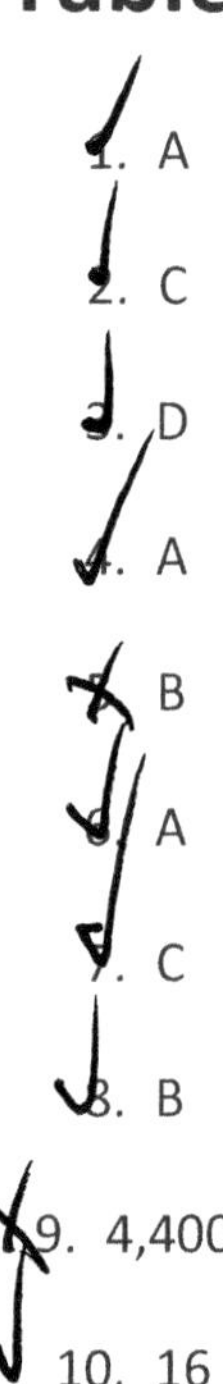

1. A
2. C
3. D
4. A
5. B
6. A
7. C
8. B
9. 4,400
10. 16

Charts & Tables Explanations

1. **A.** This question is surprisingly basic - many Charts and Tables questions are - although there is also the chance for a Misreading **Careless Error**. We're asked "what fraction the of *parakeets* prefer seeds rather than pellets." Make sure to focus *only* on the row that gives data on parakeets. We should first set up a fraction (review the lessons on **Ratios and Proportions** and/or **Probability** for a review of "Word Fractions"):

$$\frac{\text{parakeets that prefer seeds}}{\text{all parakeets}}$$

Notice that this is *not* the ratio of "all seeds to all pellets." Instead, it is a fraction of *some* parakeets taken out of *all* parakeets. We should only use data for parakeets. You can tell by the wording "what fraction of parakeets..."

Now plug in values to our Word Fraction:

$$\frac{\text{parakeets that prefer seeds}}{\text{all parakeets}}$$
$$=\frac{9}{23}$$

This fraction cannot be simplified any further, and we have our final answer, Choice A.

2. **C.** This question gives us a bar chart with an unlabeled vertical axis and asks us to correctly label that axis. The answer choices are all the same except for the units of dollars they use: are we working in tens of dollars? Hundreds? Thousands?

The best way to answer this would be to read off and total up all five bars of the bar chart first. It looks like the five bars have a sum of values, from left to right, of:

$$4.5+9+2.5+7+6.5$$
$$=29.5$$

So, if the *bar chart* shows a total of "29.5" and the *word problem* makes clear that the total value is \$29,500, we need an answer choice that correctly converts 29.5 into 29,500.

What value would multiply 29.5 to get 29,500? If you want, you could set it up as an Algebra equation:

$$29.5x = 29{,}500$$
$$\frac{29.5x}{29.5} = \frac{29{,}500}{29.5}$$
$$x = 1{,}000$$

Or, you could simply count the zeros on your fingers, or even just use basic logic. No matter what method you use, just be sure you know that the bars must be showing *thousands* of dollars, or Choice C.

3. **D.** This question refers to an upcoming lesson on **Percents**, but it's a very simple percent and we can probably handle it right now.

The question asks for a category or "cell" of the table that represents 25% of *all* survey respondents - or ".25" in decimal form. Let's set up a Word Fraction to get clear on our ideas:

$$\frac{\text{Number in Chosen Category}}{\text{All Survey Respondents}}$$

We also know that the total number of *all* survey respondents is 61, as given in the table, so let's update our fraction:

$$\frac{\text{Number in Chosen Category}}{61}$$

The next thing to do is test all four answer choices by plugging in each cell value. Remember, we're looking for the category that gets us closest to a value of .25 when we plug it into our fraction.

First, Choice A:

$$\frac{\text{Males who prefer Sweatshirts}}{61}$$
$$=\frac{9}{61}$$
$$=.098$$

Not very close to ".25", is it? The answer probably isn't Choice A.

Now let's test Choice B:

$$\frac{\text{Males who prefer Ponchos}}{61}$$
$$=\frac{4}{61}$$
$$=.147$$

The result also isn't very close to our target value of ".25".

So, let's keep looking with Choice C:

$$\frac{\text{Females who prefer Jackets}}{61}$$
$$=\frac{7}{61}$$
$$=.115$$

Still not very close to the desired value of ".25". There's one last hope in Choice D:

$$\frac{\text{Females who prefer Sweaters}}{61}$$
$$=\frac{15}{61}$$
$$=.246$$

Finally! An answer choice that gets us close to a value of ".25". This is our best answer choice.

4. **A.** Despite the "information overload" of this long word problem and bar chart, the work is actually extremely simple. According to the word problem, we're looking for any paint types that "showed a decrease in vibrance *every year*." What would that look like on the chart? Simply put, it would be a bar that continually drops from left to right.

Let's investigate each of the three Paint Types. Type A drops from Year 1 to Year 2, but then holds steady from Year 2 to Year 3. Therefore, Type A does *not* fit the requirements.

Type B drops from Year 1 to Year 2, then again from Year 2 to Year 3. In fact, if we look at each year, Type B continually drops from each year to the next. So, Type B *does* fit the requirements of the question.

Finally, let's investigate Type C. It drops from Year 1 to Year 2, but does *not* drop from Year 2 to Year 3. It does not fit the question.

Although all the Paint Types drop over time, neither Type A nor Type C drop *each* year the way that Type B does. Our answer must be Choice A, which only includes Paint Type B.

5. **B.** This question refers back to the previous bar chart and also makes use of a basic ratio (review the lesson on **Ratios and Proportions** if you need help on this topic). Let's set up a Word Fraction to get our ideas straight first:

$$\frac{\text{Total Vibrance in Year 5}}{\text{Total Vibrance in Year 1}}$$

Since we're using the *total* of each year, we'll need to estimate the value of each of the three columns, both for Year 1 and Year 5. I'll input those numbers below:

$$\frac{\text{Total Vibrance in Year 5}}{\text{Total Vibrance in Year 1}}$$
$$=\frac{7+2+7}{10+10+10}$$

And now clean up and simplify:

$$=\frac{7+2+7}{10+10+10}$$
$$=\frac{16}{30}$$
$$=\frac{8}{15}$$

This fraction can be read as a ratio of "8 to 15." If you got Choice A, make sure you didn't misread and accidentally use Year 6 instead of Year 5!

6. **A.** This question asks us for the proportion of children who preferred baseball. It doesn't necessarily require a full review of the lesson on **Ratios and Proportions**, but it would still be wise to set up a Word Fraction to get our ideas clear before continuing:

$$\frac{\text{Children who prefer Baseball}}{\text{All Children}}$$

Make sure you're clear: we're selecting children who prefer baseball out of *all* total children. Adults are completely disregarded. Now plug in the correct values from the table:

$$\frac{\text{Children who prefer Baseball}}{\text{All Children}}$$
$$=\frac{18}{75}$$

And reduce the fraction for your final answer:

$$\frac{18}{75}$$
$$=\frac{6}{25}$$

7. **C.** Not a very hard question. First locate the point of the vehicle with the highest speed. Our *y*-axis or vertical axis shows top speed, so the highest value for this axis is found at Point D.

Now we just have to read off the Horsepower of Point D. I suggest using your pencil to mark a straight vertical line downwards from Point D until you meet the "Horsepower" axis. This will just help improve your accuracy a bit, and is a good policy to follow any time you're reading coordinates off a chart or table on the SAT.

Once we've done that, we find that Point D has about 560 horsepower. If you're getting something different, be sure to double-check your tick marks, which are going up by 20 per tick.

8. **B.** This question is slightly different from others in this chapter, only because we're required to set up an Algebra equation based on the data from the chart.

First, let's get our basic setup. We'll use variables p, f, and c to represent the grams of each of the three macronutrients. Each of these variables will be multiplied by the number of calories per gram of each macronutrient. There are 4 calories per gram of protein, so we'll use $4p$. There are 9 calories per gram of fat, so we'll use $9f$. And there are 4 calories per gram of carbohydrates, so we'll use $4c$.

The total calories from each of the macronutrients will then be added up and set equal to 1200, the total calories in the cheeseburger:

$$1200 = 4p + 9f + 4c$$

Make sure you understand how we created the equation above before continuing. Now use **Basic Algebra 1** to isolate c, so that our final answer looks like the answer choices:

$$\begin{aligned} 1200 &= 4p + 9f + 4c \\ -4p &\quad -4p \\ 1200 - 4p &= 9f + 4c \\ -9f &\quad -9f \\ 1200 - 4p - 9f &= 4c \\ \frac{1200 - 4p - 9f}{4} &= \frac{4c}{4} \\ 300 - p - \frac{9}{4}f &= c \end{aligned}$$

This equation is identical to Choice B.

9. **4,400.** Well, right away we can tell that this question is trying to hit us with information overload. A huge word problem, plus not one but *two* charts and graphs?

Don't freak out. Let's give ourselves a chance to try and put our ideas in order first.

We're being asked the *difference* between two heights. Info on one of those heights, "10,000 feet at 30 seconds," is easier to find, because it comes straight from the table. We can read it straight off the table: at 30 seconds, it will be a height of 3,600.

The other height comes from a higher parachute, dropped at 12,000 feet after 15 seconds, but we should assume "that the rate of descent" is the same, according to the question.

It's not found directly on the table, but if you think about it, all we have to do is adjust the height values in our table by +2,000 feet. The rate of descent is exactly the same - only the starting point of 12,000 feet is higher than the original starting height of 10,000 feet.

So, locate the cell on the table for "15 seconds," which gives a height of 6,000 feet *when dropped from 10,000 feet*. We just need to adjust upwards by +2000 feet, which would give a height of 8,000 feet.

Now we know our two heights are 8,000 feet and 3,600 feet. To finish the question, give them what the want: the *difference* between these two heights, which is a simple subtraction problem:

$$8{,}000 - 3{,}600 = 4{,}400$$

And, we're done! There will be a 4,400 foot difference between the two heights. Notice that we didn't even need to use the graph for anything - only the table.

10. **16.** This question is very similar to Pretest Question 2 in this lesson, as well as Practice Question 9 before it. A lot of the word problem and combined table / chart is simply there to cause information overload.

Let's get clear on what we need. We're asked for the *difference* between two amounts of salt. The first amount of salt comes from ".7 liters at 10 minutes." The second amount comes from ".3 liters at 30 minutes."

Let's find each value on its own. At 10 minutes, the table shows that the concentration of salt is 60 mg/liter. Note that we are using .7 liters of solution, so we have to multiply the concentration by the volume of solution to determine exactly how many grams of salt are present (this is a variation on **Unit Conversion** and is identical to the steps we followed in Pretest Question 2 from this lesson):

$$60\frac{\text{mg}}{\text{liter}}(.7\text{ liters})$$
$$=42\text{ mg}$$

Then at 30 minutes, the table shows that the concentration is 86.4 mg/liter. We are using .3 liters of this solution, so again multiply concentration by volume:

$$86.4\frac{\text{mg}}{\text{liter}}(.3\text{ liters})$$
$$=25.92\text{ mg}$$

Now, simply use subtraction to find the difference between the two masses of sale:

$$42-25.92=16.08$$

The question asks us to round to the nearest whole number, so our final answer will be 16.

Again, like the previous question, notice that we didn't need the graph at all - just the table. The test is just trying to hit us with information overload to freak us out and make us give up before we've even really tried our best - a common tactic the SAT uses on Charts & Tables questions.

Lesson 13: Percents

Percentages

- 4% of Whole Test
- 1.5% of No-Calculator Section
- 5.3% of Calculator Section

Prerequisites

- Basic Algebra 1
- Algebra 1 Word Problems
- Ratios & Proportions

This lesson on Percents will cover one of the most important topics on the SAT Math, and one that causes a lot of problems for students of every level.

We'll calculate percents and percent change, and set up Algebra equations with percents so we can work "backwards" to find unknown starting values (such as an item's price before discounts and taxes). We'll learn how to avoid the mistakes most students make. We'll also learn how to quickly and easily apply multiple percent changes that frequently stump other students.

Expect to see a lot of word problems in this topic. The majority of percent problems have some kind of real-world story behind them.

A related topic, Percent Growth and Decay, will be covered in more detail later in the lesson on **Exponential Growth**.

Percents Quick Reference

- To calculate a basic percentage, use "Desired over Total."
- To calculate Percent Growth, use "Change over Original"
- Stop trying to use addition and subtraction to calculate percents - this is the mark of an amateur. Start using "Percent Multipliers" instead.
- Understand the metaphor of "Sea Level" to determine your Percent Multipliers.
- If there is more than one Percent Multiplier, they can easily be strung together in sequence.
- Keep track of the difference between a percent and its corresponding decimal version.
- Clean Algebra setups are essential in harder Percent problems.
- Percent Increase and Percent Decrease are *not* mirror images of each other.

Intro to Basic Percents: Desired Over Total

Let's start with the most basic of all Percent calculations.

To find a simple percentage that one value represents to another, we use a "Desired over Total" setup and multiply by 100%.

$$\frac{\text{Desired}}{\text{Total}}(100\%)$$

For example, if Sarah saves $25 out of every $100 she earns at her job, we could calculate the percentage of her income that she saves. Our input for "Desired" would be her savings of $25, and our input for "Total" would be her total income of $100:

$$\begin{aligned}&\frac{\text{Desired}}{\text{Total}}(100\%)\\&=\frac{\$25}{\$100}(100\%)\\&=.25(100\%)\\&=25\%\end{aligned}$$

In this case, we can calculate that Sarah is saving 25% of her income.

Try to solve the following question on your own, using the same technique. You can use a calculator:

> **Terrence's monthly paycheck is $2,000. If he spends $75 of every paycheck on popcorn, what percent of his monthly paycheck does he spend on popcorn?**

In this case, $75 is our "desired" value and $2,000 is our "total" value. Set up the fraction and evaluate it:

$$\begin{aligned}&\frac{\text{Desired}}{\text{Total}}(100\%)\\&=\frac{\$75}{\$2{,}000}(100\%)\\&=.0375(100\%)\\&=3.75\%\end{aligned}$$

We can see that Terrence spends **3.75%** of his monthly paycheck on popcorn.

The "Desired over Total" fraction can be used whenever you need to calculate what percentage *A* represents of *B*, where *A* and *B* are any two values.

"Sea Level" is 100%

Here I'm going to introduce one of the most valuable concepts I've ever discovered for teaching Percents to my students. This single concept has helped hundreds of students overcome their confusion surrounding Percents - especially Percent Increase and Percent Decrease.

This teaching concept is based around the metaphor of "Sea Level." Imagine sea level as a place where everything is flat - where everything is at a normal baseline. It's just a basic starting point we use for comparisons - it's not high, it's not low. Sea Level is just nice and flat and "normal."

Let's refer to Sea Level as "100%". When you have 100% of something, you have its basic starting amount. A nice, flat, baseline level - 100% of something is "Sea Level."

Keep in mind the decimal equivalent of 100% is just "1". So if you have, say, 100% of a sweater, you just have 1 complete sweater.

Now, picture the sea rising and falling from this level of 100%. If we *increase* by 20%, we will rise up from 100% Sea Level to a new percent of 120%. If we *fall* by 30%, we will drop from 100% Sea Level to a new level of 70%.

Everything starts from sea level. Rise or fall from this baseline of 100%. Increase by 60%? That's a new level of 160%. If we decrease by 5%? That's 95%. Make sense?

I'm going to refer to the "Sea Level" concept frequently throughout my explanations of the practice problems, so please make sure you understand it before you go on.

Decimals and "Percent Multipliers"

Now, time to understand another concept. Percents and Decimals are just different versions of each other. Multiply a decimal by 100 to get the equivalent percent. Divide a percent by 100 to get the equivalent decimal.

20% is equivalent to the decimal ".2"

150% is equivalent to the decimal "1.5"

The decimal 1.3 is equivalent to 130%

The decimal .08 is equivalent to 8%

This is an extremely easy conversion - you just move the decimal point two spots to the right or two spots to the left to convert between percentages and their equivalent decimal values.

We can use this to create what I call a "Percent Multiplier."

See, Percents don't work based on addition or subtraction. They work based on *multiplication*.

In fact, the very words "80% *of*" indicate that we're working with multiplication. In the lesson on **Algebra 1 Word Problems**, I pointed out that the word "of" means "multiply."

So, combine everything I've said and you'll get the idea of the "Percent Multiplier."

If you want to increase the value of x by 30%....

First, rise by 30% from the "sea level" of 100% to the higher value of 130%.

Then Convert 130% into the decimal version, 1.3.

Then multiply x by 1.3 (this value of 1.3 is what I call the "Percent Multiplier").

The value of $1.3x$ is equivalent to "a 30% increase in x."

No addition or subtraction involved. Just multiplication.

Now, riddle me this:

What is the percent multiplier for an increase of 22%?

We rise from 100% "sea level" by 22%. That puts us at 122%, a Percent Multiplier of **1.22**.

What is the percent multiplier for a decrease of 45%?

We drop from 100% "sea level" by 45%. That puts us at 55%, a Percent Multiplier of **.55**.

If a car increases from speed s by 82%, what is the new speed in terms of s?

We rise from 100% sea level by 82%. That puts us at 182%. Convert to a decimal of 1.82. Now apply this Percent Multiplier to s to get $1.82s$.

If a plane decreases in speed n by 37%, what is the new speed in terms of n?

We drop from 100% sea level by 37%, which puts us at 63%. Convert to a decimal of .63. Now apply this Percent Multiplier to get $.63n$.

I will refer to the concept of "Percent Multipliers" (or simply "Multipliers") frequently throughout the lesson and practice problems, so make sure you understand it.

Using Multiple Percent Multipliers

You can even combine multiple Percent Multipliers in an unbroken chain. This is a lifesaver in multi-step Percent questions.

For example, what if a sweater of price c is on sale for 20% off, but we have to pay an 8% sales tax on the final price?

First, the discount is a drop of 20% from "sea level" of 100%, which puts us at 80%. That's a decimal of .8, so we can apply that Percent Multiplier to c and get $.8c$.

The tax is an increase of 8% from "sea level" of 100%, which puts us at 108%. That's a decimal of 1.08, so we can apply that Percent Multiplier to $.8c$ and get $(1.08)(.8c)$. If you want, you can even simplify that by multiplying 1.08 and .8 to get a cumulative final Percent Multiplier of $.864c$.

Now try Multiple Percent Multipliers yourself:

> A certain toy originally costs $20. The store puts it on sale for 10% off. There is also a 5% sales tax on the purchase. What is the price paid for the toy?

A 10% discount is a *drop* from sea level, which puts us at 90%. That's a decimal Percent Multiplier of .9.

A 5% sales tax is an *increase* from sea level, which puts us at 105%. That's a decimal Percent Multiplier of 1.05.

Multiply the original price of $20 by these two Percent Multipliers to get the final price paid:

$$\$20(.90)(1.05) = \$18.90$$

The final price after discount and tax will be **$18.90**.

Do you see yet how easy and powerful this approach can be? Use the "sea level" metaphor to find your Percent Multiplier(s) and then just string them together. It's a much better approach than most of my students were taught in school.

Percent Change

Let's move on to a different but related topic: "Percent Change."

Percent Change is calculated using a "Change over Original" setup. This is used to calculate values for "percent increase" and "percent decrease".

$$\frac{\text{Change}}{\text{Original Value}}(100\%)$$

For example, if Tim's original weight was 150 pounds, and he lost 15 pounds through training and exercise, we could calculate the percent decrease in his weight. The Change was 15 pounds, and the Original value was 150 pounds:

$$\begin{aligned} &\frac{\text{Change}}{\text{Original Value}}(100\%) \\ &= \frac{15 \text{ pounds}}{150 \text{ pounds}}(100\%) \\ &= .1(100\%) \\ &= 10\% \end{aligned}$$

Tim lost 10% of his weight by dropping 15 pounds from his original 150 pounds, leaving him at a new weight of 135 pounds.

Try calculating Percent Change yourself on the following example. You can use a calculator:

> Ferris has a car that is originally worth $20,000. After he has an accident and dings the mirror on a trash can, the value of the car decreases to $14,000. What is the percent decrease in the value of the car after the accident?

Let's figure this out using our Percent Change setup of "Change over Original." The Change is $\$20{,}000 - \$14{,}000$ or a decrease of $6,000. The Original value was $20,000. Plug them into the fraction:

$$\frac{\text{Change}}{\text{Original}}$$
$$= \frac{\$6{,}000}{\$20{,}000}$$

Now reduce the fraction to a decimal:

$$= \frac{\$6{,}000}{\$20{,}000}$$
$$= .3$$

Keep in mind that a decimal value of ".3" is equivalent to 30%, because we have to multiply decimals by 100% to convert them into final percentages. We've calculated that the car experienced a **30% decrease in value** after the accident.

Percent Increase & Decrease Are NOT Mirror Images

Here is a good time to note something *very* important: "Percent Increase" and "Percent Decrease" are *not* mirror images of each other.

For example, if we return to Tim's weight loss example, and increase Tim's new weight of 135 pounds by 10%, we will *not* return to his original weight of 150 pounds. Check it out:

$$135(1.1)$$
$$= 148.5$$

Obviously 148.5 pounds is not the same as his original weight of 150 pounds. What is going on here? Didn't he lose 10% of his weight, then gain back 10% of his weight? Shouldn't he return to his original weight of 150 pounds?

Well, no. This phenomenon occurs because percent is a *multiplier* of other values; the result of a Percent Change depends entirely on what value you apply it to.

10% of a larger number (such as 150 pounds) is *not* equivalent to 10% of a smaller number (such as 135 pounds). A 10% loss is *not* exactly offset by a 10% gain. It's just not how Percent Change works, which confuses many students at first.

If an investor loses 20% of his cash on a risky investment, he *cannot* recover all of his lost cash simply by making a 20% gain on his next investment. The 20% he lost was 20% of a *larger* amount of money, and his following 20% gain is applied to the *smaller* amount of money that he was left with.

To recover back to his original amount of cash, this investor would have to make a *25%* return on his next investment. First lose 20%, then gain 25%, and he would only break even to his original amount of money.

That's why I say that "Percent Increase" and "Percent Decrease" are *not* mirror images of each other.

Let me prove this for you, if you're not convinced. We'll start with \$1,000. Now hit that with a loss of 20%. We'll decrease from 100% "sea level" by 20%, which puts us at 80% or a decimal multiplier of .80:

$$\$1{,}000(.80) = \$800$$

If, after this loss of 20% in value, we simply try to *increase* by 20%, we will *not* end up where we started. Test it out - a 20% increase would be a 1.2 multiplier (increase of 20% from "sea level" of 100%):

$$\$800(1.2) = \$960$$

See? After decreasing by 20%, then increasing by 20%, we've only recovered to \$960 of our original \$1,000.

To calculate the *actual* percentage gain we need, we could use this setup:

$$\$800(x) = \$1{,}000$$

When we solve this equation, we'll discover the Percent Multiplier that we need to go from \$800 back to \$1,000. Now solve it:

$$\frac{\$800(x)}{\$800} = \frac{\$1{,}000}{\$800}$$
$$x = 1.25$$

We need a 1.25 multiplier, which is a 25% increase. Lose 20% of your cash, and you'll have to make a gain of 25% just to break even.

It's very important to understand why this is. Make sure to reread and study this section of the lesson carefully if you still don't understand *exactly* what I mean when I say "Percent Increase and Percent Decrease are *not* mirror images of each other."

Pretest Question #1

Let's take a look at our first Pretest question on this topic. Try it yourself if you got it wrong the first time.

(CALCULATOR) A company lowered the price of a product, Item A, from \$12 to \$9. To make up for this, the company raised the price of Item B by a percent equal to twice the percent decrease on Item A. There is also a 10% shipping charge on all products the company sells. The original price of Item B was \$18. What is the new price of Item B *after* shipping?

(A) \$14.85

(B) \$22.50

(C) \$27.00

(D) \$29.70

The first step in this problem is to calculate the Percent Change of Item A. There's a decrease from $12 to $9, which is a difference of $\$12-\$9$ or $3.

Now we'll use the "Change over Original" setup to find the percent change, with our original value as $12, the price of Item A before the discount:

$$\frac{\text{Change}}{\text{Original Value}}$$
$$=\frac{\$12-\$9}{\$12}$$
$$=\frac{\$3}{\$12}$$
$$=.25$$

Remember that this is a *decimal* value, and we want a percent, so multiply by 100% to get a 25% decrease in Item A's price.

Now, the word problem says that Item B was *increased* by twice that percent. Double 25% and we get a 50% price increase on Item B.

The original price of Item B was $18. A 50% increase from "sea level" would give 150% or a multiplier of "1.5". Take the original price and multiply by 1.5 to get the new price:

$$\$18(1.5)=\$27$$

The new price of Item B is $27, but we still need to apply our shipping cost of 10%. This will *increase* the cost of the item, so we'll rise from 100% "sea level" to 110% or a multiplier of "1.1".

Take the $27 price and multiply by 1.1 to get our final cost of Item B after shipping:

$$\$27(1.1)=\$29.70$$

Our final cost is $29.70, and our answer is **Choice D**.

Using Algebra with Percents

The more advanced Percent problems on the SAT test require the use of a clean Algebra setup before they can be answered.

For example, what if we encountered a question like the following:

> A man paid $15 for an umbrella. This umbrella was on sale for 25% off its normal price. What is the normal price of the umbrella?

This is a simple example, but it's important to get it right. Try it yourself!

First, we need to notice something critical. There are *two* prices in this question: the price paid of $15, and the *unknown* original price. The question gives us the actual price paid, but we must make up our own variable to use as a placeholder for the original price. Let's keep things simple and use x to represent the unknown original price.

Now focus on the percents. A 25% discount is a *decrease* from sea level that lowers us from 100% down to 75%. This can be written as the decimal multiplier of .75 and applied to the original unknown price *x*. This can be expressed as:

$$.75x = \$15$$

This is a simple but clean Algebra with Percents setup. And, it can be easily solved using **Basic Algebra 1** to discover the original price of the umbrella before the discount.

$$\begin{aligned} .75x &= \$15 \\ \frac{.75x}{.75} &= \frac{\$15}{.75} \\ x &= \$20 \end{aligned}$$

The original price x of the umbrella must have been $20 before the discount.

For my last trick, we're going to combine *all* of the proceeding concepts into a single practice question:

> A boy paid $10.80 for a certain candle. This candle was on sale for 40% off its normal price, and there was a 20% sales tax on the purchase. What was the normal price of the candle?

First, we'll need to make up our own variable for the *original* price of the candle. Never lose track of the difference between "price paid" and "original price." Let's use x again to represent the original price.

Now, this candle is on sale for 40% off. That's a decrease of 40% from 100% sea level, and puts us at 60%. We will write this as a decimal multiplier of .6.

Furthermore, there's a sales tax of 20%. This will increase the cost from 100% sea level, and puts us at 120%. We can write this as a decimal multiplier of 1.2.

Put it all together by stringing together the two multipliers and applying them both to the original price x. Set this all equal to the actual price paid of $10.80:

$$(.6)(1.2)x = \$10.80$$

Make sure you understand how we created the equation above. It says to *decrease* the cost by 40% but to *increase* that by 20% tax. Simplified, our setup gives:

$$.72x = \$10$$

Again, this is *not* the same as just multiplying by 80%. Do you see the difference? We are at 72% of the original cost - not 80%! We can't just add "-40%" to "+20%" to get a combined "-20%" or 80% of the original value. This will not give an accurate result, and is based on a fundamental misunderstanding of how percents work.

Remember, Percent Increase and Percent Decrease are *not* mirror images of each other! A decrease in percent is *not* balanced out by an equivalent increase in percent. We *must* use multiplication when working with multiple percent changes - they *cannot* be simply added to or subtracted from each other.

Now solve the equation for the original price x using **Basic Algebra**:

$$.72x = \$10.80$$
$$\frac{.72x}{.72} = \frac{\$10.80}{.72}$$
$$x = \$15$$

The original price of the candle must have been **$15** before the discount and sales tax were applied.

Pretest Question #2

Let's take a look at another Pretest question. Try it yourself before you look at my explanation below the question.

> (CALCULATOR) FREE RESPONSE: An aeronautical engineering team has been developing new test aircraft. From Version 1 to Version 2 of the aircraft, the maximum speed increases by 60%. From Version 2 to Version 3, the maximum speed increases by 50%. From Version 3 to Version 4, the maximum speed decreases by 25%. From Version 4 to Version 5, the maximum speed increases by 5%. If Version 5 has a maximum speed of 378 miles per hour, what was the maximum speed of Version 1?

In this question, we start from an unknown speed (Version 1), which we should represent with a placeholder variable such as x. This starting speed undergoes four percent changes - some increasing, some decreasing.

Each of the percent changes will need a Percent Multiplier for us to complete our Algebra setup.

An increase of 60% from Version 1 to Version 2 can be written as a 1.6 multiplier (starting at 100% "sea level" and increasing by 60% to 160% or 1.6).

An increase of 50% from Version 2 to Version 3 is written as a 1.5 multiplier.

A decrease of 25% from Version 3 to Version 4 is a .75 multiplier (starting at 100% "sea level" and *decreasing* by 25% to 75% or .75).

The final increase of 5% from Version 4 to Version 5 can be written as a 1.05 multiplier.

Remember that it's completely fine to string together a series of Percent Multipliers. We can take our original speed x and just multiply it by all four multiplier values, like this:

$$x(1.6)(1.5)(.75)(1.05)$$

And, we know the *final* speed of Version 5 that results after all these changes is 378 miles per hour. So, we can set our expression above equal to the final speed:

$$x(1.6)(1.5)(.75)(1.05) = 378$$

Now simplify the multipliers:

$$1.89x = 378$$

For those of us keeping score, we can clearly see that there was a cumulative 89% increase in top speed from Version 1 to Version 5 (the cumulative 1.89 multiplier reveals that we've increased 89% from the original maximum speed of Version 1).

To finish the question, just divide both sides using **Basic Algebra 1**:

$$\frac{1.89x}{1.89} = \frac{378}{1.89}$$
$$x = 200$$

Now we have our original speed x of **200 miles per hour** for Version 1, and we're done.

Review & Encouragement

Whenever you find a Percent problem on the SAT, the setups for Percents ("Desired over Total") and Percent Change ("Change over Original") should leap immediately to your mind.

To truly master Percent problems, you must also clearly understand the concept of Percent Multipliers. Whether you're increasing or decreasing in value, remember the metaphor of rising or falling from a "sea level" baseline of 100%.

Keep track of your percents and decimals. In the midst of a busy question, it can be hard to remember if you're working with the decimal value ".5" or the percent "0.5%".

Don't forget to keep track of "original values" vs. "final values." The word problems on the SAT aren't exactly friendly about reminding you. You may have to make up your own variables to hold the place of unknown values given in the word problem.

Percents appear very frequently on the SAT Math test, and most students have a lot of confusion in their heads surrounding this topic. Use this lesson and the following practice problems to resolve any confusion and develop rock-solid confidence on this valuable topic.

Percents Practice Questions

YOU MAY USE YOUR CALCULATOR ON ALL OF THE FOLLOWING QUESTIONS.

1. A packing company has two types of boxes. The smaller boxes contain 30% packing materials by mass and the larger boxes contain 40% of packing materials by mass. 50 smaller boxes and 30 larger boxes contain a total of 120 kilograms of packing materials. Which equation models this relationship, where x is the mass of a single smaller box and y is the mass of a single larger box?

(A) $15x+12y=120$

(B) $12x+15y=120$

(C) $150x+120y=120$

(D) $120x+150y=120$

2. A computer chip was on sale for 35% off its original price. If the price paid for the chip was $162.50, what was the original price of the computer chip, rounded to the nearest dollar? (Assume there is no sales tax.)

(A) $57

(B) $106

(C) $219

(D) $250

3. Ian bought a pair of motorcycle gloves that cost $81.00 after an 8% sales tax was added. What was the price of the motorcycle gloves before the sales tax was added?

(A) $73.00

(B) $74.52

(C) $75.00

(D) $80.52

4. FREE RESPONSE: Christian would owe $23,000 in taxes at the end of the year without any tax deductions. This year, Christian is eligible for tax deductions that reduce the amount of taxes he owes by $5,060. If these tax deductions reduce the taxes Christian owes this year by $n\%$, what is the value of n?

5. In 2016 the number of cars in Country A and Country B were equal. From 2008 to 2016, the number of cars in Country A increased by 30% and the number of cars in Country B decreased by 20%. If the number of cars in Country A was 280,000 in 2008, what was the number of cars in Country B in 2008, rounded to the nearest whole number?

(A) 172,308

(B) 303,333

(C) 436,800

(D) 455,000

6. An internet provider company charges $52.50 per month for basic internet services. After a rate increase, the monthly charge increases to $64.05. To the nearest percent, by what percent did the monthly rate increase?

(A) 12%

(B) 18%

(C) 22%

(D) 24%

7. Horatio is studying bees and honey production. He notices that Type X bees produced 75% more honey than Type Y bees did. Based on Horatio's observation, if the Type X bees produced 525 grams of honey, approximately how many grams did the Type Y bees produce?

(A) 131

(B) 300

(C) 394

(D) 450

8. FREE RESPONSE: A music festival in Austin featured 80 artists in 2004. The number of artists featured in 2004 was 25% greater than in 2003. The number of artists was x more in 2004 than in 2003. What is the value of x?

9. FREE RESPONSE: A storekeeper is selling candles. On Monday, the storekeeper permanently reduces the price of all candles by 20% discount to encourage business. Once business has increased, the storekeeper finds that one brand of candles is very popular, and on Wednesday she increases the price of this brand of candles by 50% of its discounted price. She then feels a final adjustment is needed, and on Friday she reduces the new price of this brand of candles by 5%. If the sale price of one of these candles on Friday is $6.27 before tax is added, what was the original sticker price *before* the discount was offered on Monday? (Disregard the dollar sign when entering your final answer.)

Year	Users
2010	2,940
2011	3,381

10. FREE RESPONSE: A startup company was growing the user base of their software product. The user base of the software in 2010 and 2011 is shown in the table above. The percent increase in users from 2010 to 2011 was three times the percent increase from 2009 to 2010. How many users did the software have in 2009?

11. FREE RESPONSE: How many liters of a 50% saline solution must be added to 5 liters of a 10% saline solution to obtain a 25% saline solution?

12. FREE RESPONSE: A school is forming a team for a new sport that they've just created. Of the students on the team, 6% play goalie, 24% play defense, 50% play midfield, and the remaining 10 students play forward. How many more students play midfield than play goalie?

13. Jeffery bought a new scooter at a store that gave a 40% discount of its original price. The total amount he paid was p dollars, including an 5% sales tax on the discounted price. Which of the following represents the original price of the scooter in terms of p?

(A) $.65p$

(B) $\frac{p}{.65}$

(C) $(.6)(1.05)p$

(D) $\frac{p}{(.6)(1.05)}$

14. FREE RESPONSE: A certain river in 2012 has an average flow of x liters per hour. In 2013, this flow has grown by 10%. From 2013 to 2014, the flow grows again by 22%. From 2014 to 2015, the flow decreases by 13%. If the average flow in 2015 is 4,000 liters per hour, what was the average flow in 2012? (Round your answer to the nearest whole number.)

15. Christian and Yanik each ordered a plate of a sushi at a restaurant. The price of Christian's sushi was d dollars, and the price of Yanik's sushi was $6 more than the price of Christian's sushi. If Christian and Yanik split the cost of the sushi plates evenly and each paid a 20% tip, which of the following expressions represents the amount, in dollars, each of them paid? (Assume there is no sales tax.)

(A) $.5d + 1.2$

(B) $1.2d + 3.6$

(C) $2.4d + 1.2$

(D) $2.4d + 7.2$

Percents Answers

1. A
2. D
3. C
4. 22
5. D
6. C
7. B
8. 16
9. 5.5 or 5.50
10. 2,800
11. 3
12. 22
13. D
14. 3426
15. B

Percents Explanations

1. **A.** We need to set up an algebra equation based on percents. There are 50 small boxes of mass x and 30 larger boxes of mass y, and we'll be totaling these masses. We can get started with that:

$$50x + 30y$$

We need to include our "Percent Multipliers". The percent multiplier for x will be .30 to represent 30% packing materials in the smaller boxes. The percent multiplier for y will be .40 to represent 40% packing materials in the larger boxes. Let's add this in:

$$(.30)50x + (.40)30y$$

We know that the total mass of all packing materials is 120 kilograms. So, we can set our equation equal to 120:

$$(.30)50x + (.40)30y = 120$$

Now we can clean up and simplify to get our final form of the equation:

$$(.30)50x + (.40)30y = 120$$
$$15x + 12y = 120$$

This equation corresponds to Choice A.

2. **D.** The first issue here is that there is no variable given in the question to be a placeholder for the *original* price. Whenever you're working with "original price" vs. "discounted price", you need to keep careful track of which price is which. The question will not always make this clear, so oftentimes it's up to us. Let's use a new variable, x, to represent the *original* price before the discount.

Now, we need to apply a "percent multiplier" to our original price to account for the discount.

Use the concept of "Sea Level". We're *decreasing* the original price by 35%, so drop down from the original "sea level" or original price of 100% by subtracting 35%. That gives you $100\% - 35\%$ or 65% of the original price. Our percent multiplier should be ".65" and applied to the original price x to show that it's being discounted by 35%. This will be set equal to \$162.50, the price actually paid after discount:

$$.65x = \$162.50$$

Make sure you understand this setup with total clarity. We made a new variable, x, to hold the place of the *original* price before discount. Then we applied a percent multiplier of ".65" to the original price, which represents a *discount* of 35% from the original price. Finally, we set that equal to the *price paid* of \$162.50.

Now it's just a matter of using **Basic Algebra 1** to solve our setup for the original price x:

$$.65x = \$162.50$$
$$\frac{.65x}{.65} = \frac{\$162.50}{.65}$$
$$x = \$250$$

This is how we find the original price of the computer chip was \$250.

3. **C.** As in Question 2, we'll need to create our own variable x to represent the *original* price of the gloves *before* tax was added. Also, never be misled by the wording "sales tax was *added*." It's much better to imagine percent as a *multiplier*. Thinking of percents as "adding or subtracting" usually leads students into confusion.

Our percent multiplier for the tax will be "1.08". Imagine starting at "sea level" or base price of 100% and then *rising* by an 8% sales tax to get to 108% of the sticker price. This can be written as the decimal "1.08". This 1.08 multiplier will be applied to the original price x. We set this equal to \$81, the price paid at checkout:

$$x(1.08) = \$81$$

Make sure you understand exactly how we created this setup; the rest is easy, but you need to be perfectly clear on how we got to this point.

Now we can easily solve for x, the original price of the gloves before tax, using **Basic Algebra 1**:

$$x(1.08) = \$81$$
$$\frac{x(1.08)}{1.08} = \frac{\$81}{1.08}$$
$$x = \$75$$

Of course, the price paid of \$81 is *more* than the sticker price of \$75, since tax makes items cost *more* than their original sticker prices.

4. **22.** This is a simple "Percent Change" calculation question. There is a *change* (a decrease of $5,060) being compared to an *original value* (the $23,000 that Christian originally owed). All we have to do is use "Change over Original" to calculate the percent change.

$$\frac{\text{Change}}{\text{Original}}$$
$$= \frac{\$5,060}{\$23,000}$$

Now just finish the division in the fraction:

$$= \frac{\$5,060}{\$23,000}$$
$$= .22$$

Remember - ".22" is a *decimal* but we are supposed to enter our answer as a *percentage*, as required by the word problem. We must multiply by 100% to convert a decimal into a percent. Our final answer should be "22."

Now set these equal to each other, since the number of cars in 2016 were *equal*. Here's our setup:

$$1.3A_{2008} = .8B_{2008}$$

We still haven't used the value "280,000" that was given for the number of cars in Country A in 2008. Go ahead and plug this in, then solve for the remaining variable of B_{2008}:

$$1.3A_{2008} = .8B_{2008}$$
$$1.3(280,000) = .8B_{2008}$$
$$364,000 = .8B_{2008}$$
$$\frac{364,000}{.8} = \frac{.8B_{2008}}{.8}$$
$$455,000 = B_{2008}$$

And we're done! In 2008, Country B must have had 455,000 cars.

Make sure you understand how we set up this problem and solved it before continuing.

5. **D.** This is one of the more challenging questions we've seen in this set so far. It's not hard if you're confident in your percent skills, but if you've been struggling through this lesson, this question might seem tough.

We'll need some variables to be placeholders for the number of cars in Country A and Country B. Let's use the variable A_{2008} to represent the number of cars in Country A in 2008. Logical, right? And, we'll need a variable like B_{2008} to represent Country B's total cars in 2008.

Both of these countries experienced a percent change from 2008 to 2016. We'll use a different Percent Multiplier on both variables to show the change in each country from 2008 to 2016.

City A *increased* by 30% from 2008. That's a 1.30 multiplier (a 30% *increase* in "sea level" from 100% to 130%, written as the decimal 1.30). We can write this as $(1.3)A_{2008} = A_{2016}$, which will give the number of cars in Country A from 2008 to 2016.

Repeat a similar process for Country B, which will have a multiplier of .80 (a 20% *decrease* in "sea level" from 100% to 80%, written as the decimal .80). We can write this as $(.8)B_{2008} = B_{2016}$, which will give the number of cars in Country B by the year 2016.

6. **C.** This is a basic Percent Change problem. All we need to do is use the "Change over Original" fraction that we studied in the lesson.

First, we need the "Change" value. What's the difference between the old and new rates? That's a simple subtraction problem between the old and new rates:

$$64.05 - 52.50 = 11.55$$

The difference between the old and new rate is $11.55. Now we can use that to set up a "Change over Original" fraction, compared to the original rate of $52.50:

$$\frac{\text{Change}}{\text{Original}}$$
$$= \frac{\$11.55}{\$52.50}$$

And now evaluate the division in the fraction:

$$\frac{\$11.55}{\$52.50}$$
$$= .22$$

Remember that ".22" is the *decimal* form and we need to multiply by 100% to get our final answer of 22%.

7. **B.** This question is similar to a simplified version of Question 5. The most important step in this entire problem is getting your setup right.

We'll need the correct Percent Multiplier first. That will be "1.75," which comes from *rising* from 100% "sea level" by an increase of 75%.

Now, I'm going to show your our setup and then go over exactly what it means:

$$X = 1.75Y$$

Notice that since Type *X* bees produced *more* honey, we need to put our 1.75 multiplier on the side of *Y*.

Think carefully about this: we're supposed to have a balanced equation; both sides must be *equal*. But if Type *Y* bees produces *less* honey than Type *X*, then we need to "help" Type *Y* by boosting it up with a percent multiplier that makes the Type *Y* side "heavier."

You can literally think of it as a balance scale: on its own, Type *X* honey is "heavier" than Type *Y*, so the side with Type *Y* needs "help" from a 1.75 multiplier to make it bigger and balance the scales out so the equation is equal on both sides.

As long as you understand this setup, it's easy to finish the work. Just plug in 525 for *X* as given in the word problem and solve for *Y*.

$$X = 1.75Y$$
$$525 = 1.75Y$$
$$\frac{525}{1.75} = \frac{1.75Y}{1.75}$$
$$300 = Y$$

And we're done. The Type *Y* bees must have produced 300 grams of honey.

It's vitally important that every step of the setup in this question makes *perfect* sense to you. Do not continue onto harder questions until you're perfectly clear on this question.

8. **16.** This question focuses on using a percent multiplier to create an Algebra setup. Starting in year 2003, there is a 25% *increase* to reach the value of year 2004.

How do we write the "percent multiplier" for a 25% increase? You probably know by now: we *increase* from 100% "sea level" by 25%, reaching 125%, which can be written as the decimal "1.25".

Apply this multiplier of 1.25 to the side for year 2003 to increase to the value for year 2004. (I'll use the new variable N as a placeholder for the number of artists in each year).

$$(1.25)N_{2003} = N_{2004}$$

Now plug in the given value of "80" for N_{2004}, as given in the word problem. Then solve for N_{2003}:

$$(1.25)N_{2003} = N_{2004}$$
$$(1.25)N_{2003} = 80$$
$$\frac{(1.25)N_{2003}}{1.25} = \frac{80}{1.25}$$
$$N_{2003} = 64$$

We've shown that there were 64 artists at the festival in 2003. Almost done - now use subtraction to find the difference between 2003 and 2004, as requested by the word problem:

$$N_{2004} - N_{2003} = x$$
$$80 - 64 = x$$
$$16 = x$$

9. **5.5** or **5.50**. Wow - this word question can seem really overwhelming at first. We're tracking an original price through a series of percent-based price changes. There are *three* of these changes: a reduction by 20%, an increase of 50%, and a reduction of 5%.

However, it's *not* hard if you know how to use Percent Multipliers. We'll need three of them: one for each price change.

The first reduction of 20% can be written as ".8", which represents a 20% *decrease* from 100% "sea level."

The price increase of 50% can be written as "1.5", which represents a 50% *increase* from "sea level."

The final reduction of 5% can be written as ".95", which represents a 5% *decrease* from "sea level."

As we've learned, one enormous advantage of the Percent Multiplier approach is that you can just multiply any percent multiplier with another (or even a third or fourth one) to combine the percent changes with no need for additional calculations. No "adding or subtracting" involved!

I'll use the new variable x as a placeholder for the "original sticker price" *before* any of the price changes started taking place.

Here's my setup with the three percent multipliers in place:

$$(.8)(1.5)(.95)x = \$6.27$$

From here it's actually quite easy to finish the problem - just **Basic Algebra 1** from here on out. First I'll use multiplication to simplify and combine all three percent multipliers on the left side of the equation.

$$(.8)(1.5)(.95)x = \$6.27$$
$$1.14x = \$6.27$$

Side note: if you're paying attention, you'll notice we can now see that after *all three* price changes are made, the end result is that the original price of the candle x has increased by a total of 14%. I'm reading this final change from the simplified Percent Multiplier "1.14" that results when all three individual Percent Multipliers are combined together with multiplication - the resulting multiplier of 1.14 signifies a final increase of 14% above the original 100% "sea level" starting cost.

Now let's finish the question and find the value of x using our everyday **Basic Algebra 1** skills:

$$1.14x = \$6.27$$
$$\frac{1.14x}{1.14} = \frac{\$6.27}{1.14}$$
$$x = \$5.50$$

Like so many Percent problems on the SAT Math, it's *not* the math itself that's hard - it's the setup *before* you do the math that counts.

10. **2,800.** The first step is to find the Percent Change between the two years given in the table. We will use "Change over Original" to calculate this Percent Change.

To do so, we'll need the difference between users in 2010 and 2011, which we can easily find by subtracting the two values given in the chart. (I'll use the variable U throughout the problem as a placeholder for the number of users in any given year):

$$U_{2011} - U_{2011}$$
$$= 3{,}381 - 2{,}940$$
$$= 441$$

So, there is a difference of 441 users between years 2011 and 2010. Let's plug that as the "Change" between years and use the earlier year, 2010, as our "Original" value:

$$\frac{\text{Change}}{\text{Original}}$$
$$= \frac{U_{2011} - U_{2010}}{U_{2010}}$$
$$= \frac{441}{2940}$$

Now simplify the fraction to get a decimal:

$$= \frac{441}{2940}$$
$$= .15$$

Remember that this value of .15 is a *decimal* and must be multiplied by 100% to find the percentage. That gives us a 15% increase from users in 2010 to 2011.

OK, now the word problem indicates that this 15% increase is "three times the percent increase from 2009 to 2010":

$$15\% = 3(\text{Increase from 2009 to 2010})$$

We can just divide both sides by 3 to get the increase from 2009 to 2010:

$$15\% = 3(\text{Increase from 2009 to 2010})$$
$$\frac{15\%}{3} = \frac{3(\text{Increase from 2009 to 2010})}{3}$$
$$5\% = \text{Increase from 2009 to 2010}$$

Making progress. Now we need another Algebra setup to handle the increase from 2009 to 2010. Remember that an increase of 5% would be a Percent Multiplier of "1.05":

$$(1.05)U_{2009} = U_{2010}$$

Remember that the 2009 side needs "help" to balance the scales and make it equal to 2010, which had a bigger user base. We're *increasing* the users in 2009 by 5% (a multiplier of 1.05) so they equal the higher number of users in 2010.

Now we can plug in the value for users in 2010 as given by the table:

$$(1.05)U_{2009} = 2{,}940$$

Now solve for the target value of users in 2009:

$$\frac{(1.05)U_{2009}}{1.05} = \frac{2{,}940}{1.05}$$
$$U_{2009} = 2{,}800$$

And we're finished. It was a bit of an ordeal, but we've found the original number of users in 2009, which was 2,800.

As with the other questions in this Practice set, make *absolutely sure* you understand every step of this problem - without hesitation - before you move on to anything more advanced.

11. **3.** This question tests both our Percent skills *and* our Algebra setup skills.

Salinity is a measure of concentration, which is calculated with "Amount of Salt" over "Volume of Liquid":

$$\frac{\text{Amount of Salt}}{\text{Volume of Liquid}}$$

To calculate salinity of our final mixture, we need to know the *total* salt and divide it by the *total* volume of our end product. Also, this fraction must be equal to .25 or "25% salinity", according to the question:

$$\frac{\text{Total Salt}}{\text{Total Volume of Liquid}} = .25$$

The total of both Salt and Liquid can be broken out into the two solutions that are being mixed. There is a 50% solution and a 10% solution:

$$\frac{\text{Salt from } 50\% + \text{Salt from } 10\%}{\text{Volume from } 50\% + \text{Volume from } 10\%} = .25$$

To calculate the total amount of salt (on top of the fraction), we must multiply the volume of each solution by its percent salinity (my abbreviation "Conc" means "Concentration" or "Percent Salinity"):

$$\frac{(\text{Conc}_{50})(\text{Vol}_{50}) + (\text{Conc}_{10})(\text{Vol}_{10})}{\text{Vol}_{50} + \text{Vol}_{10}} = .25$$

Now let's plug in all the values we know from the question. I'll use the new variable x to represent the unknown volume of 50% saline solution that we are trying to find. We'll use the decimal multiplier ".50" for 50% salinity and the decimal ".10" for 10% salinity:

$$\frac{(.50)(x) + (.10)(5)}{x+5} = .25$$

Do you understand how we've gotten this far? The setup is absolutely crucial. The rest of the problem is easy. As usual with Percent problems, the setup is the hardest part by far.

From here, it's just a sequence of **Basic Algebra 1** steps that takes us to the end value for x:

$$\frac{(.50)(x) + (.10)(5)}{x+5} = .25$$

$$\frac{.5x + .5}{x+5} = .25$$

$$(x+5)\frac{.5x + .5}{x+5} = .25(x+5)$$

$$.5x + .5 = .25x + 1.25$$

$$-.5 \qquad -.5$$

$$.5x = .25x + .75$$

$$-.25x \; -.25x$$

$$.25x = .75$$

$$\frac{.25x}{.25} = \frac{.75}{.25}$$

$$x = 3$$

Now we know the value of x, the liters of of 50% saline solution, is 3. And we're done!

12. **22.** OK, to solve this question our first step is to total up all the percentages we already have:

$$6\% + 24\% + 50\% = 80\%$$

This tells us that 80% of the team is composed of goalies, defense, and midfield. Out of the 100% total of the entire team, that leaves $100\% - 80\%$ or 20% of the team, to play forward.

We also know from the question that there are 10 students playing forward. In other words, we know that 10 students compose 20% of the team. We can set this up using a new variable x to represent the total number of players on the team:

$$(.20)(x) = 10 \text{ forwards}$$

Note that we're using a ".20" decimal multiplier on x, representing 20% of the entire team, to produce the correct number of 10 forwards.

Now use **Basic Algebra 1** to solve for x, the total number of players on the team:

$$(.20)(x) = 10 \text{ forwards}$$

$$.2x = 10$$

$$\frac{.2x}{.2} = \frac{10}{.2}$$

$$x = 50$$

The entire team must have had 50 players on it.

Now that we know this, we can determine exactly how many students played goalie and midfield:

$$\begin{aligned}&(50\%\text{ played midfield})(50\text{ total players})\\&=(.5)(50)\\&=25\text{ midfielders}\end{aligned}$$

$$\begin{aligned}&(6\%\text{ played goalie})(50\text{ total players})\\&=(.06)(50)\\&=3\text{ goalies}\end{aligned}$$

To finish the word problem, we simply use subtraction to find the difference between the number of midfielders and the number of goalies:

$$\begin{aligned}&\text{midfielders}-\text{goalies}\\&=25-3\\&=22\end{aligned}$$

And now we know our final answer - there were 22 more midfielders than goalies.

13. **D.** Questions like this one - where you can't hide from the Algebra setup - tend to really throw off students who aren't confident in their Percent skills. Those students tend to "trust their instincts," rather than properly setting up an Algebra equation, and as a result, they fall for traps the SAT has deliberately left in the answer choices.

Let's not fall for those. We'll do a beautiful Algebra and Percent setup before we make our final judgement.

First, let's ignore the tax for a moment and just focus on the discount itself. Remember how important it is to keep track of *original* price vs. the *final* price. The question gives the variable p as the final price paid - don't forget that we need to represent the *original* price, too. We'll need a new variable, like x, to represent the original price.

Then, we'll use a percent multiplier of ".60" to represent the discount of 40%. Consider that we've dropped from 100% original "sea level" by a discount of 40%, which lowers us to 60% of the original price. We can represent that as a multiplier of ".60".

Put it together into a clean setup that multiplies the original price by the discount multiplier of .60 to equal the price paid after discount, p (but note, we haven't included tax yet):

$$(.60)x=p$$

Good so far. Now it's time to include tax.

The tax of 5% will *increase* our "Sea Level" from 100% to 105%, which can be written as a decimal of 1.05.

Remember, the wonderful thing about the Percent Multiplier approach is that we can just include several Percent Multipliers strung together in a single setup without the need for further work.

Now include the tax multiplier into the setup from before:

$$(1.05)(.60)x=p$$

The setup is clean. From here all we have to do is use **Basic Algebra 1** to solve for the value of x, the original price:

$$\frac{(1.05)(.60)x}{1.05}=\frac{p}{1.05}$$

$$.6x=\frac{p}{1.05}$$

$$\frac{.6x}{.6}=\frac{(\frac{p}{1.05})}{.6}$$

$$x=\frac{p}{(1.05)(.6)}$$

And we're done. Match it to the correct answer, Choice D.

14. **3,426.** OK, so here we have another "chain" of percent multipliers all strung together. Viewed in this light, the question actually isn't very difficult.

Let's start with the *original* flow back in year 2012. We'll use a new variable, x, as a placeholder for this flow.

From 2012 to 2013, this flow grew 10%. An *increase* of 10% will be a 1.1 multiplier (go up from 100% "sea level" by 10%).

From 2013 to 2014, the flow grew 22%. An *increase* of 22% will be a 1.22 multiplier (go up from 100% "sea level" by 22%).

From 2014 to 2015, the flow *decreased* by 13%. This will be a .87 multiplier (go *down* from 100% "sea level" by 13%).

The final flow in 2015 is given as 4,000. We can set all this up with a string of percent multipliers that start with the original flow x back in 2012:

$$x(1.1)(1.22)(.87)=4{,}000$$

Make absolutely sure that this setup is crystal-clear to you. It's the only hard part of the problem. Don't rush it.

From here, it's just **Basic Algebra 1** to combine like terms and solve for x:

$$x(1.1)(1.22)(.87) = 4{,}000$$
$$\frac{1.16754x}{1.16754} = \frac{4{,}000}{1.16754}$$
$$x = 3{,}426.006...$$

If you're interested in keeping track, you can see that after all three increases/decreases in flow, there was a cumulative 16.7% increase from 2012 to 2015 (I'm rounding from the final combined multiplier of 1.16754). It's not mandatory to notice this, but if you've truly learned your Percents in this lesson, my meaning should be clear.

Anyway, the question says the final answer is rounded to the nearest whole number, which is a value of 3,426 liters per hour back in 2012.

15. **B.** Like Question 13, this type of question tends to catch students who aren't confident in using Percents to set up an Algebra equation. "Trusting our instincts" or making an educated guess usually ends poorly on these questions.

Instead, let's gradually build up a complete equation while keeping clear on each step.

First, we need to total the cost of the dishes before tip.

Christian's dish costs d dollars, and Yanik's dish is \$6 more, or $d+6$.

Add the two dishes together to get the total cost of the meal before tip:

$$\begin{aligned}&(\text{Christian's Meal}) + (\text{Yanik's Meal})\\&= (d) + (d+6)\\&= 2d+6\end{aligned}$$

Now, the easiest thing to do is apply the complete tip *before* splitting the bill in half. If you understand that Percents work via *multiplication*, then it shouldn't matter if we multiply everything *before* we divide by 2 or *after* we divide by 2 to split the meal. The total tip will be the same. (If this is confusing, imagine being the waiter taking this order: whether the two customers tip 20% on each meal, or 20% on the combined price of the meal, you, the waiter, will receive the same total amount in tips).

OK, so let's apply the tip of 20% now. It's a tip, so the cost of the meal *increases*. That means we'll rise by 20% from baseline 100% "sea level," giving 120% or a 1.2 multiplier:

$$1.2(2d+6) = \text{total cost with tip}$$

Why don't we distribute right now so we don't have to do it later:

$$1.2(2d+6) = \text{total cost with tip}$$
$$2.4d+7.2 = \text{total cost with tip}$$

Now we just have to split the total cost equally between the two friends. Just divide everything by 2:

$$2.4d+7.2 = \text{total cost with tip}$$
$$\frac{2.4d+7.2}{2} = \text{total cost with tip (per person)}$$

And finally, clean up the fraction for our final answer:

$$\frac{2.4d+7.2}{2} = \text{total cost with tip (per person)}$$
$$1.2d+3.6 = \text{total cost with tip (per person)}$$

This final equation matches to Choice B, and we're done!

Lesson 14: Exponents & Roots

Percentages

- 1.4% of Whole Test
- 3.5% of No-Calculator Section
- 0.3% of Calculator Section

Prerequisites

- Basic Algebra 1
- Fractions

In this lesson, we'll learn everything there is to know about Exponents & Roots on the SAT Math test. There is quite a lot to understand, and all of it is essential knowledge, especially in the No-Calculator section.

To succeed in this topic at the higher levels, you must also be confident in working with **Fractions** (e.g. adding, subtracting, multiplying, and dividing Fractions)

Note: In this lesson we will *not* cover **Exponential Growth**, which combines Exponents with **Percentages**. It will be covered thoroughly in the next lesson.

Exponents & Roots Quick Reference

- Basic Exponents are used to multiply a number *times itself*.
- A "Base" is the number underneath an exponent. A "Power" is the number of times the base multiplies itself. In x^n, the x is the base and the n is the power.
- Any number (except zero itself) raised to the "0 power" is equal to 1.
- Multiplying two numbers with the same bases *adds* their exponents.
- Dividing two numbers with the same bases *subtracts* the bottom exponent from the top exponent.
- Raising an exponent to another exponent *multiplies* the two exponents together.
- Roots are the reverse of Powers. They can also be written as Fractional Exponents, with the Root on the *bottom* ("denominator") of the Fractional Exponent.
- Some Roots can be simplified by breaking them down into smaller factors with a Factor Tree and looking for "matched sets" of factors to remove from the root.
- Fractional Exponents can combine a Power and a Root into a single exponent. The Power is on top ("numerator") and the Root is on bottom ("denominator"). In $x^{\frac{a}{b}}$, a is the power and b is the root.
- Negative Exponents *invert* ("flip upside-down") the base they are attached to: if they're on top of a fraction, they move to the bottom; if on bottom, they move to the top. For example, $x^{-3} = \frac{1}{x^3}$.
- "Combo Exponents" combine a Fractional Exponent with a Negative Exponent, combining a Power, a Root, and an Inversion into a single exponent.
- Exponents, Roots, Fractional Exponents, and Exponent Rules can be combined with **Basic Algebra 1** to solve for unknown values.

What Are Exponents?

An exponent, in its most basic form, is a notation that means to multiply a certain number (which we call the "Base") *times itself* a certain number of times (which we call the "Power").

For example, 4^2 can be read "four to the power of two" or "four squared," and it means 4×4, or "four times itself twice", and equals a value of 16. The base is 4 and the power is 2.

Or, 2^6 means "two times itself six times," and can be read "two to the sixth power" or just "two to the sixth." It would equal $2 \times 2 \times 2 \times 2 \times 2 \times 2$, and equals a value of 64. The base is 2 and the power is 6.

Exponents have a variety of uses. The most basic use is simply to save time and space that would have been spent multiplying the same number *times itself* more than once.

Compare this to the basic concept of multiplication, which itself is just a quicker way of writing the same addition problem more than once.

For example, 4×3 literally means "four times three," or to "add 3 to itself four times":

$$4 \times 3 = 3 + 3 + 3 + 3 = 12$$

5×6 means "five times six" or to "add 6 to itself five times":

$$6 + 6 + 6 + 6 + 6 = 30$$

Just like these *multiplication* examples are a way of saving space when notating the same *addition* problem more than once, *Exponents* are a way of saving space when notating the same *multiplication* problem more than once.

Ask yourself the following practice questions:

How do we read 5^3 out loud? What does it mean? What does it equal?

How do we read 1^8 out loud? What does it mean? What does it equal?

How do we read 2^4 out loud? What does it mean? What does it equal?

5^3 is read "five to the third" or "five cubed." It means to multiply 5 times itself three times. It equals $5 \times 5 \times 5$ or a final value of 125. The base is 5 and the power is 3.

1^8 is read "one to the eighth power" or "one to the eighth." It means to multiply 1 times itself eight times. It equals $1 \times 1 \times 1 \times 1 \times 1 \times 1 \times 1 \times 1$ or a final value of 1. The base is 1 and the power is 8.

2^4 is read "two to the fourth power" or "two to the fourth." It means to multiply 2 times itself four times. It equals $2 \times 2 \times 2 \times 2$ or a final value of 16. The base is 2 and the power is 4.

Raising to the Power of 0

Any Base number raised to the 0 power will have a value of 1. It's beyond the scope of this book to explain why that is - just try to remember it.

$$5^0 = 1$$

$$289{,}421^0 = 1$$

$$36^0 = 1$$

Note: the only exception to the zero-power rule is the base number "0" itself. If you raise 0 to the zero power, you will get an "undefined" result, meaning it does not exist.

$$0^0 = \text{undefined}$$

Again, for the SAT Math, the *reasons* behind this don't matter, so it's better to just memorize it.

Essential Exponent Rule #1: Multiplying Same Bases

Now, it's time to understand the essential rules of working with Exponents. These rules are *critical* to success in this lesson - and whenever you encounter Exponents on the SAT test.

Remember, a "base" is what we call the number *underneath* an exponent, and a "power" is what we call the exponent itself. For example, 7^3 has a base of 7 and a power of 3.

The first Exponent Rule occurs when we multiply the *same* bases. When this happens, we add our exponents. For example:

$$3^2 \times 3^4 = 3^{2+4} = 3^6$$

Here are a few more examples of multiplying same bases and adding their exponents:

$$2^{11} \times 2^4 = 2^{15}$$

$$(x^9)(x^3) = x^{12}$$

$$(3^2)(3^2)(3^2) = 3^{2+2+2} = 3^6$$

Why does this work? It's actually quite easy to understand if you think about what's happening "under the hood." I'll show you.

Let's take the example of $(n^3)(n^2)$.

This could be read out loud as "n to the third times n to the second."

If we expand it out, it could be written as:

$$(n \times n \times n)(n \times n)$$

Now count up all your n's. There are *five* of them being multiplied together - which we could write as n^5 if we wanted to save some time and space.

$$\text{See? } (n^3)(n^2) = n^{3+2} = n^5.$$

Remember, when we multiply same bases, their exponents will *add* to each other.

Try these simple practice problems for yourself and fill in the final exponent:

$$(x^3)(x^5) = x^?$$

$$(t^7)(t^4) = t^?$$

$$(n^2)(n^3)(n^4) = n^?$$

And the answers are below:

$$(x^3)(x^5) = x^8$$

$$(t^7)(t^4) = t^{11}$$

$$(n^2)(n^3)(n^4) = n^9$$

We just added the exponents for each example, because we're multiplying numbers with the same bases.

Essential Exponent Rule #2: Dividing Same Bases

The reverse of multiplication is, of course, division.

When you are *dividing* the same bases, you *subtract* their exponents. The bottom exponent is subtracted from the top exponent. This is the reverse of *multiplying* the same bases and *adding* their exponents.

For example:

$$\frac{6^5}{6^2} = 6^{5-2} = 6^3$$

Here are a few more examples of dividing same bases and subtracting their exponents:

$$\frac{15^{13}}{15^{2}} = 15^{11}$$

$$\frac{x^{9}}{x^{4}} = x^{5}$$

$$\frac{7^{3}}{7^{8}} = 7^{3-8} = 7^{-5}$$

In each of these cases, we divided the same bases and subtracted the bottom exponent from the top exponent.

Notice in that last example that you *can* end up with “Negative Exponents” as a result of these division problems. We will explain Negative Exponents before this lesson is finished, but don’t worry about what it means for now. Just understand *why* we’re subtracting the exponent of 8 from the exponent of 3 and getting a final exponent of -5.

Just as with Exponent Multiplication, it’s not hard to understand why this rule works. It’s good to grasp the reasons behind these rules, because I promise it will help you remember them and feel more confident in them.

Let’s take the example of $\frac{t^{5}}{t^{3}}$.

This could be read out loud as “t to the fifth divided by t to the third.”

If we expand it out, it could be written as:

$$\frac{t \times t \times t \times t \times t}{t \times t \times t}$$

Now notice: according to the rules of **Fractions**, each of the three t’s on bottom of the fraction could cancel out one of the t’s on top. If we cancel three t’s from both top and bottom, we will be left only with $t \times t$, or t^{2}.

See? $\frac{t^{5}}{t^{3}}$ must equal t^{5-3} or t^{2}. Whenever we divide the same bases, we *subtract* the bottom exponent from the top exponent.

Try these simple practice problems for yourself and fill in the final exponent:

$$\frac{c^{7}}{c^{4}} = c^{?}$$

$$\frac{5^{10}}{5^{3}} = 5^{?}$$

$$\frac{n^{2}}{n^{2}} = n^{?}$$

And the answers are below:

$$\frac{c^7}{c^4} = c^3$$

$$\frac{5^{10}}{5^3} = 5^7$$

$$\frac{n^2}{n^2} = n^{2-2} = n^0 = 1$$

We just subtracted the exponents for each example, because we're dividing numbers with the same bases.

Essential Exponent Rule #3: Exponents on Exponents

The third and final essential exponent rule happens when we raise one exponent to another exponent. For example:

$$(3^2)^5 = 3^{2\times5} = 3^{10}$$

Here are a few more examples of raising exponents to other exponents:

$$(2^3)^3 = 2^9$$

$$x^{3^4} = x^{12}$$

$$((n^2)^3)^4 = n^{2\times3\times4} = n^{24}$$

Just as with Exponent Multiplication and Division before this, there is a very good reason for why this rule works, and it's helpful to understand why, because it will help you remember the rule and apply it more quickly and confidently.

Let's take the example of $(n^2)^4$. This could be read out loud as "n squared to the fourth."

This can be expanded out and written as:

$$(n\times n)\times(n\times n)\times(n\times n)\times(n\times n)$$

Count up all your n's and how many do you get? There are 8 total... which is just 2×4, or the two exponents multiplied.

Whenever we raise an exponent to another exponent, we *multiply* the two exponents together.

Try these simple practice problems for yourself and fill in the final exponent:

$$(c^7)^4 = c^?$$

$$(5^{10})^3 = 5^?$$

$$((n^4)^2)^3 = n^?$$

And the answers are below:

$$(c^7)^4 = c^{28}$$

$$(5^{10})^3 = 5^{30}$$

$$((n^4)^2)^3 = n^{4\times2\times3} = n^{24}$$

We just multiplied the exponents for each example, because we're raising exponents by exponents.

Rules of Roots

"Roots" are the reverse of Powers. For example, if $x^5 = 32$, then $\sqrt[5]{32} = x$.

Or, the value of $\sqrt[4]{16} = 2$, because $2^4 = 16$.

We can read the examples above out loud as "the fifth root of 32 is x" and "the fourth root of 16 is 2."

Observe how Roots and Powers are the reverse of each other in the two examples above.

Of course, we're probably most familiar with the "square root," which could also be called "the second root": $\sqrt{25} = 5$ could be read as "the square root of 25 is 5."

If you wanted, you could write it as $\sqrt[2]{25}$, but usually we don't bother writing the $\sqrt[2]{\ }$ when we use square roots; we just write $\sqrt{\ }$ and the fact that it's a square root or "second root" is automatically understood by everyone who sees it.

We could also go the opposite direction in this example by using Powers, because $5^2 = 25$. Again, Roots and Powers are the reverse of each other.

Roots are like parentheses - they apply to everything underneath them. For example:

$$\sqrt[3]{8xy} = (\sqrt[3]{8})(\sqrt[3]{x})(\sqrt[3]{y})$$

I call the work I've done above "Splitting the Root." This is when I split the *group* of terms underneath a root into *individual* terms, with the same root being applied to each of those individual terms.

Try it yourself. Split the root below:

$$\sqrt[6]{2ab}$$

Your answer should look like this:

$$\sqrt[6]{2ab} = (\sqrt[6]{2})(\sqrt[6]{a})(\sqrt[6]{b})$$

Simplifying Roots

Sometimes we will need to "Simplify a Root." It's easier to show an example of what I mean, then explain:

For example, $\sqrt[3]{54}$ is *not* a common sight in most math textbooks. There's a reason for that: it can be broken down and written in a simplified form.

We will break it down into its smallest factors - the numbers that can multiply together to equal 54. Usually I would show this with a "Factor Tree," which you may have encountered in school classes. However, I can't type that on my computer, so I'll show it with a "Factor Pyramid" instead:

$$\begin{aligned} &54 \\ &= (2)(27) \\ &= (2)(3)(9) \\ &= (2)(3)(3)(3) \end{aligned}$$

Notice how I divided out small Prime Numbers like "2" and "3" and broke the number 54 down into its smallest factors. If you multiply out $2 \times 3 \times 3 \times 3$ again, you will return to the starting point of 54.

Here's why that's particularly interesting: we are doing $\sqrt[3]{\ }$ of 54, and when we factored out 54, we found *three* multiples of 3 lurking within it. That means that the matched set of three 3s can "escape" the $\sqrt[3]{\ }$ sign, like this:

$$3\sqrt[3]{2}$$

Note that the poor factor of 2 didn't have a group of three friends to escape the $\sqrt[3]{\ }$ with, so 2 has to remain "trapped" under the root symbol.

Also note that when the three 3's escape the $\sqrt[3]{\ }$, only *one* of them manages to escape. That's because $\sqrt[3]{3 \times 3 \times 3} = \sqrt[3]{27} = 3$. Again, this is just a form of breaking down and simplifying a root.

OK, you might "get it" already, or you might still be confused. Either way, it's OK. Let's try another practice example. This time we'll use $\sqrt[5]{96}$. Try simplifying it on your own first.

Rewrite $\sqrt[5]{96}$ in its most reduced form.

The explanation is on the next page.

Break 96 down with a "Factor Tree" or "Factor Pyramid" into its smallest Prime Factors:

$$\begin{aligned}
&96 \\
&=(2)(48) \\
&=(2)(2)(24) \\
&=(2)(2)(2)(12) \\
&=(2)(2)(2)(2)(6) \\
&=(2)(2)(2)(2)(2)(3)
\end{aligned}$$

Feel free to check that if you multiply all the numbers in the bottom row of my Factor Pyramid, you will return to the original value of 96.

Now, this time we're underneath a $\sqrt[5]{}$ symbol. That means only matched sets of *five* of the same factor can escape from underneath. Luckily, we have five factors of 2, so they can all escape. But remember, when all five of them escape, only one of them gets out - *and* the factor of 3 doesn't have any friends to escape with, so it remains trapped underneath.

Here's the result: $2\sqrt[5]{3}$.

Try it yourself on the following practice example:

Rewrite $\sqrt[4]{6{,}480}$ in its most simplified form.

OK, don't hate me for making you do a giant Factor Pyramid, but I needed a question like this to help us grasp the concept. Let's start factoring:

$$\begin{aligned}
&6{,}480 \\
&=(2)(3{,}240) \\
&=(2)(2)(1{,}620) \\
&=(2)(2)(2)(810) \\
&=(2)(2)(2)(2)(405) \\
&=(2)(2)(2)(2)(5)(81) \\
&=(2)(2)(2)(2)(5)(3)(27) \\
&=(2)(2)(2)(2)(5)(3)(3)(9) \\
&=(2)(2)(2)(2)(5)(3)(3)(3)(3)
\end{aligned}$$

Now, we're using a $\sqrt[4]{}$ symbol, so only numbers in matched sets of *four* can escape. Count them up: we have four 2's *and* four 3's! *Both* of these numbers can escape the $\sqrt[4]{}$. Remember, only *one* of each factor from each group of four will escape from the $\sqrt[4]{}$, and the lonesome factor of 5 will be left behind.

This can be written as $(2)(3)\sqrt[4]{5}$ and then simplified to $6\sqrt[4]{5}$. If you test it in your calculator (and you should!) then you will be convinced that $6\sqrt[4]{5}=\sqrt[4]{6{,}480}$

Hopefully now you have a better grasp of how to simplify Roots. If necessary, review this subsection and the three examples until it makes perfect sense.

Writing Roots as Exponents

There's another very important thing to know about Roots: they can be rewritten as "Fractional Exponents." For example:

$$\sqrt{x} = x^{\frac{1}{2}}$$

$$\sqrt[3]{n} = n^{\frac{1}{3}}$$

$$\sqrt[37]{t} = t^{\frac{1}{37}}$$

Notice how each Root can be turned into a Fractional Exponent by taking the Root, putting it on the *bottom* of a fraction, and using that fraction as an exponent upon the base.

Try these simple practice problems for yourself and fill in the final exponent:

$$\sqrt[7]{5} = 5^{?}$$

$$\sqrt[12]{x} = x^{?}$$

$$\sqrt[304]{n} = n^{?}$$

And the answers are below:

$$\sqrt[7]{5} = 5^{\frac{1}{7}}$$

$$\sqrt[12]{x} = x^{\frac{1}{12}}$$

$$\sqrt[304]{n} = n^{\frac{1}{304}}$$

We just took the Root from each example and turned it into a Fractional Exponent with the Root on the bottom of the Fractional Exponent.

Fractional Exponents

Let's talk more about Fractional Exponents. It's possible to raise a Base by a Fractional Exponent, which can include *both* a Power and a Root in a single exponent.

We know that $x^{\frac{1}{3}} = \sqrt[3]{x}$. Well guess what: $x^{\frac{2}{3}} = \sqrt[3]{x^2}$!

See, our Fractional Exponent can provide both a Power (2) and a Root (3) in the same exponent!

The Power goes on top of the Fractional Exponent. That makes sense, since Powers usually make numbers "bigger." The Root goes on bottom of the Fractional Exponent. That also makes sense, since Roots usually make numbers "smaller."

Here are a few more examples of Fractional Exponents with both Powers and Roots:

$$t^{\frac{3}{5}} = \sqrt[5]{t^3}$$

$$m^{\frac{9}{2}} = \sqrt[2]{m^9}$$

$$5^{\frac{7}{3}} = \sqrt[3]{5^7}$$

Make sure you see how we can convert back and forth between the two sides of each of the examples above.

Now try a few simple practice problems for yourself:

$$7^{\frac{4}{3}} = ?$$

$$n^{\frac{13}{8}} = ?$$

$$d^{\frac{2}{11}} = ?$$

And the answers are below:

$$7^{\frac{4}{3}} = \sqrt[3]{7^4}$$

$$n^{\frac{13}{8}} = \sqrt[8]{n^{13}}$$

$$d^{\frac{2}{11}} = \sqrt[11]{d^2}$$

We just took the Power from the top of each Fractional Exponent and the Root from the bottom of each Fractional Exponent.

Fractional Exponents can be reduced like any other **Fraction**. For example:

$$b^{\frac{6}{3}} = b^2$$

$$x^{\frac{15}{12}} = x^{\frac{5}{4}}$$

$$n^{\frac{6}{18}} = n^{\frac{1}{3}}$$

And, all the Basic Rules of Exponents *still apply* to Fractional Exponents.

Multiplying same bases will add the exponents, like always:

$$(x^{\frac{3}{2}})(x^{\frac{5}{2}}) = x^{\frac{3}{2}+\frac{5}{2}} = x^{\frac{8}{2}} = x^4$$

Dividing same bases will subtract the exponents, like always:

$$\frac{n^{\frac{5}{3}}}{n^{\frac{1}{3}}} = n^{\frac{5}{3}-\frac{1}{3}} = n^{\frac{4}{3}}$$

Raising exponents to other exponents will multiply the exponents, like always:

$$(y^{\frac{5}{6}})^{\frac{2}{3}} = y^{(\frac{5}{6})(\frac{2}{3})} = y^{\frac{10}{18}} = y^{\frac{5}{9}}$$

The Exponent Rules operate exactly the same way they did with simpler-looking exponents.

These type of Fractional Exponent situations often seem *very* complicated to novices but *very* simple to those of us who understand the basic rules of Exponents and feel confident working with Fractions.

Also notice that whenever I can reduce the Fractions in the examples above, I go ahead and do so. It's good to simplify fractions whenever possible, because it makes your numbers smaller and simpler.

Now try a few more practice problems for yourself:

$$(x^{\frac{5}{3}})(x^{\frac{7}{3}}) = ?$$

$$\frac{n^{\frac{7}{9}}}{n^{\frac{2}{9}}} = ?$$

$$(d^{\frac{1}{5}})^{\frac{3}{4}} = ?$$

And the answers are below:

$$(x^{\frac{5}{3}})(x^{\frac{7}{3}}) = x^{\frac{5}{3}+\frac{7}{3}} = x^{\frac{12}{3}} = x^4$$

$$\frac{n^{\frac{7}{9}}}{n^{\frac{2}{9}}} = n^{\frac{7}{9}-\frac{2}{9}} = n^{\frac{5}{9}}$$

$$(d^{\frac{1}{5}})^{\frac{3}{4}} = d^{(\frac{1}{5})(\frac{3}{4})} = d^{\frac{3}{20}}$$

We just applied the Basic Rules of Exponents to each of the Fractional Exponent examples, and simplified our Fractions whenever possible.

Negative Exponents

It is possible to raise a base to a Negative Exponent. A negative exponent means to "invert", or flip upside down, the base number it is applied to. For example:

$$3^{-2} = \frac{1}{3^2}$$

We can read this out loud as "three to the power of negative two." Notice that 3 is still raised to the power of 2, but it is also *inverted*, or flipped upside down to the bottom of a fraction.

Here are a few more simple examples to help you get the idea:

$$7^{-4} = \frac{1}{7^4}$$

$$x^{-1} = \frac{1}{x^1} = \frac{1}{x}$$

$$(3n)^{-4} = \frac{1}{(3n)^4}$$

Negative Exponents are *not* the same as "negative numbers" - not at all! The purpose of a negative exponent is simply to invert its base, either from top to bottom, or bottom to top.

If a Negative Exponent is present on the *bottom* of a fraction, it will invert its base to the *top* of the fraction. Here are some examples:

$$\frac{1}{x^{-5}} = x^5$$

$$\frac{15}{t^{-1}} = 15t^1 = 15t$$

$$\frac{4}{x^2 y^{-3}} = \frac{4y^3}{x^2}$$

Notice how each base with a negative exponent is moved from the bottom to the top of the fraction. To "invert" is simply to flip from top to bottom, or from bottom to top. This is the sole purpose of all negative exponents.

If a fraction has multiple negative exponents, they each should be inverted from top to bottom or bottom to top, depending on where they started. Here are some examples:

$$\frac{x^{-2}}{y^{-3}} = \frac{y^3}{x^2}$$

$$\frac{5^{-7}}{n^{-4}} = \frac{n^4}{5^7}$$

$$\frac{x^{-6} y^8}{x^{11} y^{-2}} = \frac{y^8 y^2}{x^{11} x^6}$$

Are you starting to get the idea of Negative Exponents? Good, if so. And if not, review the examples until you can see exactly what's happening in each of them. Track how the negative exponents *invert* their base to the top or bottom of the fraction. It's actually a very simple concept once you grasp it.

Try these simple practice problems for yourself:

$$c^{-7} = ?$$

$$\frac{x^{-3}}{y^{-5}} = ?$$

$$\frac{5n^3 t^{-4}}{2^{-3} x^5 y^{-8}} = ?$$

And the answers are below:

$$c^{-7} = \frac{1}{c^7}$$

$$\frac{x^{-3}}{y^{-5}} = \frac{y^5}{x^3}$$

$$\frac{5n^3t^{-4}}{2^{-3}x^5y^{-8}} = \frac{(2^3)5n^3y^8}{t^4x^5}$$

We simply inverted the bases for each negative exponent in the examples, because the only thing a negative exponent does is invert the base it's attached to.

Combo Exponents

A "Combo Exponent" is what I call an exponent that combines a Power, a Root, and a Negative Exponent all into a single exponent. For example:

$$2^{-\frac{5}{3}} = \frac{1}{\sqrt[3]{2^5}}$$

Notice: the base of 2 is *raised* to the 5th power, which comes from the top of the Fractional Exponent; that is then put under a 3rd root, which comes from the bottom of the Fractional Exponent; all of that is then *inverted* because of the negative exponent.

There is *nothing new to see here* - it's just a combination of Powers, Roots, and Inversion all in a single, tidy package.

Try these simple practice problems for yourself:

$$c^{-\frac{7}{3}} = ?$$

$$6^{-\frac{2}{3}} = ?$$

$$\frac{2}{x^{-\frac{4}{9}}} = ?$$

And the answers are below:

$$c^{-\frac{7}{3}} = \frac{1}{\sqrt[3]{c^7}}$$

$$6^{-\frac{2}{3}} = \frac{1}{\sqrt[3]{6^2}}$$

$$\frac{2}{x^{-\frac{4}{9}}} = 2\sqrt[9]{x^4}$$

We applied the Combo Exponent for each of the examples. We raised to the Power, then applied the Root, then Inverted any negative exponents from top to bottom, or from bottom to top.

Combo Exponents still follow the exact same Essential Exponent Rules as any other exponents do.

Don't forget that Combo and Fractional Exponents can be reduced exactly like all **Fractions** can be. For example, $x^{-\frac{8}{4}}$ can be reduced to x^{-2} and then written as $\frac{1}{x^2}$.

Try simplifying the following practice examples:

$$\frac{(c^{-\frac{2}{3}})^{-\frac{3}{2}}}{c^{-\frac{1}{3}}} = ?$$

$$\left(\frac{x^{-\frac{3}{5}}}{x^{-\frac{5}{3}}}\right)^{\frac{1}{2}} = ?$$

And the answer steps are below:

$$\frac{(c^{-\frac{2}{3}})^{-\frac{3}{2}}}{c^{-\frac{1}{3}}} = \frac{c^{(-\frac{2}{3})(-\frac{3}{2})}}{c^{-\frac{1}{3}}} = \frac{c^{\frac{6}{6}}}{c^{-\frac{1}{3}}} = c^{\frac{6}{6}-(-\frac{1}{3})} = c^{\frac{6}{6}-(-\frac{2}{6})} = c^{\frac{6}{6}+\frac{2}{6}} = c^{\frac{8}{6}} = c^{\frac{4}{3}}$$

$$\left(\frac{x^{-\frac{3}{5}}}{x^{-\frac{5}{3}}}\right)^{\frac{1}{2}} = \frac{x^{(\frac{-3}{5})(\frac{1}{2})}}{x^{(-\frac{5}{3})(\frac{1}{2})}} = \frac{x^{\frac{-3}{10}}}{x^{-\frac{5}{6}}} = x^{\frac{-3}{10}-(-\frac{5}{6})} = x^{\frac{-3}{10}+\frac{5}{6}} = x^{\frac{-9}{30}+\frac{25}{30}} = x^{\frac{16}{30}} = x^{\frac{8}{15}}$$

Obviously these two practice examples are more complex than most of the previous examples. However, we are applying all the same rules as before. It just takes more steps to work through each set of maneuvers carefully, while also testing our knowledge of **Fractions**.

Try to make sure you understand these two practice examples perfectly before continuing.

Pretest Question #1

Let's take a look at our first Pretest question on this topic. Try it yourself if you got it wrong the first time.

The expression $\dfrac{\sqrt{\left(x^{\frac{7}{3}}y^{-\frac{4}{3}}\right)}}{\sqrt[3]{\left(x^{-\frac{3}{2}}y^{\frac{5}{2}}\right)}}$, where $x>1$ and $y>1$, is equivalent to which of the following?

(A) $\dfrac{1}{\sqrt[36]{(xy)^{11}}}$

(B) $\dfrac{\sqrt[3]{x^5}}{\sqrt{y^3}}$

(C) $\dfrac{\sqrt{y^3}}{\sqrt[3]{x^5}}$

(D) $\sqrt[36]{(xy)^{11}}$

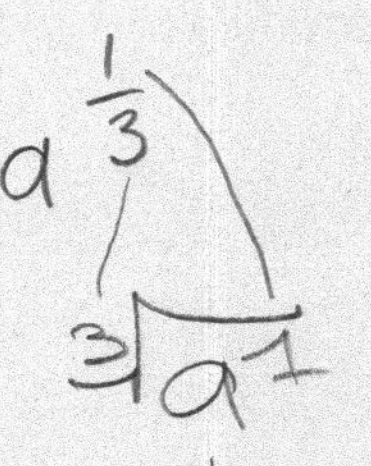

OK, this question presents us with a huge variety of Roots and Combo Exponents. We'll also test our essential Exponent Rules.

I would start by rewriting the Roots as Fractional Exponents on top and bottom of the fraction, like this:

$$\frac{\left(x^{\frac{7}{3}}y^{-\frac{4}{3}}\right)^{\frac{1}{2}}}{\left(x^{-\frac{3}{2}}y^{\frac{5}{2}}\right)^{\frac{1}{3}}}$$

Then, I would distribute the exponents to the parentheses on both top and bottom. Remember, raising an exponent to another exponent will *multiply* the exponents:

$$\frac{x^{(\frac{7}{3})(\frac{1}{2})}y^{(-\frac{4}{3})(\frac{1}{2})}}{x^{(-\frac{3}{2})(\frac{1}{3})}y^{(\frac{5}{2})(\frac{1}{3})}}$$

Now multiply all the Fractional Exponents:

$$\frac{x^{\frac{7}{6}}y^{-\frac{4}{6}}}{x^{-\frac{3}{6}}y^{\frac{5}{6}}}$$

Next, I would invert all of my Negative exponents. Top goes to bottom, and bottom goes to top:

$$\frac{x^{\frac{7}{6}}x^{\frac{3}{6}}}{y^{\frac{5}{6}}y^{\frac{4}{6}}}$$

At this point, we have some "same bases" multiplying each other, so their exponents will *add* to each other.

$$\frac{x^{\frac{7}{6}+\frac{3}{6}}}{y^{\frac{5}{6}+\frac{4}{6}}}$$

Luckily, the fractions conveniently have the same denominators already, so I won't need to use any special **Fraction** tricks to get common denominators. I can just add right away:

$$\frac{x^{\frac{10}{6}}}{y^{\frac{9}{6}}}$$

The fractions in the exponents can be simplified:

$$\frac{x^{\frac{5}{3}}}{y^{\frac{3}{2}}}$$

Finally, we should rewrite this using Root symbols in order to match the format of the Answer Choices:

$$\frac{\sqrt[3]{x^5}}{\sqrt{y^3}}$$

This puts us at **Choice B**, and we've definitely proven our skill at manipulating Combo Exponents with the Essential Exponent Rules.

Algebra with Exponents

We can also use the Rules of Exponents to solve Algebra equations.

For example, what if we were given this question:

If $x^{\frac{3}{2}} = 5$, what is the value of x?

We can apply our Exponent Rules to solve this question quite easily.

We want to solve for the value of x - in other words, we want to get x^1 by itself. That means we have to find a way to "strip" or remove the exponent $\frac{3}{2}$ from x.

What's the easiest way to remove that exponent from x? Well, we know the rule that raising an exponent to another exponent will *multiply* the two exponents. So, what exponent could we use to multiply the existing exponent of $\frac{3}{2}$ to transform it to an exponent of "1"?

Or in other words, if $\frac{3}{2}n = 1$, what is the value of n?

Well, just multiply both sides by the reciprocal fraction (this requires a solid understanding of **Fractions**):

$$\frac{3}{2}n = 1$$
$$(\frac{2}{3})\frac{3}{2}n = 1(\frac{2}{3})$$
$$n = \frac{2}{3}$$

Now we know that we can raise both sides of our original equation by the exponent $\frac{2}{3}$ to get x by itself:

$$x^{\frac{3}{2}} = 5$$
$$(x^{\frac{3}{2}})^{\frac{2}{3}} = (5)^{\frac{2}{3}}$$
$$x^{\frac{6}{6}} = 5^{\frac{2}{3}}$$
$$x^1 = 5^{\frac{2}{3}}$$
$$x = \sqrt[3]{5^2}$$

Since we raised *both sides* by the same exponent, the equation remains balanced. Now you know how Rules of Exponents can add an exciting and useful tool to your Algebra-Solving toolkit.

Try it yourself on the following practice example:

If $\frac{n^4}{n^2} = 4$, what is the value of n?

In this case, the first thing to do is simplify the fraction with the same base on top and bottom. When we divide the same base, we *subtract* the bottom exponent from the top exponent:

$$n^{4-2} = 4$$
$$n^2 = 4$$

Now we can raise both sides to the power of $\frac{1}{2}$ (or just square root both sides, if that's how you prefer to think about it):

$$(n^2)^{\frac{1}{2}} = (4)^{\frac{1}{2}}$$
$$n^{\frac{2}{2}} = 4^{\frac{1}{2}}$$
$$n^1 = \sqrt{4}$$
$$n = 2$$

Admittedly, these last two examples have been fairly simple, but you can still start to see how the same basic Exponent Rules and Fractional Exponents that we've been studying throughout the lesson can be used to solve Algebra problems.

Pretest Question #2

Let's take a look at another Pretest question. Try it yourself before you look at my explanation below the question.

If $x - 4y = 7$, what is the value of $\frac{2^x}{16^y}$?

(A) $(\frac{1}{8})^3$

(B) $8^{\frac{7}{4}}$

(C) 2^7

(D) The value cannot be determined from the information given.

This question will test a bit of our Algebra with Exponents skills.

It also requires a special insight before you can make any real progress. That insight is the fact that "16" can be rewritten as 2^4.

$$\frac{2^x}{(2^4)^y}$$

Notice the benefit we gain - our fraction now has the same base of "2" on top and bottom, which really opens up our opportunities to use the Essential Exponent Rules.

But, before we continue, how were you even supposed to think of this move?

Well, the hints are subtle, but they're there. It's obviously a question involving Exponents - the Question and Answer Choices give this away. And, we know it's easier to work with Exponents when the bases are the same. Furthermore, 16 is a power of 2. So, take a chance and rewrite "16" as 2^4, and now we can make real progress.

Returning to our rewritten fraction, the next step is to simplify the bottom of the fraction. We know that exponents raised to other exponents will multiply:

$$\frac{2^x}{(2^4)^y}$$

$$= \frac{2^x}{2^{4y}}$$

Then, when we divide the same bases, we subtract the bottom exponent from the top exponent:

$$\frac{2^x}{2^{4y}}$$
$$= 2^{x-4y}$$

Now, stop for a moment and return to the original question. They *gave* us the fact that $x - 4y = 7$. We can substitute it directly!

$$\text{If } 2^{x-4y} \text{ and } x - 4y = 7$$
$$\text{Then } 2^{x-4y} = 2^7$$

And we're done. **Choice C** gives us the option 2^7.

Review & Encouragement

I know that this chapter contains a lot of info - but it's vital that you understand each subsection of the lesson.

I've found something interesting about these rules - the more clearly you understand them, the easier they are to remember. It stops seeming like a pile of disconnected "rules" and becomes a cohesive set of a few concepts that all make sense together. Try to reach this level of understanding yourself. Even if it takes hard work to get there, it will make the whole topic seem quite easy when you reach it.

Furthermore, when you see Exponent & Root questions on the SAT Test, they frequently mix together many of the subtopics from this lesson into a single question. If there's even *one* part of this lesson that you don't understand, the test will probably catch you out with a trap and you'll get the question wrong.

The writers of the SAT Math invest a lot of effort into making "trick" answers for Exponent-based questions. They know all of the common mistakes and misunderstandings that students have in this topic, and they *will* exploit those misunderstandings any time they can.

All of the rules of Exponents and Roots work together like a well-oiled machine. When it finally "clicks" for you, it will feel amazing! Now give your full attention to the set of practice problems to test your mastery of this topic.

Exponents & Roots Practice Questions

NO CALCULATOR IS PERMITTED ON ANY OF THESE QUESTIONS.

1. Which of the following is equal to $x^{\frac{4}{3}}$, for all values of x?

   (A) $\sqrt[3]{x^{\frac{1}{4}}}$

   (B) $\sqrt[3]{x^4}$

   (C) $\sqrt[4]{x^3}$

   (D) $\sqrt[4]{x^{\frac{1}{3}}}$

$$\sqrt{16x^2}$$

2. If $x > 0$, which of the following is equivalent to the given expression?

   (A) $4x$

   (B) $4x^2$

   (C) $8x$

   (D) $32x$

3. FREE RESPONSE: If $q^{\frac{g}{3}} = 81$ for positive integers q and g, what is one possible value of g?

4. Which of the following is equivalent to $4^{\frac{3}{4}}$?

   (A) 3

   (B) $2\sqrt{2}$

   (C) $2\sqrt[4]{4}$

   (D) $4\sqrt[3]{4}$

5. If $n^{-\frac{2}{3}} = p$, where $n > 0$, what is n in terms of p?

   (A) $-\frac{2}{3}p$

   (B) $-\sqrt[3]{p^2}$

   (C) $\frac{1}{\sqrt{p^3}}$

   (D) $-\frac{1}{\sqrt[3]{p^2}}$

6. FREE RESPONSE: If $x^{\frac{n-3}{2.5}} = 1$ and $x \neq 0$, what is the value of n?

7. The expression $\dfrac{x^{-\frac{3}{2}}y^{\frac{2}{3}}}{x^{\frac{3}{2}}y^{-\frac{1}{6}}}$, where $x>1$ and $y>1$, is equivalent to which of the following?

(A) $\sqrt{y}$

(B) $\dfrac{\sqrt[6]{y^5}}{x^3}$

(C) $\dfrac{\sqrt[9]{y}}{\sqrt[4]{x^9}}$

(D) $\dfrac{\sqrt[4]{xy}}{xy}$

8. If $2x+y=4$, what is the value of $\dfrac{9^x}{3^{-y}}$?

(A) 3^4

(B) 3^8

(C) 9^6

(D) The value cannot be determined from the information given.

9. If $\dfrac{n^{b^4}}{n^{a^4}}=n^{24}$ and $a+b=4$, what is the value of $a-b$?

(A) -6

(B) -4

(C) 4

(D) 6

10. The expression $\dfrac{(x^{\frac{2}{3}}y^{-\frac{3}{2}})^{-\frac{5}{6}}}{\sqrt[3]{x^{-\frac{3}{2}}y^{\frac{5}{6}}}}$, where $x\neq 0$ and $y\neq 0$, is equivalent to which of the following?

(A) $\sqrt[9]{\dfrac{1}{x^6y^{22}}}$

(B) $\dfrac{1}{x^{\frac{13}{24}}y^{\frac{5}{4}}}$

(C) $(xy)^{\frac{55}{12}}$

(D) $\sqrt[36]{\dfrac{y^{35}}{x^2}}$

Exponents & Roots Answers

1. B
2. A
3. 1, 2, 3, 4, 6, or 12
4. C
5. C
6. 3
7. B
8. A
9. A
10. D

Exponents & Roots Explanations

1. **B.** This is a simple question that asks us to translate a Fractional Exponent. Remember that with Fractional Exponents, the numerator (top number of the fraction) gives the *power*, and the denominator (bottom number of the fraction) gives the *root*.

Therefore, the *power* of the fractional exponent on x is 4, and the *root* is 3. This can be written in words as "the third root of x to the fourth power" - or expressed in mathematical symbols as $\sqrt[3]{x^4}$, Choice B.

2. **A.** In this question, we need to simplify the original expression. Remember that roots are like parentheses - they apply to everything underneath them. If we wanted to, we could "split the root" and rewrite this as:

$$(\sqrt{16})(\sqrt{x^2})$$

This can be simplified in two parts. The square root of 16 is 4:

$$\sqrt{16} = 4$$

The square root of x^2 is x.

$$\sqrt{x^2} = x$$

Therefore, $\sqrt{16x^2} = 4x$. Can you see how the square root distributed to both the "16" and the "x^2"?

Our final answer can be written as $4x$, or Choice A.

3. **1, 2, 3, 4, 6, or 12.** There are a variety of possible answers to this question, but the process for finding them all is essentially the same. I will not explain every possible answer, but you should be able to figure out the rest from my work below.

First, remember that an "integer" is a *whole number*, including all negative and positive whole numbers, and 0.

Now, consider some possibilities. We'll choose a possible integer value for q, and then determine what integer g would need to be in order to make the equality $q^{\frac{g}{3}} = 81$ be true.

First, and most simply, q could just equal 81 already.

Then, because $81^1 = 81$, the exponent $\frac{g}{3}$ would have to equal 1. And if $\frac{g}{3} = 1$, then $g = 3$ according to **Basic Algebra 1**.

Or, consider that q could equal 9. Then, since $9^2 = 81$, we could surmise that the exponent $\frac{g}{3} = 2$, and $g = 6$.

Now we've found two possible solutions for the value of g - it could equal 3, or 6.

What if $q = 3$? The value of 3^4 is 81. So, the exponent $\frac{g}{3}$ would have to equal 4, and $g = 12$.

If you feel like getting more creative, you could also consider more complex fractional exponents. For example, $27^{\frac{4}{3}} = 81$, so g could also be 4. However, I would discourage making this question any harder than it needs to be. The easiest and fastest answers are just $g = 3$ and $g = 6$.

4. **C.** This question asks us to simplify a Fractional Exponent as far as we can.

Remember that the top number of the Fractional Exponent is the *power* and the bottom number is the *root*. We could say $4^{\frac{3}{4}}$ means "the fourth root of 4 to the third power", or we could rewrite it with math symbols as $\sqrt[4]{4^3}$.

But, this form doesn't look like any of our answer choices. What gives?

Well, our rewriting isn't finished yet. Let's explore what happens if we calculate the value of 4^3:

$$4^3 = (4)(4)(4) = 64$$

So, we could rewrite $\sqrt[4]{4^3}$ as "$\sqrt[4]{64}$". But it *still* doesn't look any of the answers.

What we need to do next is break down 64 into smaller factors and look for any specific factors that appear four times to match with the fourth root or $\sqrt[4]{}$ symbol.

Many students call this technique a "factor tree," although I can't make it look like a "tree" on my computer, so I'll use a "Factor Pyramid" - but the idea is exactly the same. I'm breaking down 64 into its smallest factors. Follow each line as I break the number down into smaller and smaller multiples:

$$\begin{aligned}&64\\&=2\times32\\&=2\times2\times16\\&=2\times2\times2\times8\\&=2\times2\times2\times2\times4\\&=2\times2\times2\times2\times2\times2\end{aligned}$$

Let's look at the very bottom row, where we've broken out 64 into the smallest possible factors that make it up.

Do we have a group of four of the same number? Yes - we have a group of four "2s" that we can pull out from under the $\sqrt[4]{\ }$ symbol.

If we pull a matched set of four 2s out from a *fourth root*, they will "escape" as only a single 2. It takes four of the same number to escape a fourth root, and together, all four values will escape as one, like this: $2\sqrt[4]{\ }$

But, that doesn't account for all of the factors in our "tree": we found that 64 is made of *six* factors of 2. Four of them have escaped to the outside of the $\sqrt[4]{\ }$, but the remaining two factors of 2 are still trapped under the root symbol.

We could write this situation as $2\sqrt[4]{2\times2}$.

Or, more simply, as $2\sqrt[4]{4}$ - Choice C.

5. **C.** This question gives us an Algebra equation based on Fractional Exponents. We'll use the Rules of Exponents to solve it.

We start with the given equation $n^{-\frac{2}{3}}=p$ and we're trying to get n by itself. This can be accomplished in a single maneuver, if you know the Rules of Exponents.

Think - if we want n by itself, we could also say we want n^1, since n^1 is just "n".

So, how can we remove the exponent of $-\frac{2}{3}$ from n?

Well, according to the Rules of Exponents, if we raise one exponent to another exponent, then the two exponents will *multiply* each other.

So, what number x would we multiply $-\frac{2}{3}$ by to result in a final exponent of "1"?

You can set this up with Algebra (or just solve it in your head if you're strong with **Fractions**):

$$-\frac{2}{3}(x)=1$$

And then solve by multiplying both sides by $-\frac{3}{2}$ using **Basic Algebra 1**:

$$-\frac{2}{3}(x)=1$$

$$(-\frac{3}{2})(-\frac{2}{3})(x)=(1)(-\frac{3}{2})$$

$$x=-\frac{3}{2}$$

What we've learned is that if we raise $n^{-\frac{2}{3}}$ to the $-\frac{3}{2}$ power, the exponents will multiply each other and cancel to "1".

So, return to the original equation:

$$n^{-\frac{2}{3}}=p$$

And raise both sides to the $-\frac{3}{2}$ power to cancel the $-\frac{2}{3}$ exponent on n:

$$(n^{-\frac{2}{3}})^{-\frac{3}{2}}=(p)^{-\frac{3}{2}}$$

And then simplify:

$$n^1=p^{-\frac{3}{2}}$$

$$n=p^{-\frac{3}{2}}$$

And we're done! Or are we?

Our math is correct, but unfortunately our written form doesn't quite match the answer choices. We'll need to use the rules of Combo Exponents to rewrite $n=p^{-\frac{3}{2}}$ using roots, powers, and an inverse fraction.

First, apply the negative exponent to make an inverse fraction:

$$n=\frac{1}{p^{\frac{3}{2}}}$$

Then, change the "2" on bottom of the Fractional Exponent into a square root (or "second root") symbol:

$$n = \frac{1}{\sqrt{p^3}}$$

And *now* we're done. We've gotten n by itself by using the Rules of Exponents and raising an exponent to another exponent; then we've rewritten our answer according to Fractional Exponent rules. Our answer is Choice C.

6. **3.** To solve this question, remember that *any* number (other than zero) raised to the "zero power" will equal "1." It's safe to assume, then, that x is being raised to the zero power.

Therefore, the entire exponent on x must be equal to "0".

Set this up:

$$\frac{n-3}{2.5} = 0$$

And then solve for n using **Basic Algebra 1**.

$$\frac{n-3}{2.5} = 0$$
$$(2.5)\frac{n-3}{2.5} = 0(2.5)$$
$$n - 3 = 0$$
$$+3 + 3$$
$$n = 3$$

And we're done. The value of n must be 3.

7. **B.** Wow, the original form looks like an absolute *mess* when we first see it:

$$\frac{x^{-\frac{3}{2}}y^{\frac{2}{3}}}{x^{\frac{3}{2}}y^{-\frac{1}{6}}}$$

Let's start by moving the negative exponents from top to bottom, and vice-versa. Remember that negative exponents represent *inverse*, so any number with a negative exponent on top of the fraction will move to the bottom of the fraction, and any number with a negative exponent on the bottom of a fraction will move to the top.

After inverting our negative exponents, it looks like this:

$$\frac{y^{\frac{1}{6}}y^{\frac{2}{3}}}{x^{\frac{3}{2}}x^{\frac{3}{2}}}$$

OK, it already looks a tiny bit better.

Furthermore, we know from Exponent Rules that if the *bases* are the same and we multiply two bases, their exponents will *add* to each other. We have the same bases on top (both y) and the same bases on bottom (both x). So, let's apply this rule to simplify further:

$$\frac{y^{\frac{1}{6}+\frac{2}{3}}}{x^{\frac{3}{2}+\frac{3}{2}}}$$

Now we turn to the rules of **Fractions** to simplify even more. First I need to get common denominators on top of the fraction, so I'll convert $\frac{2}{3}$ to $\frac{4}{6}$:

$$\frac{y^{\frac{1}{6}+\frac{4}{6}}}{x^{\frac{3}{2}+\frac{3}{2}}}$$

Now add the fractions to each other:

$$\frac{y^{\frac{5}{6}}}{x^{\frac{6}{2}}}$$

It's looking better and better. Also note that the fraction $\frac{6}{2}$ on bottom can be simplified to "3":

$$\frac{y^{\frac{5}{6}}}{x^{3}}$$

Last but not least, we need to match the styles of the answer choices. That requires rewriting our Fractional Exponent $y^{\frac{5}{6}}$ as $\sqrt[6]{y^5}$. At this point, you should feel very comfortable making this rewrite.

And, we're done! Our final expression is $\frac{\sqrt[6]{y^5}}{x^3}$, or Choice B.

8. **A.** A lot of students are tempted to pick "cannot be determined" because they're not sure how to handle this problem. But, there *is* a solution, and we're going to find it. The process is very similar to Pretest Question 2 from this lesson.

Here's the trick with this question. You have to notice that "9" can be rewritten as 3^2.

How are you supposed to think of this in the first place?

Well, the clues are subtle, but all around you. Clearly, the question is about Exponents. And, both "3" and "9" are related by powers of exponents. So, take a chance and replace the "9" on top with 3^2.

$$\frac{(3^2)^x}{3^{-y}}$$

Look at what an advantage this gains us: now the fraction is an Exponent Division problem with the same base of 3 on top and bottom.

We can clean the top up using the Rules of Exponents, because raising an exponent to another exponent will multiply the two exponents together:

$$\frac{3^{2x}}{3^{-y}}$$

And, according to the Rules of Exponents, we can divide same-base numbers by subtracting the bottom exponent from the top exponent, like this:

$$\frac{3^{2x}}{3^{-y}}$$
$$= 3^{2x-(-y)}$$

And that can be cleaned up:

$$= 3^{2x-(-y)}$$
$$= 3^{2x+y}$$

Now check it out: our new exponent of $2x+y$ is the same equation given in the beginning of the question! We were given that $2x+y=4$. So, we can do a direct substitution:

$$\text{If } 2x+y=4$$
$$\text{Then } 3^{2x+y}=3^4$$

And we're done - this is the same as Answer Choice A. Personally, I find this to be a very satisfying question - probably because of how many people are tempted to give up on it too soon.

9. **A.** This is a combined Exponent Algebra and **Systems of Equations** question (we'll cover Systems of Equations in more detail in a later lesson, so just follow along with my work for now).

First, clean up and simplify the exponents in the original expression. We know from Rules of Exponents that exponents raised to exponents will multiply the exponents together:

$$\frac{n^{b^4}}{n^{a^4}} = n^{24}$$
$$\frac{n^{4b}}{n^{4a}} = n^{24}$$

And, when we divide same-base numbers, we can subtract the bottom exponent from the top exponent:

$$\frac{n^{4b}}{n^{4a}} = n^{24}$$
$$n^{4b-4a} = n^{24}$$

If the bases are the same on both sides, then the exponents must be equal to each other as well:

$$4b-4a=24$$

Divide both sides by 4 to simplify:

$$\frac{4b-4a}{4} = \frac{24}{4}$$
$$b-a=6$$

It looks *much* better than it did at first, doesn't it?

From here, we'll use **Systems of Equations** (covered in a future lesson) and **Basic Algebra 1** to finish the question. I'll show my steps below:

$$a+b=4$$
$$b-a=6$$

First, I'll Isolate a from the first equation on top:

$$a+b=4$$
$$-b \quad -b$$
$$a=4-b$$

Now I'll Substitute this for a into the second equation on bottom:

$$b-a=6$$
$$b-(4-b)=6$$

Now simplify and solve for b:

$$b-(4-b)=6$$
$$b-4+b=6$$
$$2b-4=6$$
$$+4\ +4$$
$$2b=10$$
$$\frac{2b}{2}=\frac{10}{2}$$
$$b=5$$

Now that we know the value of b is 5, we can plug it back into either of our two original equations:

$$a+b=4$$
$$a+5=4$$
$$-5\ -5$$
$$a=-1$$

Armed with the values of a and b, we can finish the final question, which asked for the value of $a-b$.

$$a-b=?$$
$$(-1)-(5)=?$$
$$-1-5=-6$$

Our final result, -6, corresponds to Choice A, and we're done.

10. **D.** This question is basically a bigger version of Pretest Question 1 and Practice Question 7. It will test all of our Exponent Rules to the limit. In fact, it's slightly more difficult than any Exponent question you're likely to see on the SAT test, so if you can handle this, you can handle just about anything.

First, we should focus on cleaning the exponents up as much as possible.

The first step is to look at the top of the fraction, where there are exponents being raised by other exponents. The $-\frac{5}{6}$ exponent will distribute to both x and y terms inside the parentheses, and exponents raised to other exponents will multiply each other.

Let's do this now:

$$\frac{(x^{\frac{2}{3}}y^{-\frac{3}{2}})^{-\frac{5}{6}}}{\sqrt[3]{x^{-\frac{3}{2}}y^{\frac{5}{6}}}}$$

$$=\frac{x^{(\frac{2}{3})(-\frac{5}{6})}y^{(-\frac{3}{2})(-\frac{5}{6})}}{\sqrt[3]{x^{-\frac{3}{2}}y^{\frac{5}{6}}}}$$

$$=\frac{x^{-\frac{10}{18}}y^{\frac{15}{12}}}{\sqrt[3]{x^{-\frac{3}{2}}y^{\frac{5}{6}}}}$$

Also, we should simplify the fractional exponents on the top of the fraction just to make things simpler:

$$\frac{x^{-\frac{10}{18}}y^{\frac{15}{12}}}{\sqrt[3]{x^{-\frac{3}{2}}y^{\frac{5}{6}}}}$$

$$=\frac{x^{-\frac{5}{9}}y^{\frac{5}{4}}}{\sqrt[3]{x^{-\frac{3}{2}}y^{\frac{5}{6}}}}$$

OK, the top of the fraction is looking a little better. Now let's turn our attention to the bottom of the fraction.

We know by now that the root $\sqrt[3]{n}$ can be rewritten as the fractional exponent $n^{\frac{1}{3}}$. Let's use that on the bottom of the fraction:

$$\frac{x^{-\frac{5}{9}}y^{\frac{5}{4}}}{\sqrt[3]{x^{-\frac{3}{2}}y^{\frac{5}{6}}}}$$

$$=\frac{x^{-\frac{5}{9}}y^{\frac{5}{4}}}{(x^{-\frac{3}{2}}y^{\frac{5}{6}})^{\frac{1}{3}}}$$

Now, on the bottom just as we did with the top of the fraction, we'll distribute the exponent to the x and y terms. Raising one an exponent to another exponent will multiply the exponents. Stay focused on the bottom of the fraction only. Here's what we get:

$$\frac{x^{-\frac{5}{9}}y^{\frac{5}{4}}}{(x^{-\frac{3}{2}}y^{\frac{5}{6}})^{\frac{1}{3}}}$$

$$=\frac{x^{-\frac{5}{9}}y^{\frac{5}{4}}}{x^{(-\frac{3}{2})(\frac{1}{3})}y^{(\frac{5}{6})(\frac{1}{3})}}$$

Now finish multiplying the Fractional Exponents. We're still only working on the bottom half of the main fraction:

$$\frac{x^{-\frac{5}{9}}y^{\frac{5}{4}}}{x^{(-\frac{3}{2})(\frac{1}{3})}y^{(\frac{5}{6})(\frac{1}{3})}}$$

$$=\frac{x^{-\frac{5}{9}}y^{\frac{5}{4}}}{x^{-\frac{3}{6}}y^{\frac{5}{18}}}$$

We can also clean up the bottom by simplifying one of the Fractional Exponents:

$$\frac{x^{-\frac{5}{9}}y^{\frac{5}{4}}}{x^{-\frac{3}{6}}y^{\frac{5}{18}}}$$

$$=\frac{x^{-\frac{5}{9}}y^{\frac{5}{4}}}{x^{-\frac{1}{2}}y^{\frac{5}{18}}}$$

It's still a pretty busy fraction, but if you compare it to our original expression from the question, it's actually looking much simpler.

OK, now let's turn our attention to the negative exponents. We know that negative exponents are *inverses*, so any negative exponents on bottom of the fraction will jump to the top, and any on top will jump to the bottom. Let's take care of this now:

$$\frac{x^{-\frac{5}{9}}y^{\frac{5}{4}}}{x^{-\frac{1}{2}}y^{\frac{5}{18}}}$$

$$=\frac{x^{\frac{1}{2}}y^{\frac{5}{4}}}{x^{\frac{5}{9}}y^{\frac{5}{18}}}$$

So now what's wrong? It still doesn't look like the final Answer Choices.

Well, remember that if you divide two same-base numbers, you can subtract the bottom exponent from the top exponent. We have x bases on top and bottom. Let's subtract the bottom exponent from the top exponent on the x's:

$$\frac{x^{\frac{1}{2}}y^{\frac{5}{4}}}{x^{\frac{5}{9}}y^{\frac{5}{18}}}$$

$$=(x^{(\frac{1}{2}-\frac{5}{9})})\frac{y^{\frac{5}{4}}}{y^{\frac{5}{18}}}$$

Now repeat the same move of subtracting exponents for the y bases:

$$(x^{(\frac{1}{2}-\frac{5}{9})})\frac{y^{\frac{5}{4}}}{y^{\frac{5}{18}}}$$

$$=(x^{(\frac{1}{2}-\frac{5}{9})})(y^{\frac{5}{4}-\frac{5}{18}})$$

OK, now we switch over to **Fractions** - we need to get the same denominators and then complete the subtraction:

$$(x^{(\frac{1}{2}-\frac{5}{9})})(y^{\frac{5}{4}-\frac{5}{18}})$$

$$=(x^{(\frac{9}{18}-\frac{10}{18})})(y^{\frac{45}{36}-\frac{10}{36}})$$

$$=x^{-\frac{1}{18}}y^{\frac{35}{36}}$$

Finally, let's use our knowledge of Combo Exponents to rewrite this. First, take care of the negative exponent on x by putting it on the bottom of a fraction:

$$x^{-\frac{1}{18}}y^{\frac{35}{36}}$$

$$=\frac{y^{\frac{35}{36}}}{x^{\frac{1}{18}}}$$

Wait - there's one more thing to notice. We can rewrite the denominator (bottom) of the exponent on x (which is 18) to match the denominator of the exponent on y (which is 36):

$$\frac{y^{\frac{35}{36}}}{x^{\frac{1}{18}}}$$

$$=\frac{y^{\frac{35}{36}}}{x^{\frac{2}{36}}}$$

Then rewrite the Fractional Exponents as roots:

$$\frac{y^{\frac{35}{36}}}{x^{\frac{2}{36}}}$$

$$=\frac{\sqrt[36]{y^{35}}}{\sqrt[36]{x^{2}}}$$

And now put everything under the umbrella of a single $\sqrt[36]{}$ symbol:

$$\frac{\sqrt[36]{y^{35}}}{\sqrt[36]{x^{2}}}$$

$$=\sqrt[36]{\frac{y^{35}}{x^{2}}}$$

This matches Choice D, and we're done. Quite a journey! If you understand exactly how we solved this question, then you should be confident going into any exponent question on the SAT Math test.

Lesson 15: Exponential Growth & Decay

Percentages

- 2.8% of Whole Test
- 1% of No-Calculator Section
- 3.7% of Calculator Section

Prerequisites

- Percents
- Exponents & Roots
- Linear Equations
- Algebra 1 Word Problems
- Basic Algebra 1

In this lesson we'll cover the topic of Exponential Growth & Decay (or "Exponential Change") which is a hybrid topic that mixes **Percents**, **Exponents**, and **Word Problems**.

Exponential Change is frequently contrasted with Linear Change, which we studied in two previous lessons on **Linear Equations**. Whereas Linear Change is a *constant* rate of change, Exponential Change is a *changing* rate of change, based on a percentage increase or decrease from the current value. As the current value grows or shrinks, so does the rate of change.

This topic is more commonly found in the Calculator Section than the Non-Calculator section, probably because calculating Exponential Change is often tedious without a calculator.

Exponential Growth & Decay Quick Reference

- Exponential Change is often contrasted with Linear Change. Whereas Linear Change is a *constant* rate of change, Exponential Change is a *changing* (increasing or decreasing) rate of change.
- Exponential Growth gets faster and faster as time goes on. Exponential Decay gets slower and slower as time goes on. In contrast, Linear Growth and Linear Decrease happen at the same pace regardless of how much time has passed.
- It's easier to *understand* Exponential Growth & Decay than it is to *memorize* the formula for it.
- Many SAT questions will simply ask us to *identify* Exponential Growth (or Decay) in a word problem or an equation - usually contrasting it with Linear Change.
- Harder questions require us to *set up* an Exponential Growth or Decay equation and possibly even solve for an unknown value with Algebra.
- Keep an eye out for tricky "time values," such as dividing years into quarter-years, or working in decade-long intervals rather than in 1-year intervals.

What is Exponential Growth?

So, what is "Exponential Growth" exactly?

Exponential Growth occurs when a starting value continually increases by a constant *percentage* of its current value. As the current value grows larger, so does its rate of growth, because a percentage of a *bigger* number is more than the *same* percentage of a *smaller* number.

For example, 20% of $100 is $20, but 20% of $120 is $24. The percentage stays the same at 20%, but the *result* of that percentage is bigger for the higher starting value - contrast the two increases of $20 vs $24.

I often think of Exponential Growth as a combination of **Percents** and **Exponents**. It is a constant percent change happening over and over again. We've seen how percents are a form of multiplication, and exponents multiply the same number more than once.

Before you continue in this lesson, be absolutely sure that you understand the content and practice problems in the lesson on **Percents**. You will not get very far in this lesson without that foundational understanding. It would also be wise to review the lesson on **Exponents & Roots**.

Now again, with Exponential Growth, although the Percent Growth stays the same, it's acting upon a bigger and bigger number over time, which leads to faster and faster growth - something we could call a "Feedback Effect."

For example, let's look at a table that shows what would happen if we invested $1,000 into an investment that gave a consistent 10% return each year for the next five years:

Year	Value	Growth
0	$1,000	n/a
1	$1,100	$100
2	$1,210	$110
3	$1,331	$121
4	$1,464.10	$133.10
5	$1,610.51	$146.41

The most important thing to observe is that the year-to-year growth *itself* is growing faster and faster - first +$100, then +$110, then +$121, onwards and upwards.

This is the "feedback effect" of the account's growth being fed back into the account each year, and then the new, higher account value growing by another 10% the next year. This new growth is returned into the account, then even more growth occurs the next year as the 10% growth multiplies against an even *higher* account value.

The longer this investment continues, the *faster* it will grow each year. This is a classic and common example of Exponential Growth in the real world.

In contrast, an account with *Linear* Growth would just plod along at "+$100, +$100, +$100" and so forth.

Exponential Growth situations can be found in populations of growing cities - as the population grows, more babies are born, who then have babies of their own, causing the city to grow faster and have more babies than the previous generation...

Or in bacterial colonies and cell growth - as the cells divide and double in count, then *more* cells will divide and double in the next growth phase, and then *even more* cells will double themselves in the phase after that.

We can identify Exponential Growth situations in word problems about values that increase by *multiplication* or by *percentages* or by "doubling each time period," etc.

Again, contrast this with Linear Growth situations that increase by *adding* or "increase at a *constant* rate."

When written as equations, we can identify Exponential Growth by the presence of exponential variables acting on numbers *greater than* 1.00 - for example:

$$250(1.3)^t$$

The Exponential Growth equation above represents a starting value of "250" that increases by 30% for each additional time period (the "Percent Multiplier" of 1.3 takes care of this 30% growth, as we learned in the lesson on **Percents**). Notice the t acting as an exponential variable on this Exponential Growth equation. This equation will grow faster and faster as time goes on.

Contrast the previous Exponential Growth equation with a Linear Growth Equation, such as:

$$250 + 1.3t$$

The Linear Growth equation above represents a starting value of "250" that increases by a constant "1.3" for each additional time period. Notice the lack of any exponential variables in this Linear Growth equation. The value of this equation will grow at the same rate regardless of how much time has passed.

What is Exponential Decay?

In contrast to Exponential Growth, what is Exponential Decay?

Exponential Decay occurs when a starting value continually *decreases* by a constant percentage of its current value. As the current value grows smaller, so does its rate of decrease, because a percentage of a *smaller* number is less than the *same* percentage of a *bigger* number.

Exponential Decay also has a "feedback effect" like Exponential Growth does, but it's the *opposite* effect. Whereas Exponential Growth gets faster and faster over time, Exponential Decay has its biggest decrease immediately, then the decrease gets smaller and smaller for each successive time period.

This makes sense if you think about it, because although the Percent Multiplier stays the same, it's acting upon a *smaller and smaller* number in each time period, causing the rate of decrease to get *smaller* over time.

For example, let's look at a table that shows what would happen if we invested $1,000 into a bad investment that gave a consistent 10% loss each year for the next five years:

Year	Value	Decrease
0	$1,000	n/a
1	$900	$100
2	$810	$90
3	$729	$81
4	$656.10	$72.90
5	$590.49	$65.61

The most important thing to observe is that the year-to-year decrease *itself* is falling slower and slower - first -$100, then -$90, then -$81, a smaller and smaller decrease each year.

This is the "feedback effect" of the account's losses being deducted from the account each year, and then the new, lower account value declining by another 10% the next year. These new losses are deducted from the account, then even less decrease occurs the next year as the 10% loss multiplies against an even *lower* account value.

The longer this bad investment continues, the *slower* it will lose money each year. This is a classic and common example of Exponential Decay in the real world.

Something else interesting about this situation is that the account can technically never reach a value of $0! If you let this bad investment continue forever, it could reach an infinitely tiny value - imagine something like $.00000001 dollars, or even less - but it's mathematically unable to Exponentially Decay to a value of 0.

In contrast, an account with *Linear* Decrease would just plod along at "-$100, -$100, -$100" year after year until the account reached $0.

Examples of Exponential Decay can be found in cities with *declining* populations, in the half-lives of radioactive materials, or in concentrations of medicine in the bloodstream as the medicine is processed and removed from the body.

When written as equations, we can identify Exponential Decay by the presence of exponential variables acting on numbers *less than* 1.00 - for example:

$$300(.75)^t$$

The Exponential Decay equation above represents a starting value of "300" that decreases by 25% for each additional time period (the "Percent Multiplier" of .75 takes care of this 25% decrease, as we learned in the lesson on **Percents**). Notice the t acting as an exponential variable on this Exponential Decay equation. This equation will decrease more and more slowly as time goes on.

Contrast the Exponential Decay equation above with a Linear Decrease Equation, such as:

$$300-.75t$$

The Linear Decrease equation above represents a starting value of "300" that decreases by a constant ".75" for each additional time period. Notice the lack of any exponential variables in this Linear Decrease equation. The value of this equation will fall at the same rate regardless of how much time has passed.

Calculating Exponential Growth & Decay

Let me make something perfectly clear: there *is* a formula for Exponential Growth & Decay, and I *do not* have it memorized.

That's right - I do *not* have the following formula committed to memory. To be honest, I find it more of a burden than a blessing. I'll explain why in just a few moments.

Nevertheless, I will give you the formula for Exponential Change below:

$$n(\frac{100\pm x}{100})^t$$

In the equation above, n represents the initial value, x represents the percent growth or decrease (add for growth, or subtract for decrease), and t represents the number of time periods for which the exponential change occurs.

For example, if an initial investment of $200 grew at 3% of its current value each year for 6 years, we could plug in the values as:

$$200(\frac{100+3}{100})^6$$

Or, if an initial investment of $300 lost its value at a rate of 5% of its current value each year for 11 years, we could express this as:

$$300(\frac{100-5}{100})^{11}$$

So how do I get away with not memorizing this formula, yet still end up with a perfect score on the SAT Math test?

Well, I prefer to just *think* about the math behind Exponential Change, rather than *memorize* some formula that I might forget on test day!

So *think* about it. We have a starting value. That value is multiplied by a Percent Multiplier to increase or decrease it. The same Percent Multiplier happens over and over with each passing time period, and I can use an exponent to show how many times that happens.

My own "formula," if you could even call it that, is more like this:

$$\text{(Starting Value)(Percent Multiplier)}^{\text{number of changes}}$$

No formula needed. Just a clear understanding of the situation and the math behind it.

Now try the following two practice problems. Whether you simply use the formula for Exponential Change, or you follow an approach like my own, just make sure you *understand* what you're doing, and not just plugging in numbers like a monkey with a typewriter!

> A certain bacterial colony begins with 400 organisms and grows at a rate of 80% of its current population each week for 7 weeks. Write an expression that represents the current population of this colony.

In this case, we have our initial value of 400. The Percent Multiplier is "1.80" to account for 80% growth. The 80% growth happens 7 times. My expression is:

$$400(1.80)^7$$

Let's try another:

> A used car currently has a value of $8,000. The value decreases by 10% of its current value each year for a period of 9 years. Write an expression that represents the current value of this car.

In this case, we have an initial value of $8000. The Percent Multiplier is ".90" to account for the 10% decrease. The 10% decrease happens 9 times. My expression is:

$$8000(.90)^9$$

Again, make sure you *understand* the equation. Don't just plug numbers in blindly.

Unusual Time Periods

Occasionally we encounter Exponential Functions that don't operate according to "simple" time periods.

For example, consider a company that begins with 20 employees and grows its number of employees by 15% every *quarter* year as part of a question that asks about the growth of the company *each year*.

The equation would look like this, where t is the number of *complete* years:

$$20(1.15)^{4t}$$

Notice what's changed: the exponent has a value of $4t$, not just "t" this time. That's because for each individual year, the exponential growth happens *four* times. There are four quarters per year. So, in one year, this company would experience 15% growth *four* times.

This is an example of what I mean by "unusual time periods" as part of Exponential Growth & Decay. Luckily, we don't see this much on the SAT test, which makes our lives simpler - but we *do* see it occasionally.

Try writing an expression for the situation below:

An certain house has an initial value of $500,000. The value of this house appreciates in value by 40% every 15 years. Write an expression that represents the value of this house after t years.

Make sure you've tried your own hand first. My answer and explanation are below:

$$500{,}000(1.40)^{\frac{t}{15}}$$

Note the initial value ($500,000) times the Percent Multiplier (1.40 represents the 40% growth).

Most importantly, note the exponent $\frac{t}{15}$. This is the "tough part." We have to divide the years by 15, because the Percent Multiplier only occurs *once* for every *fifteen* years that pass.

Check that if you plug in "15" for t, you'll get $\frac{15}{15}$, which reduces to "1." In other words, the Percent Multiplier of 1.40 will only happen *once* after a total of 15 years have passed.

Try one more practice example of unusual time periods for Exponential Change:

A colony of cells begins with an initial population of 2,500 cells. The cells die at a rate of 33% of their current population every five days. Write an expression that represents the current number of cells on day d of the experiment.

Make sure to try it on your own first, then check my answer and explanation below.

$$2{,}500(.67)^{\frac{d}{5}}$$

We start with an initial value (2500 cells) times a Percent Multiplier (.67 represents the 33% decrease - again, we studied this in the lesson on **Percents**). The trickiest element is the exponent $\frac{d}{5}$. We have to divide the total days by 5, because the Percent Multiplier only occurs *once* for every *five* days that pass.

Check that if you plug in "5" for d, you'll get $\frac{5}{5}$, which reduces to "1." So, the Percent Multiplier of .67 will only happen *once* after a total of 5 days have passed.

Get the idea? Good. Keep an eye out for unusual time periods hidden in the word problems on the SAT. You might also want to watch out for any unusual exponents in the answer choices, since they can also be a subtle reminder to double-check the time periods involved in the question.

Graphing Exponential Growth & Decay

What do Exponential Growth & Decay *look* like?

When graphed on the *xy*-plane, both Exponential functions have a distinctive shape. These are not straight lines like Linear Equations are - these exponential functions are *curves.*

Below is a picture of an **Exponential Growth** curve. Notice how the slope of the line gets steeper and steeper as it moves from left to right. The rate of growth is increasing faster and faster.

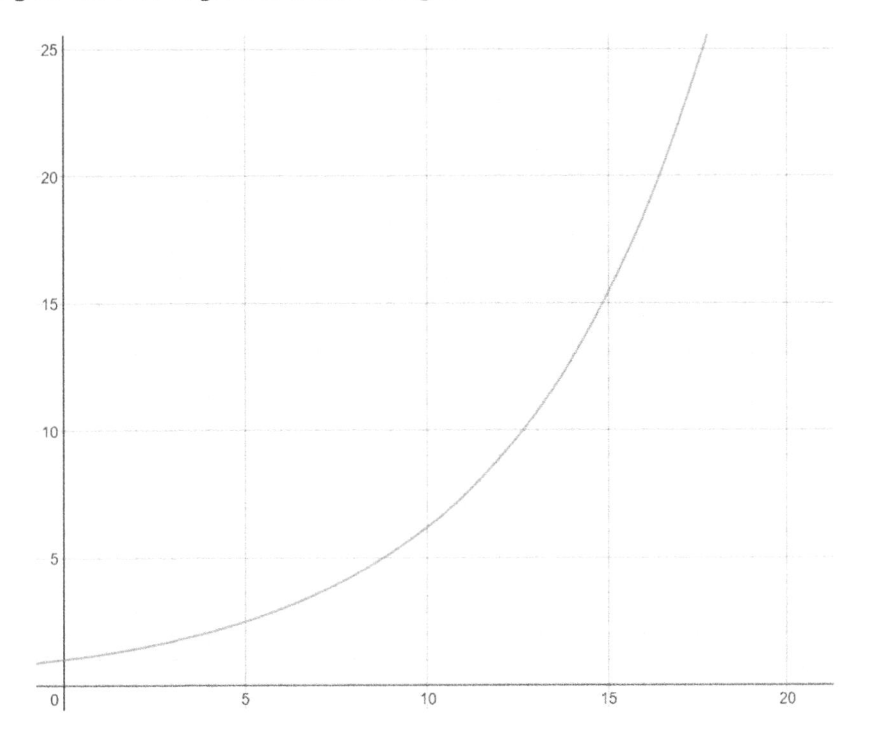

This curve will continue to rise more and more steeply over time.

And below is a picture of an **Exponential Decay** curve.

Notice how the slope of the line gets shallower and shallower as it moves to the right. The rate of decrease is falling slower and slower. Also notice that the line can never reach 0.

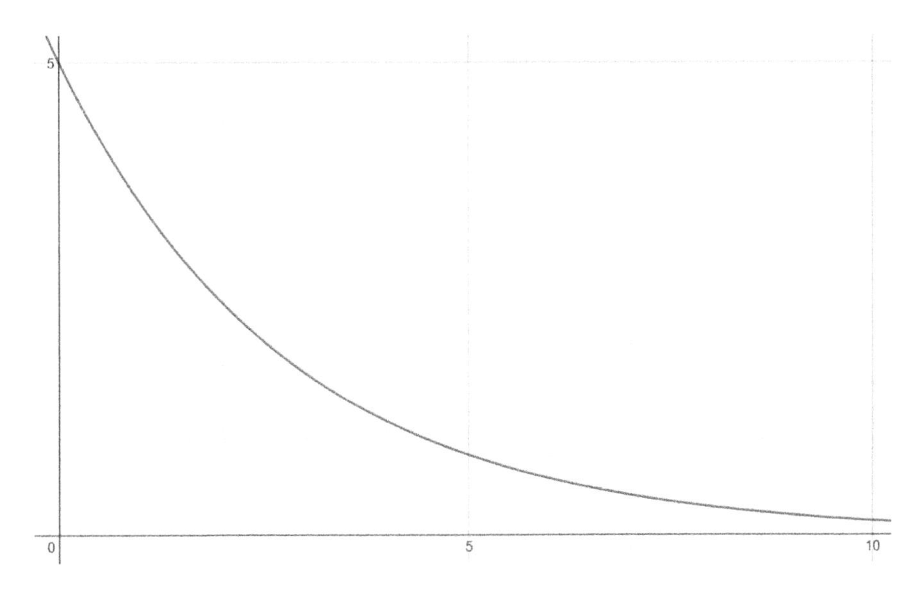

The curve will become so shallow over time that it is almost flat horizontal - but never *quite* flat - and it will almost reach zero - but never *quite* reach 0.

Pretest Question #1

Let's take a look at our first Pretest question on this topic. Try it yourself if you got it wrong the first time.

A certain town's population is currently 30,000 and is forecast to decrease by 10% every six years for the next 60 years. Which of the following expressions represents the population of this town after t years for the next 60 years?

(A) $30{,}000(.1)^{t}$

(B) $30{,}000(.9)^{6t}$

(C) $30{,}000(.9)^{\frac{t}{6}}$

(D) $30{,}000(.1)^{\frac{t}{6}}$

This is a common type of Exponential Change question found on the SAT. We are given a word problem that describes an Exponential Decay and are asked to match it to the correct formula.

Frankly, I find it easier to write my *own* expression and then match it to the answer choices after I'm finished.

We have an initial value of 30,000 people:

$$30{,}000$$

This initial value decreases by 10%, which is a .90 Multiplier (again, review the lesson on **Percents** if this is *in any way* confusing to you).

$$30{,}000(.90)$$

The .90 Multiplier needs a "time exponent" so the percent decrease can happen continually over time. However, the decrease only happens once every *six* years, so we need to divide the total years t by 6:

$$30{,}000(.90)^{\frac{t}{6}}$$

My completed expression clearly matches **Choice C**.

There's absolutely nothing new in this Pretest question that we haven't already explored in the lesson. Be sure you understand it with confidence before moving on!

Algebra and Exponential Change

As with so many topics on the SAT Math test, there are always ways to mix **Basic Algebra 1** into almost any other concept - including Exponential Change.

Try the following example question. You can use your calculator:

> Terry made an investment of d dollars that increased in value by 10% each year for 3 years. If the investment was worth $10,914.20 at the end of this three-year period, what was the value of d?

In this question, we'll first need to set up an Exponential Growth equation with the given information, and use the variable d for our initial value.

We start with d dollars, which multiplies by a Percent Multiplier of 1.10 each year, for 3 years. This result of this equation is set equal to the final value of the investment:

$$d(1.10)^3 = \$10{,}914.20$$

Now solve for the value of d using **Basic Algebra 1**. The first step is to evaluate the value of $(1.10)^3$, which your calculator will tell you is 1.331:

$$d(1.331) = \$10{,}914.20$$

Now divide both sides by 1.331 (again using your calculator) to find the value of d, the initial investment:

$$\frac{d(1.331)}{1.331} = \frac{\$10{,}914.20}{1.331}$$
$$d = \$8{,}200$$

So, we've managed to work backwards and calculate the initial value - first by setting up the Exponential Growth in an equation, then using algebra to solve for our unknown value.

Pretest Question #2

Let's take a look at another Pretest question. Try it yourself before you look at my explanation below the question.

> (CALCULATOR) FREE RESPONSE: Alex made an initial investment at the beginning of the year in 2008. The value of this investment increased by 40% per year for exactly three years, at which point the investment was worth $6,860. What was the value of Alex's initial investment?

The first step is to set up an Exponential Growth equation. We start with an initial *unknown* investment value that I'll call d dollars:

$$d$$

This unknown value increases by 40%, which is a 1.40 Multiplier:

$$d(1.40)$$

The 1.40 Multiplier needs to happen once per year. I'll use the variable t for years (real creative, I know!):

$$d(1.40)^t$$

We know that this growth continues for 3 years, so plug that in for t:

$$d(1.40)^3$$

And after three years, it has a final value of $6,860. Set them equal:

$$d(1.40)^3 = 6{,}860$$

Now we'll solve for d. First, use your calculator to evaluate $(1.40)^3$ and find that it equals 2.744:

$$d(2.744) = 6{,}860$$

Now use **Basic Algebra 1** and divide both sides by 2.744 to isolate d:

$$\frac{d(2.744)}{2.744} = \frac{6{,}860}{2.744}$$
$$d = 2{,}500$$

The initial investment must have been **$2,500**. This is an example of how an Exponential Change setup can be combined with Basic Algebra to find unknown values in a word problem.

Review & Encouragement

Exponential Change is another one of those math topics that I see students struggle with every day in SAT tutoring. I often notice that they are *so close* to getting the right answer figured out, but they're just *a little* confused about the fine details, and that's enough to get the whole problem wrong because of traps the SAT has purposefully laid for them.

Most of all, I feel like I always see these same students trying to *remember* "the formula" for Exponential Change, instead of just trying to *understand* the underlying concepts. If I had a dollar for every time I've seen this happen, I could invest in a fast-growing stock at 10% annual gains...

But I'm not kidding when I say I don't even have the Exponential Change formula memorized. Once you understand the concepts it's built from, you won't need to memorize it either.

Exponential Growth & Decay Practice Questions

YOU MAY USE A CALCULATOR FOR ALL OF THE FOLLOWING PRACTICE PROBLEMS.

1. When a certain concert hall was opened, its first performance had an audience of 10 people. For each of the next 7 performances, the audience at each performance doubled over the previous performance. If $a(p)$ is a function for the audience size at each concert p performances after opening the concert hall, which of the following statements best describes the function a?

(A) The function a is a decreasing linear function.

(B) The function a is an increasing linear function.

(C) The function a is a decreasing exponential function

(D) The function a is an increasing exponential function.

2. James is choosing a bank account into which he will make an initial deposit of \$350. Of the following four types of bank account, which choice would yield exponential growth of the money in the account?

(A) At the end of each year, \$35 is added to the value of the account.

(B) At the end of each year, 3.5% of the original deposit is added to the value of the account.

(C) At the end of each year, 3.5% of the current value is added to the value of the account.

(D) At the end of each year, 3.5% of the original deposit and \$35 is added to the value of the account.

Time (Years)	Value (Dollars)
0	8100
2	2700
4	900
6	300
8	100

3. The value of an investment over time is shown in the table above. Which of the following best describes the relationship between time and the value of the investment?

(A) Exponential Decay

(B) Exponential Growth

(C) Increasing Linear

(D) Decreasing Linear

4. Dan modeled the growth of a colony of bacterial organisms he was studying in Biology class. He estimated the colony began with 440 bacterial organisms and had a 3% daily increase in the number of bacterial organisms every day thereafter. Which of the following functions models $B(d)$, the number of bacterial organisms in the colony t days after the first day that Dan began tracking the growth of the colony?

(A) $B(d) = 440^{1.03t}$

(B) $B(d) = 440t^{1.03}$

(C) $B(d) = 440(1.3)^t$

(D) $B(d) = 440(1.03)^t$

5. FREE RESPONSE: Christian purchased a guitar that had a value of $2000 at the time of purchase. Each year, the value of the guitar is estimated to increase 20% over its value the previous year. The estimated value of the guitar, in dollars, 2 years after purchase can be represented by the expression $2000b$, where b is a constant. What is the value of b?

$$U = 325(1.03)^n$$

6. The equation above models the number of users, U, of a recording studio n years after the studio opens. Which of the following equations models the number of users of the recording studio q quarter years after the studio opens?

(A) $U = 325(1.0075)^{4q}$

(B) $U = 325(1.03)^{\frac{q}{4}}$

(C) $U = 325(1.03)^{4q}$

(D) $U = 325(1.12)^q$

7. A radioactive substance decays at an annual rate of 35%. If the initial amount of the substance is 1.3 kilograms, which of the following functions f models the remaining amount of the substance, in kilograms, t years later?

(A) $f(t) = 1.3(.65)^t$

(B) $f(t) = 1.3(.35)^t$

(C) $f(t) = .35(1.3)^t$

(D) $f(t) = .65(1.3)^t$

$$2{,}000(1+\frac{n}{1{,}200})^{12}$$

8. The expression above gives the amount of money, in dollars, generated in a year by a $2,000 deposit in a savings account that pays an annual interest rate of $n\%$, compounded monthly. Which of the following expressions shows how much additional money is generated in one year at an interest rate of 6% than an interest rate of 2%?

(A) $2{,}000(1+\frac{6-2}{1{,}200})^{12}$

(B) $2{,}000(1+\frac{\frac{6}{2}}{1{,}200})^{12}$

(C) $2{,}000(1+\frac{6}{1{,}200})^{12}-2{,}000(1+\frac{2}{1{,}200})^{12}$

(D) $\dfrac{2{,}000(1+\frac{6}{1{,}200})^{12}}{2{,}000(1+\frac{2}{1{,}200})^{12}}$

9. A student starts a business that resells limited-edition shoes online. The student begins with one pair of shoes in stock at the start of Month 1. At the end of each month, the student takes the profits and reinvests them to buy twice as many shoes to sell in the following month. For the next 12 months, the student sells out their entire stock of shoes before the end of the month. Which of the following expressions represents S, the number of individual shoes the student has in stock at the start of month t?

(A) $S=1(2)^{t-1}$

(B) $S=2(2)^{t-1}$

(C) $S=1(2)^{t}$

(D) $S=2(2)^{t}$

10. An urban planner estimates that, starting from the present day, the population of a certain city will decrease by 5% every 15 years. If the present population of the city is 80,000, which of the following expressions represents the urban planner's estimate of the population t years from now?

(A) $80{,}000(.05)^{15t}$

(B) $80{,}000(.95)^{15t}$

(C) $80{,}000(.95)^{\frac{t}{15}}$

(D) $80{,}000(.05)^{\frac{t}{15}}$

11. FREE RESPONSE: Ian made an investment of d dollars on January 1st, 2015. The value of his investment increased by 50% each year until Ian's investment was worth $5,568.75 on January 1st, 2019. What is the value of d?

12. FREE RESPONSE: The population of water bugs in a certain lake is 15,000 on January 1st, 2010. A scientist believes the population will decrease by 14% per month for the next 6 months. The scientist uses the equation $P = 15{,}000(x)^t$ to model the population, P, of water bugs in the lake t months after January 1st, 2010. What value should the scientist use for x?

13. (REFER TO QUESTION #12) To the nearest whole number, what is the predicted population of water bugs in the lake at the end of the six-month period?

Time (Minutes)	Concentration (mg/liter)
0	600
5	438
10	320
15	233
20	170

14. FREE RESPONSE: The concentration in milligrams per liter of a certain substance dissolved in water can be modeled by $C(t) = 600(n)^{\frac{t}{5}}$ as the solution evaporates, where t is the number of minutes since the substance was mixed into the water. The table above shows a set of times and concentrations for this solution. If C approximates the values in the table above, what is the value of n, rounded to the nearest tenth?

Exponential Growth & Decay Answers

1. D
2. C
3. A
4. D
5. 1.44
6. B
7. A
8. C
9. B
10. C
11. 1,100
12. .86
13. 6,069
14. .7

Exponential Growth & Decay Explanations

1. **D.** This question simply asks us to decide between a "decreasing" or "increasing" function, as well as identifying whether it is "linear" or "exponential." (Note: it also uses some basic **Functions** notation, which we'll study in-depth in a future lesson. Don't get caught up in the details of functions - focus on the bigger picture of Exponential Growth.)

The first choice is easy: the audience *doubles* after each performance, so the audience is growing over time and this is an *increasing* function, eliminating Choice A and Choice C, which are *decreasing*.

As for whether this situation is linear or exponential, consider that the audience will grow faster and faster after each performance. "Doubling" is inherently an *exponential* form of growth, because as the audience gets bigger, it grows faster. The rate of growth is not constant; each doubling acts upon a *larger* audience - first 10 people in the audience, then 20, then 40, then 80 people - so the audience would grow exponentially, from +10 between performances, then +20 people, then +40, and so forth, after each performance.

The answer must be Choice D, an "increasing exponential" function.

It it was a *linear* equation, the size of the audience would simply grow by the same constant number of people after each performance (and this number would be the "slope" of the linear function, as we studied in the two earlier Lessons on **Linear Equations**).

2. **C.** This question explicitly asks us for "exponential growth." By elimination, it can't come from Choice A, which just adds a constant $35 to the account each year (this is "linear growth" and 35 would be the slope of the function, as we covered in the earlier lessons on **Linear Equations**).

We need an answer choice that contains some sort of percentage, since Exponential Growth is based on repeatedly multiplying by a percentage. But, all three remaining answer choices involve Percents.

However, Choices B and D are trap answer choices. Notice the fine print: these two choices both take 3.5% of the *original* deposit. There is no "feedback effect" of additional interest on the growing value of the bank account; the yearly increase only comes from what was already in the account at the very beginning. Despite paying lip service to percentages - and possibly grabbing our attention for a moment - both Choices B and D are only smokescreens for what are actually *linear* growth situations.

Only **Choice C** resolves this problem and provides *exponential* growth, because a percentage of the *current* value is being added back into the account each year. As the account grows in value, so will the amount of money it earns at the end of the year. And in the following year, the bank account will have more money in it, so 3.5% of its "current value" will be applied to a *higher* value, providing the "feedback effect" that Exponential Growth is so famous for.

3. **A.** Another question of the common type on the SAT that asks us to identify Exponential vs. Linear relationships - this time from a table, instead of a word problem.

Notice first that the value of the investment is *decreasing* over time. It starts at $8,100, then falls to $2,700, and keeps falling until it reaches only $100 in value. Not a very good investment! It's certainly *decreasing* in value, which eliminates the "increasing" Choices B and C.

Now, is the change a *constant* decrease, which would signify a **Linear Equation**? No - the decrease in value is different from time period to time period. In the first two years, it loses $\$8{,}100 - \$2{,}700$ or $5,400 in value, but from years 2 to 4 it only loses $\$2{,}700 - \$900$ or $1,800 in value. Therefore, this data *cannot* represent a Linear Equation. The answer must be Choice A.

If you want some personal "bonus points" (of course, the SAT won't give these to you, but it doesn't hurt to understand better), then you could even determine the percent decrease in each 2-year period. Remember from the prior lesson on **Percentages** that Percent Change is calculated via "Difference over Original", so we could use the difference in value from Year 0 to Year 2, and divide it by the original value of $8,100:

$$\frac{\text{Difference}}{\text{Original}}$$
$$= \frac{\$8{,}100 - \$2{,}700}{\$8{,}100}$$
$$= \frac{\$5{,}400}{\$8{,}100}$$
$$= .6666...$$

We get the infinitely-repeating decimal ".6666..." Remember that this is a *decimal* result, so we must multiply by 100% to get the equivalent percentage of an approximate 66.67% decrease per 2-year period.

This investment seems to be losing value very quickly!

If you want, you can do the same Percent Change calculation for Year 2 to Year 4, or indeed, between any two years on the table. You'll find that the investment consistently decreases in value by approximately 66.67% every two years.

For mega-extra bonus points, you could write your own equation to estimate the value t years after the investment is first made. If you have a moment, I suggest you practice this right now - it will come in handy later. Give it a try!

My equation looks like this:

$$\text{Current Value} = 8100(.3333...)^{\frac{t}{2}}$$

(Since ".3333..." is an infinitely-repeating decimal, the more "3s" you enter into your calculator, the more accurate the final estimate will be.)

Notice that our "Percent Multiplier" must be ".3333...", *not* ".6666...", for reasons that were explained in the earlier lesson on **Percentages** (remember the "decrease from Sea Level" metaphor?) Also note that we have to divide t by 2, because our table gives us the decrease for an unusual time period of every *two* years.

4. **D.** This question is the first example in this set of how I do *not* think of Exponential Growth as a "formula." As I expressed in the lesson, I try to "remember" as few things as possible, because my memory is fallible. I've always hated "formulas" because they get so hard to remember - and if I remember even a small part of the equation wrong, I'm likely to miss the entire question.

So here's my thought process on a question like this one. First of all, I know how much I start with: the colony "began with 440 bacterial organisms." So, I'll make a note of my starting number, 440:

$$440$$

Next, I read that there is a "3% daily increase" in the number of these organisms. I check ahead in the problem to make sure I'm working in *days* and not something tricky like weeks or months. Luckily, the use of "days" remains consistent throughout the problem.

So, then I ask myself *how to represent a 3% increase*. If you've studied the earlier lesson on **Percentages**, you'll know I use the "Sea Level" method in which I always start at "100%" or "baseline," then I increase or decrease from "sea level" to determine my "Percent Multiplier."

This colony is *growing*, so I need to *increase* from sea level by 3%. That puts me at 103% of the original value, which can be written as the decimal "1.03." This is the multiplier that I can attach to the starting value to cause it to "increase by 3%":

$$440(1.03)$$

But, I'm still missing something.

What I've written so far will only increase the size of the colony *once*, but I need a formula that increases the population *day after day*. The 1.03 multiplier needs to happen again and again - once per day - to keep increasing the colony by another 3% each day.

In other words, I need to multiply my population by 1.03 an extra time for each day that passes in the experiment. And that's exactly what **Exponents** are for: to repeat the same multiplication more than one time.

So, I'll include an exponent on my Percent Multiplier, representing the number of times that my colony increases. I will use t as that exponent, since the increase happens once per day for as many days as the experiment goes on for:

$$440(1.03)^t$$

As you can see, I've ended up creating the same equation as given in Choice D. But it wasn't because I had a formula memorized. It's because I constructed my equation, piece by piece, by just thinking logically about the value I started with and what happened to that value over time. My understanding of Percents and Exponents helped me along.

I was using my brain to *think*, rather than to *memorize equations*. And I sincerely suggest you do the same.

5. **1.44.** My explanation for this question calls back to my previous explanation of Question 4. I can create my own exponential growth equation, free of any "memorization," and just based on logical and mathematical reasoning.

First, the guitar starts at a value of $2000. Then, this value increases by 20%. A 20% increase can be written as a multiplier of 1.20 (if you're confused review the lesson on **Percentages** with special attention to the concept of "sea level" and "Percent Multipliers").

That gives me $2000(1.2)$. However, this expression only increases the value *one* time, but it needs to increase *once per year*. As always, **Exponents** are great for writing repeated multiplications by the same number. So I'll apply an exponent to the "1.2" in my equation, allowing it to multiply again and again, once for each year t (note that I just made up the variable t to help me write my equation):

$$2000(1.2)^t$$

Now I can plug in a value of "2" for t since the question asks for the value of the guitar "2 years after purchase."

$$2000(1.2)^2$$

And, I can set this equal to the expression given in the word problem for the value after two years:

$$2000(1.2)^2 = 2000b$$

Now just solve for b using **Basic Algebra 1**:

$$\frac{2000(1.2)^2}{2000} = \frac{2000b}{2000}$$
$$(1.2)^2 = b$$
$$1.44 = b$$

And we're done - the value of b must be 1.44.

6. **B.** There's a sneaky little trick in this otherwise-simple problem, and it has to do with how we handle the "time exponent" in our Exponential Growth model.

Notice that we're given a complete equation for the Exponential Growth from the very beginning of the question, which saves us some trouble.

The only issue is that the original equation models users based on *years* after opening, but we need an answer that operates according to *quarter years*.

Of course, there are 4 quarter-years in a year. But then again, all of the answer choices use "4" in some way or another. How are we to tell which is correct?

My suggestion is to try a simple test: if we plug in a value of "4" for q quarters, the equation should end up exactly the same as if we plugged in "1" for n years. This makes sense if you think about it: a time period of "4 quarter years" is the same amount of time as "1 year."

First let's plug in "1 year" for n in the original given equation, just so we have a clear point of reference:

$$U = 325(1.03)^1$$

If you plug 4 quarters into q for Choice A, you'll get an equation that looks nothing like our reference equation: $U = 325(1.0075)^{16}$

When you plug in 4 quarters for q into Choice B and simplify the fraction $\frac{4}{4}$, you get an equation that looks exactly like our reference equation: $U = 325(1.03)^1$

When you plug in 4 quarters for q into Choice C, you get an equation with too high of an exponent compared to the original: $U = 325(1.03)^{16}$

When you plug in 4 quarters for q into Choice D, you get another equation that looks significantly different from our reference equation: $U = 325(1.12)^4$

Only Choice B matches our reference equation and gives the same result for "4 quarters" and "1 year," and Choice B is the correct answer.

Why is that? Well, for every four *quarter-years* of time that pass, our *yearly* percent growth of 3% (from 1.03) should only happen one time. So, you need to divide the number of quarter-years by 4 to get an equivalent time period in full years.

7. **A.** This is the first question so far that has asked us to model Exponential *Decay*. However, our thought process will be very similar to what we used on Question 4, where we modeled Exponential *Growth*.

As I've said, I'm not very fond of memorizing formulas. So when it comes to Exponential Growth and Decay, I prefer just to *understand* and *think about* my work, rather than work from rote memorization.

First, we have a starting value of 1.3 kilograms:

$$1.3$$

Then, we decay by 35%. If I decrease by 35% from 100% baseline or "sea level", I'll drop to 65%, which I can write as a multiplier of ".65." (If you're confused by this, review the lesson on **Percentages** and pay attention to the concepts of "Sea Level" and "Percent Multipliers").

$$1.3(.65)$$

As in Question 4, I'll need this multiplier of .65 to happen again and again, once per year. **Exponents** are used to repeat the same multiplication more than once, so I'll include an exponent for t years on my Percent Multiplier:

$$1.3(.65)^t$$

And now I've got my final equation, which will start at 1.3 kilograms and decrease by 35% each year. The matching answer is Choice A.

8. **C.** The answer choices may seem a bit overwhelming at first, but on the flip side, most of the hard work is done for us already. We can take for granted that something about $2{,}000(1+\frac{n}{1{,}200})^{12}$ is correct, since all four answer choices have this same structure.

The only real question is what is supposed to go in for n in my equation above? It must be the piece of the puzzle that accounts for the difference between 6% and 2% interest.

Let me point out that *difference* is an issue of subtraction, not division. It always has been, and it always will be. On that basis I will eliminate Choices B and D, which use division instead of subtraction.

Now let's analyze the makeup of these equations. We can compare them to this general form:

$$(\text{Initial Value})(\text{Percent Multiplier})^{\text{Time Exponent}}$$

The "2,000" must be the initial value of the deposit. The "12" must be the time exponent (this makes sense since the *annual* interest is $n\%$ but the interest compounds *monthly* and there are 12 months in a year).

The problem with Choice A is that it inputs $6-2$ for the percent multiplier - which would equal an annual interest of 4% - but that's not what we want. Instead, we want to calculate the total value at 6% interest, and separately calculate the total value at 2% interest, and then use subtraction to find the difference between these totals. That leaves us with only one option: Choice C.

For those interested in a deeper understanding, why do we divide by "1200"? Normally we'd be dividing by "100" because we're converting n percent to decimal form. But, the n percent usually applies to a *full year* and we're calculating 12 instances of *monthly* interest per year. Therefore, we have to divide the yearly interest by 12, and 1200 on the bottom of the fraction is equal to $(100\text{ percent})(12\text{ months})$. Note, however, that the SAT won't test you on the details of this concept.

9. **B.** This is a pretty standard "create the exponential growth equation" with only a couple small variations.

First of all, let's set up the basics. We start with one pair of shoes, doubling each month. That's a small point of difference from most of these questions, which have grown based on percents, but it's no big deal.

We'll just use "2" as our multiplier so the number of pairs shoes doubles each month:

$$(1\text{ pair of shoes})(2)$$

Here's a small issue we can take of right away: the word problem asks for the number of *individual shoes*, not pairs of shoes. Tricky - a cheap shot, even - but let's take care of it now by multiplying the entire equation by 2, since there are two shoes per pair:

$$(2\text{ shoes per pair})(1\text{ pair of shoes})(2)$$

The multiplier of 2 will happen once per month, which we'll represent with the exponent t.

$$(2\text{ shoes per pair})(1\text{ pair of shoes})(2)^t$$

But here we encounter another small problem - we're asked to calculate the number of shoes at the *start* of each month, not the *end* of each month.

That may seem like just a minor detail, but it's actually a big deal. For example, when we plug in "1" for t, we should get the number of shoes that the student *starts* with at the beginning of the first month. But if we plug "1" for month t into our current equation, here's what happens:

$$(2\text{ shoes per pair})(1\text{ pair of shoes})(2)^2$$
$$=(2)(1)(2)^1\text{ shoes}$$
$$=2(2)\text{ shoes}$$
$$=4\text{ shoes}$$

Our equation is giving us the wrong number! We're only supposed to have *one* pair (two shoes) at the start of the first month.

Here's what we do to compensate: *subtract* 1 from t:

$$(2\text{ shoes per pair})(1\text{ pair of shoes})(2)^{t-1}$$

Now when we plug in 1 for t, we'll get:

$$(2\text{ shoes per pair})(1\text{ pair of shoes})(2)^{1-1}$$
$$=(2)(1)(2)^0\text{ shoes}$$
$$=(2)(1)(1)\text{ shoes}$$
$$=2\text{ shoes}$$

And we'll correctly calculate that the student has *two* shoes (or one pair) at the beginning of the first month.

The "minus 1" on t lets us calculate the value *before* the growth of each month takes place.

Now we're in a position to finish the question with a cleaned-up version of our equation:

$$S = 2(2)^{t-1}$$

Just to review: the first "2" represents the fact that there are 2 individual shoes per pair. The second "2" represents the Multiplier for "doubling" the number of pairs of shoes each month. The exponent $t-1$ is the time exponent that adjusts to calculate the *beginning* of the month *before* each monthly doubling takes place. And we have our final answer - Choice B.

10. **C.** Another variation on the common "create an exponential equation" type of SAT problem we encounter on the topic of Exponential Change. If you've studied the lesson and the practice problems before this one, it should be pretty straightforward by now.

First, our initial value is 80,000. This is multiplied by .95 to represent the decrease of 5%:

$$80{,}000(.95)$$

The .95 multiplier happens more than once over time, so it needs an exponent. But be careful - this Exponential Decay only happens *once* every *fifteen years*. So, the years t must be divided by 15:

$$80{,}000(.95)^{\frac{t}{15}}$$

Check by plugging in "15" for t and notice that the decrease of 5% happens exactly once, just as it should after 15 years. Our equation is correct and complete, and we have the final answer- Choice C.

11. **1,100.** This question is an "Algebra with Exponential Change" problem. Luckily, we're completely ready to take it on!

First, set up an equation for the Exponential Growth of this investment. We start with an initial (unknown) value of d dollars. This has a 1.50 multiplier to represent the 50% growth per year. The multiplier needs a time exponent to represent the years - let's just use t for this:

$$d(1.5)^t$$

We can count the number of years the investment grew for - the difference between Year 2019 and 2015, which is 4 years of investment growth. Plug in 4 for t:

$$d(1.5)^4$$

And, the question informs us that after four years, the investment is worth $5,568.75.

$$d(1.5)^4 = 5{,}568.75$$

Now solve for d. First, use your calculator to evaluate the value of $(1.5)^4$, which equals 5.0625.

$$d(5.0625) = 5{,}568.75$$

Then use **Basic Algebra 1** to solve for d:

$$\frac{d(5.0625)}{5.0625} = \frac{5{,}568.75}{5.0625}$$
$$d = 1{,}100$$

And we're done. We've found the value of d, the initial value of the investment, was 1,100.

12. **.86.** This is a surprisingly easy question, given what we've already been through. We know all the pieces of our Exponential Decay equation - there's the 15,000 for initial value, the x for the Percent Multiplier, and the t to act as a time exponent. Also, everything is nice and simple with no weird time values - everything is calculated in terms of months.

The question is asking for x, which is the Percent Multiplier. To tell the truth, this question has as much to do with **Percentages** as it does with Exponential Growth. If we want to *decrease* by 14% each month, we need to drop from 100% "sea level" by 14%, which brings us to 86%, and we'd write that as the decimal ".86."

Our complete equation would look like this:

$$P = 15{,}000(.86)^t$$

And, we're done - the value of x, the Multiplier, must be .86.

13. **6069.** In this question, we can refer to our work on the previous Question #12. In fact, we can use the equation we've already completed for Exponential Decay on the water bug population:

$$P = 15{,}000(.86)^t$$

And all we have to do is plug in "6" for t. The equation is best evaluated by calculator, because of the repeated decimals we encounter.

$$P = 15{,}000(.86)^6$$
$$P = 15{,}000(.4045...)$$
$$P = 6{,}068.508...$$

The question tells us to round our final answer to the nearest whole number, which is 6,069 (it's a small thing, but remember that decimals ending in ".5" round *up*. Miss this detail, and you'll get the whole problem wrong!)

14. **.7.** So, this is the first Exponential Change problem we've encountered with a table in it. We need to connect the word problem with the given equation and the data in the table.

How can we plug in values from the table? Well, let's just pick a row and plug it in! Let's use the time "0" for t and the concentration "600" for $C(t)$. Remember from the lesson on **Exponents & Roots** that the value of any number raised to the 0 power is 1:

$$C(t) = 600(n)^{\frac{t}{5}}$$
$$600 = 600(n)^{\frac{0}{5}}$$
$$600 = 600(n)^0$$
$$600 = 600(1)$$
$$600 = 600$$

OK, that wasn't very useful, I'll admit. Obviously 600 equals 600. Not much use.

Let's use a different row of the table and see what happens. How about "5" for t and "438" for $C(t)$. (Note that it doesn't really matter which row you use - they should all work, other than the first row).

$$C(t) = 600(n)^{\frac{t}{5}}$$
$$438 = 600(n)^{\frac{5}{5}}$$
$$438 = 600(n)^1$$
$$438 = 600(n)$$
$$\frac{438}{600} = \frac{600(n)}{600}$$
$$.73... = n$$

Now, that is more useful. We've got a value for n. Don't forget the question asked us to round to the nearest tenth, which gives a final answer of .7. And we're done!

Lesson 16: Basic Algebra 2

Percentages

- 3.8% of Whole Test
- 6.5% of No-Calculator Section
- 2.4% of Calculator Section

Prerequisites

- Basic Algebra 1
- Advanced Algebra 1
- Linear Equations (Algebraic)

Algebra 2 is a core concept of the SAT Math test. The percentage appearance of these questions, while not the *highest* our of all our lessons, is still substantial - and several other key topics are firmly rooted in the concepts of this chapter. If you fail to understand this lesson, you will sacrifice many points on SAT Math test.

In later lessons on Algebra 2, we will explore Parabolas, the Quadratic Formula, and other more advanced topics in depth - all based upon the main concepts in this lesson.

For now, we're just trying to get the lay of the land regarding Algebra 2. This includes understanding the differences between Algebra 2 and Algebra 1 and mastering key techniques like FOILing, factoring, and finding multiple solutions to Algebra 2 equations.

Basic Algebra 2 Quick Reference

- Algebra 1 dealt with Linear Equations, single solutions, and straight lines. Algebra 2 deals with Quadratic Equations, multiple solutions, x^2 terms, parabolas, and polynomials.
- The Quadratic Equation $ax^2 + bx + c = 0$, which is the basis of Algebra 2, contrasts with the Linear Equation $y = mx + b$, which is the basis of Algebra 1.
- In Algebra 2, the four keywords "Zeros," "Roots," "Solutions," and "*x*-intercepts" all mean the same thing. These are all different words that mean essentially the same thing: the *answers* to an Algebra 2 equation.
- The two most essential skills in Basic Algebra 2 are FOILing and Factoring. Be sure you feel confident in these two skills before the end of the lesson.
- Look for "Differences of Squares" to make Factoring easier whenever possible.
- In certain types of SAT questions, we may need to check for False Roots to avoid traps. Learn the signs of these questions.

Differences Between Algebra 1 & Algebra 2

I've thought long and hard on the essential differences between Algebra 1 and Algebra 2. Here are my conclusions.

Algebra 1 equations have only *one* solution. There is only one answer that can be plugged in to solve the equation. Algebra 2 equations frequently have *two* solutions. There are usually *two* different answers that can be plugged in to solve the equation.

Algebra 1 deals with Linear Equations, which graph as straight lines. Algebra 2 deals with the Quadratic Equation, which graph as parabolas (more on **Parabolas** in a future lesson).

In Algebra 1, there are no exponents on the variables (except in rare and simple cases). In Algebra 2, there are variables with exponents (e.g. x^2).

Algebra 1's essential formula is the Linear Equation: $y = mx + b$. Algebra 2's essential formula is the Quadratic Equation: $ax^2 + bx + c = 0$.

For future reference, we can call the Quadratic Equation a "2nd order polynomial." The word "polynomial" literally means "many terms." It's called "2nd order" because of the ^{2} in "x^2".

There are also 3rd-order polynomials (based on x^3), 4th-order polynomials (based on x^4), and so forth - but our exploration of these higher-order polynomials will be limited, because they are not a major topic on the SAT Math test.

Zeros, Roots, Solutions, and x-Intercepts

"Zeros," "Roots," "x-Intercepts," and "Solutions" are four words that all represent the *exact* same thing: the answers to an $ax^2 + bx + c = 0$ Quadratic Equation.

The name "Zeros" comes from the fact that we set the Quadratic Equation equal to "0" and solve for x.

The name "Solutions" comes from the fact that we will find values of x that make the equation true.

The name "Roots" comes from... well, to be honest, I don't really know where. But it doesn't really matter. It's just another name for the same thing.

As we get into a future lesson on **Parabolas** we will explore more about what "*x*-Intercept" means. But essentially, it's the value of the *x*-coordinate when the equation crosses the *x*-axis $y = 0$. And it's just another name for "roots," "zeros," or "solutions."

Again, all four of these keywords mean essentially *the same thing*.

We must understand that these four keywords are *completely interchangeable* and the SAT Math test can use any of them whenever it wants to, depending on the moods and whims of whoever was writing the math question at that time.

FOILing

FOILing is a central technique in Algebra 2 questions. "FOIL" is an acronym that stands for multiplying the "Firsts, Outsides, Insides, Lasts" of a pair of parentheses.

We apply FOILing when we see anything of this general form:

$$(a+b)(c+d)$$

The "Firsts" are a and c, the first terms inside each of the parentheses. The "Outsides" are a and d, which are the outermost terms. The "Insides" are b and c, in the innermost terms. The "Lasts" are b and d, which are at the end of each pair of parentheses.

We use FOILing to multiply. Here's an example of FOILing in practice:

FOIL the expression $(x+2)(x-3)$.

And here we go. Watch how I multiply each pair of "Firsts, Outsides, Insides, Lasts" to accomplish this:

$$\begin{aligned}
&(x+2)(x-3)\\
&=(x)(x)+(x)(-3)+(2)(x)+(2)(-3)\\
&=x^2-3x+2x-6\\
&=x^2-x-6
\end{aligned}$$

When cleaned up, it's no coincidence that the final expression fits the Quadratic Equation form ax^2+bx+c. Again, FOILing is a central concept of Algebra 2. It is an essential technique for setting up Quadratic Equations.

Now try your own hand at FOILing the following practice examples. Be sure to reduce them down to their simplest ax^2+bx+c after FOILing.

Rewrite the expression $(n-5)(n-2)$ in the form ax^2+bx+c.

Rewrite the expression $(b+3)(b-4)$ in the form ax^2+bx+c.

Rewrite the expression $(x+2)^2$ in the form ax^2+bx+c.

I've shown my steps for FOILing and simplifying each expression below.

$$\begin{aligned}
&(n-5)(n-2)\\
&=(n)(n)+(n)(-2)+(-5)(n)+(-5)(-2)\\
&=n^2-2n-5n+10\\
&=n^2-7n+10
\end{aligned}$$

And the second example:

$$\begin{aligned}
&(b+3)(b-4)\\
&=(b)(b)+(b)(-4)+(3)(b)+(3)(-4)\\
&=b^2-4b+3b-12\\
&=b^2-b-12
\end{aligned}$$

On the final practice example, be sure you remember that "squaring" means *multiplying something times itself.*

Don't make the classic mistake of thinking that $(x+2)^2$ is equal to x^2+4. It's not. We have to expand to $(x+2)(x+2)$ and FOIL it to get the right answer.

$$\begin{aligned}
&(x+2)^2 \\
&=(x+2)(x+2) \\
&=(x)(x)+(x)(2)+(2)(x)+(2)(2) \\
&=x^2+2x+2x+4 \\
&=x^2+4x+4
\end{aligned}$$

OK - are you feeling solid on your FOILing technique? Don't progress any deeper into the lesson until you're 100% confident you understand how and when to FOIL. We'll need it for a lot of the upcoming Algebra 2 questions.

Factoring

To "factor" is to break a number or an expression down into smaller multiples.

In simple examples, we might find ourselves factoring a single number. For example, the number "60" can be broken down into smaller multiples:

$$\begin{aligned}
&60 \\
&=2\times30 \\
&=2\times2\times15 \\
&=2\times2\times3\times5
\end{aligned}$$

If we multiply $2\times2\times3\times5$, we will return to our original value of 60. We also explored this type of factoring in the previous lesson on **Exponents & Roots**, where we encountered it in "Factor Pyramids" (or "Factor Trees") - although we will take Factoring in a different direction for this lesson.

On the other hand, we could look for just *two* factors of 60 - any two numbers that multiply to 60 - in which case it would be better to create a "Factor *Table*."

1	60
2	30
3	20
4	15
5	12
6	10

Notice that each row gives a *pair* of factors that will multiply to produce a final result of 60.

Also note how I kept my table orderly by starting at "1" and its partner factor "60," and working my way up methodically, instead of jumping around and just writing down whatever factors came to mind.

This "Factor Table" approach is what we will use in Algebra 2 more than any Factor Trees or Pyramids. That's because we're more interested now in *pairs* of factors, instead of the *smallest* possible factors.

Now try creating a factor table for each of the following numbers:

Create a factor table for 12.

Create a factor table for 64.

Create a factor table for 100.

And here are my tables for each of those numbers:

12:

1	12
2	6
3	4

64:

1	64
2	32
4	16
8	8

100:

1	100
2	50
4	25
5	20
10	10

Factoring numbers into pairs is an important basic skill that we'll need for the next challenge: factoring Quadratic Equations.

For example, we will frequently need to factor something like this:

Factor x^2+5x+6 into the form $(x+a)(x+b)$.

In factored form, this expression will be:

$$(x+3)(x+2)$$

If you FOIL $(x+3)(x+2)$ back out, you will return to the original expression x^2+5x+6. Try it now to see for yourself. In essence, Factoring and FOILing are the reverse of each other.

But how did I get to my factored form?

Notice the following: "3" and "2" are a factor pair that multiply to 6. Also, 3 and 2 *add* to 5.

Whenever we're factoring an Quadratic Expression of form $ax^2+bx+c=0$, we need to identify the one pair of factors that *multiply* to c and *add* to b.

This is why writing out your Factor Tables can be so useful: you can see a list of possible number pairs, which makes it easier to methodically search for a pair that both *multiplies* and *adds* to the desired values.

What about this example?

$$\text{Factor } x^2+3x-10 \text{ into the form } (x+a)(x+b)$$

In this case we need a pair of factors that *multiply* to "-10" and *add* to "3." It's guaranteed that one of our factors must be negative, because the only way to multiply to "-10" is if *one* and *only one* of our two factors is also a negative number.

So, we can make a factor table for "10" and look for any numbers that have a *difference* of 3. The factors "5" and "2" would fit that description perfectly.

Now we have to decide which of those factors, 5 or 2, is the negative value. It's easy. Remember, the two numbers have to *add* to 3. It the 5 was negative and the 2 was positive, we would add to *negative* 3. But if the 5 is positive and the 2 is negative, we'll have a pair of numbers that successfully add to $+3$ and multiply to -10.

Therefore, the factored form of $x^2+3x-10$ is:

$$(x+5)(x-2)$$

Again, if you try FOILing this back out (and you should), you'll return to your original $x^2+3x-10$ expression, because FOILing and factoring are the reverse of each other.

Now try your hand at some practice factoring examples. Be sure to do them on your own before you look at my explanations below.

Rewrite the expression x^2+6x+8 in the form $(x+a)(x+b)$.

Rewrite the expression n^2-7x+6 in the form $(n+a)(n+b)$.

Rewrite the expression $x^2+12x-45$ in the form $(x+a)(x+b)$.

To factor x^2+6x+8, we need a pair of numbers that multiply to 8 and add to 6. The factors "4" and "2" will work.

$$x^2+6x+8=(x+4)(x+2)$$

To factor n^2-7x+6, we need a pair of numbers that multiply to 6 and add to -7. The factors "-6" and "-1" will work.

$$n^2-7x+6=(n-6)(n-1)$$

To factor $x^2+12x-45$, we need a pair of numbers that multiply to -45 and add to 12. The factors "-3" and "15" will work:

$$x^2+12x-45=(x-3)(x+15)$$

If you're feeling uncertain, it's worth the time to A) create a factor table and B) FOIL out your final factored-form answer to make absolutely sure it returns to the original expression. Even after 2 decades of practice with factoring, I still follow the two steps above when I want to be extra-sure that I'm factoring correctly.

Solving with FOILing, Factoring, & Quadratic Equations

So, now let's try putting it all together and solving common some common Basic Algebra 2 situations we'll encounter on the SAT Math test.

Here's an example of something we'd commonly see:

> If a and b are the solutions to the equation $x^2 + 6x + 28 = 4 - 4x$, what is the value of $a+b$?

The first step is to recognize the signs of a Quadratic Equation. The question states that there are *two* solutions, and the presence of the x^2 term, along with the rest of the equation, strongly suggests that this equation can be put into $ax^2 + bx + c = 0$ form.

Let's do that first, using **Basic Algebra 1** to move the equation into Quadratic form.:

$$\begin{aligned} x^2 + 6x + 28 &= 4 - 4x \\ -4 \quad & \;\; -4 \\ x^2 + 6x + 24 &= -4x \\ +4x \qquad & \quad +4x \\ x^2 + 10x + 24 &= 0 \end{aligned}$$

OK, good. Now we should try to factor the equation. Is there any pair of numbers that *multiply* to 24 and *add* to 10?

Yes, the factors $+6$ and $+4$ would fit that description. Now use them to factor the equation:

$$\begin{gathered} x^2 + 10x + 24 = 0 \\ (x+6)(x+4) = 0 \end{gathered}$$

OK, now consider this. Our equation is set equal to zero. How could we make this equation be 0 on both sides, thus creating a true and balanced solution?

The key is that *any number* times "zero" is zero. So, as long as *one* of our parentheses equals zero, then the rest of the equation will also multiply to zero, creating the balanced equation $0 = 0$.

In other words, there are two possibilities: either $(x+6) = 0$ or $(x+4) = 0$. Either way, the equation will reduce to $0 = 0$, making it "true."

So what values could we plug in for x to make each of those equations equal 0? First let's try $(x+6)=0$.

$$
\begin{aligned}
x+6&=0\\
-6\quad -6&\\
x&=-6
\end{aligned}
$$

If $x=-6$ then the whole equation is zero. Test it out in our factored form and see:

$$
\begin{aligned}
(x+6)(x+4)&=0\\
(-6+6)(-6+4)&=0\\
(0)(-2)&=0\\
0&=0
\end{aligned}
$$

Or, you could even back up to the original form of the equation in the question and make sure that it turns out true if you plug in $x=-6$. Let's check:

$$
\begin{aligned}
x^2+6x+28&=4-4x\\
(-6)^2+6(-6)+28&=4-4(-6)\\
36-36+28&=4+24\\
28&=28
\end{aligned}
$$

See? There's the *proof* that we've found a solution with $x=-6$. But we're not done yet - we still need to consider the second possibility of $(x+4)=0$:

$$
\begin{aligned}
x+4&=0\\
-4\quad -4&\\
x&=-4
\end{aligned}
$$

You could run the same tests with $x=-4$ and you will find that it *also* makes the equation true.

So, there are *two* answers to our original question.

Remember, the question called these two solutions "a" and "b", and asked that we calculate the value of $a+b$. Since we're just adding two values, it doesn't matter which solution is "a" or "b". Let's do that now:

$$
\begin{aligned}
&a+b\\
&=(-4)+(-6)\\
&=-4-6\\
&=-10
\end{aligned}
$$

And our final answer is -10.

This was an essential **Basic Algebra 2** question and workflow. Make sure you understand how we recognized the problem, used the Quadratic form, then factored, and then found *two* solutions for x in this question. It's very important - we'll be doing a lot of this sort of work on the SAT Math test.

Once you're clear on the preceding example, go ahead and try the three practice examples below.

1. If a and b are the solutions to the equation $x^2+48=16x$, what is the value of $a+b$?

2. If a is a solution to the equation $x^2-2x-2=2x+10$ and $a<0$, what is the value of a?

3. What are two solutions to the equation $100=x^2+21x$

1. First, we need to rearrange $x^2+48=16x$ into the Quadratic form $ax^2+bx+c=0$:

$$x^2+48=16x$$
$$-16x-16x$$
$$x^2-16x+48=0$$

Now factor:

$$x^2-16x+48=0$$
$$(x-12)(x-4)=0$$

Then identify your two solutions or "zeros": $x=12$ and $x=4$.

Finally, give the question what it wants: the value of the two solutions $a+b$, which is $12+4=16$. Our final answer is **16.**

2. This follows the same pattern as the previous example. First, we need to rearrange $x^2-2x-2=2x+10$ into the Quadratic form $ax^2+bx+c=0$:

$$x^2-2x-2=2x+10$$
$$-2x \quad -2x$$
$$x^2-4x-2=10$$
$$-10-10$$
$$x^2-4x-12=0$$

Now factor:

$$x^2-4x-12=0$$
$$(x-6)(x+2)=0$$

Then identify your two solutions or "zeros": $x=6$ and $x=-2$.

Finally, give the question what they want with a solution a that fits $a<0$. Only one of our solutions is less than 0. Our final answer should be -2.

3. This follows the same pattern as the previous two examples. First, we need to rearrange $100 = x^2 + 21x$ into the Quadratic form $ax^2 + bx + c = 0$:

$$\begin{aligned} 100 &= x^2 + 21x \\ -100 & \qquad\quad -100 \\ 0 &= x^2 + 21x - 100 \end{aligned}$$

Now factor:

$$\begin{aligned} 0 &= x^2 + 21x - 100 \\ 0 &= (x+25)(x-4) \end{aligned}$$

Then identify your two solutions: $x = -25$ and $x = 4$. Our final answers should be -25 and 4 since the question asked for both solutions.

Checking for False Roots

We already explored "False Roots" in the lesson on **Advanced Algebra 1**, but now we have a bit more context within which we can revisit them.

False Roots are "false solutions" that occasionally arise when solving a Quadratic Equation that began with an x under a $\sqrt{\ }$ symbol.

For example, the equation below will produce a False Root or "false solution":

$$\sqrt{3x+1} = x - 3$$

Let's solve this out to demonstrate what happens. First, square both sides to eliminate the square root:

$$\begin{aligned} (\sqrt{3x+1})^2 &= (x-3)^2 \\ 3x+1 &= (x-3)^2 \end{aligned}$$

Be sure to FOIL $(x-3)^2$. Do *not* fall for the common error of thinking this is just $x^2 - 9$ or $x^2 + 9$, which is a common mistake. To "square" something means to *multiply it by itself*, like this:

$$\begin{aligned} (x-3)^2 &= (x-3)(x-3) \\ (x-3)(x-3) &= x^2 - 6x + 9 \end{aligned}$$

Now that we've FOILed $(x-3)^2$ properly, continue the original problem:

$$3x + 1 = x^2 - 6x + 9$$

This is a Quadratic Equation, so we need to put it into $ax^2 + bx + c = 0$ form:

$$\begin{aligned} 3x+1 &= x^2 - 6x + 9 \\ -3x-1 & \qquad -3x-1 \\ 0 &= x^2 - 9x + 8 \end{aligned}$$

And now factor the right side. I'll assume you've learned to factor from the previous part of the lesson.

$$0 = x^2 - 9x + 8$$
$$0 = (x-8)(x-1)$$

This gives us the two "solutions" of $x = 8$ and $x = 1$. However, one of these is a False Solution. It is a trap, and it does NOT actually solve the equation.

The only way to know for sure, though, is to plug them each back into the original equation. We'll start by testing $x = 8$ by plugging it back in:

$$\sqrt{3x+1} = x-3$$
$$\sqrt{3(8)+1} = (8)-3$$
$$\sqrt{24+1} = 8-3$$
$$\sqrt{25} = 5$$
$$5 = 5$$

When we plug in our solution $x = 8$, the equation holds true. Great! This solution works. But what happens if we plug in $x = 1$? Let's see:

$$\sqrt{3x+1} = x-3$$
$$\sqrt{3(1)+1} = (1)-3$$
$$\sqrt{3+1} = 1-3$$
$$\sqrt{4} = -2$$
$$2 \neq -2$$

Note that we do *not* get a true equation. The square root of 4 is *positive* 2, but we have a *negative* 2 on the other side. This is the perfect example of a False Solution to a Quadratic Equation. The only way to be sure was to plug our "solutions" back into the *original* equation and verify whether they held true or not.

Luckily, we don't have to do this extra final check very often on the SAT Math, and when we do, the test provides certain "signals" that we should watch out for False Roots in our answers.

The two signals are, first of all, an Algebra 2 Quadratic Equation that begins with an x trapped under a $\sqrt{\ }$ symbol (just as we saw in this example).

And secondly, the answer choices will contain a set of answers something like the following:

(A) {1, 8}
(B) {1}
(C) {4}
(D) {8}

Can you see how the test's answers try to lure us into picking Choice A? They know that many students will solve the Algebra equation without checking their solutions.

For those of us with keener eyes, we'll recognize that the solution set {1, 8} is also broken into options of *just* {1} or *just* {8}. This is a critical clue that reminds us to test our two "solutions" before making a final choice.

In fact, it's actually *faster* to just skip all the Algebra steps and just go straight to testing the answer choices. After all, there's no benefit to doing the algebra work on this question - you'll just end up getting the same possible solutions the SAT gives you to choose from, and you'll still have to test them to be sure.

It's better in these cases to simply recognize the problem type and skip straight to testing the Answer Choices by plugging them into the equation. It works every time, and saves a substantial amount of work, while eliminating the risk of accidentally including a False Root in your answer.

That being said, try the following practice example, which does *not* offer any multiple-choice answers. Be sure to work it yourself before checking my work below.

What are all real solutions for x in the equation $\sqrt{7-2x}=x+4$?

OK, I'm not going to explain every single step - you should be able to follow most of my work by now:

$$(\sqrt{7-2x})^2=(x+4)^2$$
$$7-2x=x^2+8x+16$$
$$-7 \qquad\qquad -7$$
$$-2x=x^2+8x+9$$
$$+2x \qquad +2x$$
$$0=x^2+10x+9$$
$$0=(x+9)(x+1)$$
$$x=-9,-1$$

Now that I have the two "solutions" $x=-9$ and $x=-1$, I need to test them for False Roots by plugging each of them back into the original equation. I'll start with $x=-9$:

$$\sqrt{7-2x}=x+4$$
$$\sqrt{7-2(-9)}=(-9)+4$$
$$\sqrt{7+18}=-9+4$$
$$\sqrt{25}=-5$$
$$5\neq-5$$

Notice that plugging in my "solution" of -9 results in a false equality, so $x=-9$ must be a False Solution.

Now let's plug $x=-1$ into our original equation:

$$\sqrt{7-2x}=x+4$$
$$\sqrt{7-2(-1)}=(-1)+4$$
$$\sqrt{7+2}=-1+4$$
$$\sqrt{9}=3$$
$$3=3$$

Good, $x=-1$ returns a true equality and must be the only real solution. Remember to look for these signals and avoid False Roots on the real SAT!

Pretest Question #1

Let's take a look at our first Pretest question on Basic Algebra 2. Try it yourself before looking at my explanation below.

$$x^2 - 15 = 2(4x + 9)$$

FREE RESPONSE: If b is a solution to the equation above and $b < 0$, what is the value of $-b$?

OK, this question is essentially a "standard" Basic Algebra 2 equation. Our first goal is to rearrange it into $ax^2 + bx + c = 0$ form to prepare it for factoring, a common step in many of these problems.

We'll start off just using Distributing and **Basic Algebra 1** to maneuver our starting equation into the $ax^2 + bx + c = 0$ form:

$$\begin{aligned} x^2 - 15 &= 2(4x + 9) \\ x^2 - 15 &= 8x + 18 \\ -18 \quad &\quad -18 \\ x^2 - 33 &= 8x \\ -8x \quad &\quad -8x \\ x^2 - 8x - 33 &= 0 \end{aligned}$$

OK, this is good. We haven't even used any Algebra 2 topics yet - just Algebra 1, prepping our equation into the correct form to use our new techniques on.

Now, our next goal is to *factor* this form, if possible, into something resembling $(x + a)(x + b) = 0$.

What we need most are two numbers that will *multiply* to -33 but *add* to -8. When you're first getting handy with this, I highly suggest just writing out a complete factor table for the number you're trying to multiply. For example:

1	33
3	11

Luckily, there aren't many factors of 33, so my complete table is listed above.

Now we look for any pair of factors (in the same row) that *add* or *subtract* to our target value of 8.

In fact, we actually know (in this case) that they must have a *difference* of 8. How? Well, our two factors must *multiply* to a negative number, "-33," which means that one factor must be positive and the other must be negative. Therefore I can look for a pair of factors in the table that have a *difference* of 8.

The only candidates are "3" and "11." These must be my factors. But which should be positive and which should be negative?

Well, the two factors must add to "-8." If the 11 is positive and the 3 is negative, they'll add to *positive* 8, which I don't want. So, the 11 must be negative and the 3 must be positive.

Check it out: do -11 and 3 multiply to -33? Yes. And do -11 and 3 add to -8? Yes.

So, we have the two factors we need. The next step is to bring them back to our factored-form equation:

$$(x-11)(x+3)=0$$

Just to be sure we're correct, it can be smart to FOIL back out again and make sure we return to our original equation. So try it out. Multiply "Firsts, Outsides, Insides, Lasts":

$$\begin{aligned}(x-11)(x+3)&=0\\(x)(x)+(x)(3)+(-11)(x)+(-11)(3)&=0\\x^2+3x-11x-33&=0\\x^2-8x-33&=0\end{aligned}$$

And yes, we've properly returned to the $ax^2+bx+c=0$ form that we found in the first stages of the problem. That's a confirmation that we've factored correctly.

Almost done - now we just need to find the *solutions* to the equation $(x-11)(x+3)=0$.

Remember that if *either* of our factors (in parentheses) are equal to "0", then the entire equation will equal 0 and be "true." Zero times *any* number is zero. This is why our equation will have *two* solutions: one for $x-11=0$ and one for $x+3=0$.

Let's solve each of those by hand (when you're comfortable with these concepts you can just do this in your head - one of the rare times when I'll recommend mental math over written work, but it does save time). Anyway, for now we'll carefully show our work:

$$\begin{aligned}x-11&=0\\+11&+11\\x&=11\end{aligned}$$

$$\begin{aligned}x+3&=0\\-3&\ \ -3\\x&=-3\end{aligned}$$

Our two solutions for the values of x are "11" and "-3". However, the question says that the solution b must be *less than* 0. It must be referring only to "-3," which is less than 0.

Last but not least, our final answer should be the value of $-b$, which is $-(-3)$ or our final answer, **3.**

Difference of Squares

A "Difference of Squares" situation is a convenient special case of Factoring. It's mainly special because it allows us to save some time on a common Factoring situation, and it can also be important to recognize in certain SAT Math questions.

The most general form of a Difference of Squares is:

$$a^2 - b^2$$

The name has a literal meaning: since "difference" is subtraction, a "Difference of Squares" is subtraction of two squared values, as you can see in the format above.

A more "everyday" example of a Difference of Squares might look like this:

$$x^2 - 9$$

Notice how similar the format is to $a^2 - b^2$. "9" is a square of 3, x^2 is of course a square of x, and one of these squares is being subtracted from the other.

Now here's the cool thing: these Difference of Squares problems always factor in the same way. Here are some examples:

$$a^2 - b^2 = (a+b)(a-b)$$

$$x^2 - 9 = (x+3)(x-3)$$

$$x^2 - 25 = (x+5)(x-5)$$

$$4n^2 - 81 = (2n+9)(2n-9)$$

Do you see the pattern? The original form $a^2 - b^2$ is factored into the form $(a+b)(a-b)$. Try the following practice examples to test your skills:

Factor the expression $t^2 - 4$.

Factor the expression $49 - x^2$.

Factor the expression $9x^2 - 100$.

Each of these three examples is a Difference of Squares. They all follow the format $a^2 - b^2$.

$t^2 - 4$ can be factored as $(t+2)(t-2)$.

$49 - x^2$ can be factored as $(7+x)(7-x)$

$9x^2 - 100$ can be factored as $(3x+10)(3x-10)$.

Pretest Question #2

Let's take a look at another Pretest question. Try it yourself before you look at my explanation below the question.

> FREE RESPONSE: If $x+y=-2$ and $x-y=3$, what is the value of $(x^2-y^2)(x+y)$?

This question is similar to one on the SAT that I see a lot of students get stuck on. But it's much easier than it looks - especially if you see the opportunity of a Difference of Squares in (x^2-y^2).

We need to factor this Difference of Squares. This is easy!

$$x^2-y^2=(x+y)(x-y)$$

So, if x^2-y^2 is the same as $(x+y)(x-y)$, then we can plug into the original question. Instead of $(x^2-y^2)(x+y)$, we can write it as:

$$[(x+y)(x-y)](x+y)$$

And in that case, we can use the information given in the problem for a substitution. Since $x+y=-2$ and $x-y=3$, we can plug these values directly into our factored form:

$$\begin{aligned}&[(x+y)(x-y)](x+y)\\&=[(-2)(3)](-2)\\&=[-6](-2)\\&=12\end{aligned}$$

The value of $(x^2-y^2)(x+y)$ must be **12.**

See how we factored the Difference of Squares, then plugged in the values given in the equation? That's all it takes to solve this problem. This explanation often surprises students who felt "completely stuck" just moments before, when they suddenly see the explanation with new-found clarity. It was right under their noses, if they just look for Differences of Squares and other opportunities to factor!

Review & Encouragement

Algebra 2 isn't just an extension of Algebra 1 - it's a quantum leap. That's not in terms of *difficulty*, but in terms of the possibilities that are available to us when solving algebraic equations.

We're learning to deal with FOILing, factors, and multiple solutions to equations. Soon we'll be heading even deeper into parabolas, the Quadratic Formula, and related topics.

For now, be sure you understand the basics of the $ax^2+bx+c=0$ Quadratic form, FOILing, factoring, "false roots," Difference of Squares, and solving Quadratic Equations for multiple solutions. Complete the practice set and be sure everything makes sense to you before moving on to the next lessons.

Basic Algebra 2 Practice Questions

DO NOT USE A CALCULATOR ON ANY OF THE FOLLOWING QUESTIONS.

$$3x(x+6)+4(2-2x)=ax^2+bx+c$$

1. FREE RESPONSE: In the equation above, a, b, and c are constants. If the equation is true for all values of x, what is the value of b?

$$(nx+f)(nx-f)=25x^2-81$$

2. In the equation above, n and f are constants. Which of the following could be the value of n?

(A) -9

(B) 3

(C) 5

(D) 9

$$2x^2-8x-42=0$$

3. FREE RESPONSE: If n is a solution to the equation above and $n>0$, what is the value of n?

$$\frac{x^2+4}{x-1}=-4$$

4. What are all values of x that satisfy the equation above?

(A) -4

(B) 0

(C) 1

(D) 0 and -4

$$\sqrt{3x} = x - 6$$

5. What is the solution set for x to the equation above?

(A) $\{3, 12\}$

(B) $\{3\}$

(C) $\{6\}$

(D) $\{12\}$

$$\frac{1+x}{x} = -\frac{x}{4}$$

6. Which of the following represents all the possible values of x that satisfy the equation above?

(A) -2

(B) 0 and -2

(C) 0 and 2

(D) 2

7. Which of the following is a value of x for which the expression $\dfrac{2x+4}{x^2+6x-27}$ is undefined?

(A) -3

(B) -2

(C) 3

(D) 9

8. What is the sum of the solutions to $(x-4.4)(x+7.3)=0$?

(A) -11.7

(B) -2.9

(C) 2.9

(D) 11.7

9. FREE RESPONSE: If $x+y=6$ and $x-y=3$, what is the value of $(x-y)(x^2-y^2)$?

$$bx^3+dx^2+gx+m=0$$

10. In the equation above, b, d, g, and m are constants. If the equation has roots -6, 3, and 7, which of the following is a factor of bx^3+dx^2+gx+m?

(A) $x+6$

(B) $x+3$

(C) $x-6$

(D) $x+7$

11. The expression $\frac{1}{5}x^2-10$ can be rewritten as $\frac{1}{5}(x-n)(x+n)$, where n is a positive constant. What is the value of n?

(A) $\sqrt{5}$

(B) $5\sqrt{2}$

(C) 10

(D) 50

$$y=4x^2-b$$

12. In the equation above, b is a positive constant and the graph of the equation in the *xy*-plane is a parabola. Which of the following is an equivalent form of the equation?

(A) $y=4(x+b)(x-b)$

(B) $y=(x+\frac{b}{2})(x-\frac{b}{2})$

(C) $y=(2x+b)(2x-b)$

(D) $y=4(x+\frac{\sqrt{b}}{2})(x-\frac{\sqrt{b}}{2})$

Basic Algebra 2 Answers

1. 10
2. C
3. 7
4. D
5. D
6. A
7. C
8. B
9. 54
10. A
11. B
12. D

Basic Algebra 2 Explanations

1. **10.** This question begins with a pair of simple Distributions and then moves on to Combining Like Terms. In many ways, it's barely more advanced than the Algebra 1 questions we did earlier in this book.

First, distribute to the parentheses via multiplication:

$$3x(x+6)+4(2-2x)$$
$$=(3x)(x)+(3x)(6)+(4)(2)-(4)(2x)$$

Then clean up as much as possible and combine like terms:

$$(3x)(x)+(3x)(6)+(4)(2)-(4)(2x)$$
$$=3x^2+18x+8-8x$$
$$=3x^2+10x+8$$

Now compare our cleaned-up version of the equation to the right side of the equation.

$$3x^2+10x+8=ax^2+bx+c$$

The question asks us for the value of b. Notice that b is the coefficient of x on the right side. Therefore, the value of b will correspond to the coefficient of x on the left side of the equation (which is 10).

Therefore, the value of b must be 10.

2. **C.** Our first step is to FOIL out the left side of the equation. Remember a general rule of Algebra 2 questions on the SAT: if you're *able* to FOIL something, you probably *should* FOIL it.

Remember, "First Outside Inside Last."

$$(nx+f)(nx-f)$$
$$=(nx)(nx)+(nx)(-f)+(f)(nx)+(f)(-f)$$
$$=n^2x^2-fnx+fnx-f^2$$

Careful observers will notice from the very beginning that $(nx+f)(nx-f)$ is a "Difference of Squares." But even if you don't see that at first, it's OK - simply FOILing and cleaning up will still get you to the right place.

Now we need to clean up our work. Most importantly, notice the middle terms of $-fnx$ and $+fnx$. These two equivalent but opposite terms will cancel each other out, and we're left with the classic result of FOILing a Difference of Squares:

$$n^2x^2-f^2$$

OK, now that we've simplified the left side of the original equation as much as possible, let's bring back the right side:

$$n^2x^2-f^2=25x^2-81$$

And the question asks for the value of n. We can find n^2 as a coefficient of x^2 on the left side of the equation. The corresponding coefficient of x^2 on the right side is "25." Therefore, we can assume that n^2 has a value of 25:

$$n^2=25$$

So, square root both sides to solve for the value of n:

$$\sqrt{n^2}=\sqrt{25}$$
$$n=5$$

And we're done. The value of n is 5, which is Choice C.

3. **7.** This is a classic Algebra 2 equation. Luckily, it's already set equal to 0 (remember how important it is to set these up in the Quadratic format $ax^2+bx+c=0$).

There's also a good first move for this Algebra 2 question. Let's divide everything by 2 before proceeding. This will just help simplify by lowering the values of the numbers we're working with, and also make it easier to factor:

$$\frac{2x^2-8x-42}{2}=\frac{0}{2}$$
$$x^2-4x-21=0$$

OK, now let's see if we can factor it. Remember, we need to find a pair of numbers that *multiply* to "-21" and *add* to "-4".

One way to do this is simply make a table of factors for 21. Luckily there aren't very many:

1	21
3	7

Then look for a pair of factors that can add or subtract to "4". Only the factors "3" and "7" fit the criteria.

Finally, we need to *multiply* to -21 and *add* to -4. So, *one* of our factors must be negative. But which one? Well, since they add to -4, then the larger factor (7) must be negative.

Here's the factored version:

$$(x-7)(x+3)=0$$

And, what are the two values for x that solve this equation?

Take the two factors one at a time and set each equal to 0. Remember, if *either* of the expressions in the pair of parentheses equals "0", then the entire equation will multiply to 0 and be a true equation (because 0 times anything always $=0$).

The better you get at this, the more easily you can do it without spending so much time on your work, but we'll be safe for now:

$$\begin{array}{ll} x-7=0 & x+3=0 \\ +7\ +7 & -3\ \ -3 \\ x=7 & x=-3 \end{array}$$

The two solutions to this equation are $x=7$ and $x=-3$. Either of them will make the equation true. (Test by plugging them back in, if you want.) However, the word problem asked us to solve for a value n that is a solution where $n>0$. Since we need a positive number, that eliminates $x=-3$ and leaves us with only one choice: n must equal 7.

4. **D.** The first step in this Algebra 2 problem is to get rid of the fraction by multiplying both sides, then cleaning up:

$$\frac{x^2+4}{x-1}=-4$$

$$(x-1)\frac{x^2+4}{x-1}=-4(x-1)$$

$$x^2+4=-(4)(x)-(4)(-1)$$

$$x^2+4=-4x+4$$

Next we need to set everything equal to 0 to get the Quadratic format $ax^2+bx+c=0$.

$$\begin{array}{c} x^2+4=-4x+4 \\ -4 \qquad\quad -4 \\ x^2=-4x \\ +4x \quad +4x \\ x^2+4x=0 \end{array}$$

Believe it or not, we *are* in the correct format $ax^2+bx+c=0$. It's just that the value of c is 0.

Now, as usual, we'll factor. This time we'll just have to pull out an x from both terms:

$$x^2+4x=0$$

$$x(x+4)=0$$

And now, consider that either x or $(x+4)$ must equal 0. Either one can equal 0, because 0 times any other value will equal 0 and give a true equation. So consider both possibilities:

$$\begin{array}{lll} x+4=0 & & \\ -4\ \ -4 & \text{or} & x=0 \\ x=-4 & & \end{array}$$

This puts us at Choice D. We've found the two values of x are -4 and 0. We accomplished this by using Algebra to set up our $ax^2+bx+c=0$ equation, then factoring, and finally solving for the two possible values of x that give the two "zeros."

5. **D.** There's a lot to learn from this question - most of it based around common Careless Mistakes that students commit in Algebra 2 questions.

Remember, the first goal of Algebra 2 questions should usually be to set up an $ax^2+bx+c=0$ equation. To do so, we first need to eliminate the $\sqrt{\ }$ symbol by squaring both sides.

$$\sqrt{3x}=x-6$$

$$(\sqrt{3x})^2=(x-6)^2$$

$$3x=(x-6)(x-6)$$

Be *very* alert that you don't make one of the most famous **Careless Errors** in Algebra 2, which is to incorrectly distribute $(x-6)^2$. Just for illustration purposes, I'll now show what many students do *WRONG* on this step:

$$(x-6)^2 \neq x^2-36$$

or possibly...

$$(x-6)^2 \neq x^2+36$$

The above equations are WRONG. They happen because we forgot what *squaring* really means. To "square" is to multiply something *times itself*.

If we want to successfully square $(x-6)^2$, we must multiply $(x-6)(x-6)$.

Can you see the difference? Can you understand how easy it is to make the Careless Error above? Be sure you understand the dangers before continuing.

OK, now let's keep the Algebra train going. Next, we need to FOIL the right side of the equation:

$$3x = (x-6)(x-6)$$
$$3x = x^2 - 6x - 6x + 36$$
$$3x = x^2 - 12x + 36$$

And now, make sure we're set equal to 0.

$$3x = x^2 - 12x + 36$$
$$-3x \qquad -3x$$
$$0 = x^2 - 15x + 36$$

Next, we need to try to factor the right side of the equation. We need a pair of numbers that *multiply* to 36 and *add* to -15. As in Question 3, it can be useful to create a table of factors for "36" if you're having trouble finding such a pair.

1	36
2	18
3	12
6	6

Now look for any pair of factors that can add or subtract to "15." The only factors that fit the criteria are 3 and 12.

Since they have to *multiply* to $+36$, the two factors must be either *both* positive or *both* negative. But, since they must *add* to -15, we know that both factors must be *negative*.

Therefore, our factored equation looks like this:

$$0 = x^2 - 15x + 36$$
$$0 = (x-12)(x-3)$$

And now we can find our two "zeros," which are $x = 12$ and $x = 3$. But wait - doesn't that just put us back at the Answer Choices we were already given? Is the correct answer just Choice A?

Well, unfortunately it's not. We encountered the same issue earlier in the lesson on **Advanced Algebra 1**. What happened is that we (may) have found a *false root.* This can occasionally happen when we "square a square root."

So check this out - there's a bit of dark humor here, but we actually could have just skipped all of the Algebra we just did, and instead just plugged in the answer choices that were given to us at the beginning. This is known as "checking your roots," and we can learn to *identify* this type of question on the SAT Math test by the style of Answer Choices we're given.

As we learned in the lesson, the clues that this is a "False Roots" problem lie in two parts: first, the Algebra 2 question involving an x trapped under a $\sqrt{\ }$ sign. Second, the "sets" of answer choices that include values like {3, 12} or just {3} or just {12}.

Anyway, it's not like we did anything *wrong* by solving the Algebra equation - we just wasted a lot of time when we could have skipped straight to the next steps below.

Let's get it over with and plug in the given answer choices to test for False Roots.

First, we'll test "3" for x.

$$\sqrt{3x} = x - 6$$
$$\sqrt{3(3)} = (3) - 6$$
$$\sqrt{9} = -3$$
$$3 \neq -3$$

OK, so despite our Algebra work, which told us that x could be 3, $x = 3$ actually does *not* work as a solution, because a square root of a number can never equal a negative number. All square roots return positive values.

Now, let's test "12" for x.

$$\sqrt{3x} = x - 6$$
$$\sqrt{3(12)} = (12) - 6$$
$$\sqrt{36} = 6$$
$$6 = 6$$

OK, so $x = 12$ actually *does* work. That's a good sign.

Just to be safe, let's test the final choice of "6" for x.

$$\sqrt{3x} = x - 6$$
$$\sqrt{3(6)} = (6) - 6$$
$$\sqrt{18} \neq 0$$

OK, $x = 6$ definitely doesn't work. This isn't really surprising, since our earlier Algebra work never suggested that $x = 6$ was a possible solution.

And, we're left with only one choice - that $x = 12$ is the only possible solution, and the correct answer is Choice D.

Learn the signs of "Testing Roots" questions on the SAT, because not only do you run the risk of wasting a lot of time doing unnecessary Algebra work, you run the even greater risk of falling for a trap answer that includes a False Root in the solution set.

6. **A.** To solve this algebra equation, the first move should be to eliminate both fractions in a single move by Cross-Multiplying, which we studied in the lesson on **Advanced Algebra 1**.

$$\frac{1+x}{x} = -\frac{x}{4}$$
$$(1+x)(4) = (-x)(x)$$

Then distribute and clean up:

$$4 + 4x = -x^2$$

Now arrange it into the classic Algebra 2 solution setup of $ax^2 + bx + c = 0$:

$$4 + 4x = -x^2$$
$$+x^2 \quad +x^2$$
$$x^2 + 4x + 4 = 0$$

And now factor the right side. I'll take it for granted that you've got your factoring locked down at this point, but if you need some guidance, you can review the lesson and/or the explanations for Questions 3 and 5.

$$x^2 + 4x + 4 = 0$$
$$(x+2)(x+2) = 0$$

Our "zeros" will result from plugging in $x = -2$ and... $x = -2$. Interestingly, there only appears to be *one* solution for x, which is "-2". And, we've got our final answer, which is Choice A.

Also note that our final form could be reduced further, if we wanted to:

$$(x+2)(x+2) = 0$$
$$(x+2)^2 = 0$$

7. **C.** This problem raises an interesting question: what would make $\dfrac{2x+4}{x^2+6x-27}$ be "undefined"?

Well, any fraction divided by "0" gives an "undefined" result. So, the bottom (or "denominator") of this fraction must be set equal to zero, then solved for x.

$$x^2 + 6x - 27 = 0$$

It's already in our favorite Algebra 2 setup of $ax^2 + bx + c = 0$, so all we have to do is factor and solve for x.

I'll take for granted that you understand how to factor by this point, but if you need help on factoring still, you can review the lesson or the explanations of Question 3 and Question 5.

In factored form, this equation is:

$$(x-3)(x+9) = 0$$

And our two "zeros" or "solutions" will be found at $x = 3$ and $x = -9$.

Only one of these solutions is available as an answer choice, which is $x = 3$ and Choice C.

8. **B.** This question is extremely quick and easy as long as we understand the concepts we've used so far throughout this lesson. This equation is *already* set equal to 0 and factored. All we have to do is read off the two values of x that would give us zero. That would be $x = 4.4$ and $x = -7.3$. See how easy that is? All the work is already done for us!

All that's left is to give them "the sum of the solutions", which is just a simple addition problem:

$$4.4 + (-7.3) = -2.9$$

And we're done. The answer must be Choice B.

9. **54.** This question is quite interesting and also nearly identical to Pretest Question #2 in this lesson. It *looks* like a **System of Equations** question, but it's *actually* just based on factoring.

Pay special attention to the $x^2 - y^2$ part of the equation. This is a Difference of Squares, and like any difference of squares, it can be easily factored:

$$x^2 - y^2 = (x - y)(x + y)$$

So, we can rewrite the original equation $(x - y)(x^2 - y^2)$ as something interesting:

$$(x - y)[(x - y)(y + y)]$$

And, best of all, the question gives us the info that $x + y = 6$ and $x - y = 3$. So we can just plug those values directly in:

$$\begin{aligned}&(x - y)[(x - y)(y + y)]\\&= (3)[(3)(6)]\\&= (3)[18]\\&= 54\end{aligned}$$

And we get our final answer, 54. This question shows the value of carefully looking for opportunities to factor, and to look for a Difference of Squares in particular.

10. **A.** So, this question slightly stretches the bounds of "Basic" Algebra 2. But just barely. You see, this question is based on a *3rd-order* polynomial, not the 2nd-order polynomials we've grown used to throughout the chapter.

Just as a reminder, a 2nd-order polynomial follows the form $ax^2 + bx + c = 0$ and a 3rd-order polynomial follows $ax^3 + bx^2 + cx + d = 0$. It's called "3rd-order" because of the ^{3} exponent.

Now, with 2nd-order polynomials, we've gotten used to setting our equation equal to 0, then finding two factors - such as $(x - 3)(x + 2)$, just as an example.

3rd-order polynomials aren't really any different, they just have three factors instead of two - for example $(x - 2)(x + 4)(x - 1)$.

And, instead of having *two* zeros to find, like a 2nd-order, we find that a 3rd-order polynomial has *three* zeros. Like always, these are the solutions where plugging in a value of x results in a final value of "0."

We learned in the lesson that "roots," "solutions, and "zeros" are all names for the same concept. So when this question tells us that the 3rd-order polynomial has *three* roots of -6, 3, and 7, that means that each of these values will produce a zero when plugged in for x.

Hopefully it's clearer by now what we're looking for: an answer choice which, if -6, 3, or 7 is plugged in for x, will produce a factor with a value of 0.

The only such factor is found in Choice A, where plugging in the root "-6" will return a value of 0. You can test any other combination of given roots from the question and possible factors from the answer choices, but none will produce a value of 0 except Choice A.

11. **B.** OK, here's how we'll handle this question. First, set the two forms of the expression equal to each other:

$$\frac{1}{5}x^2 - 10 = \frac{1}{5}(x - n)(x + n)$$

Then, multiply both sides by 5 to eliminate the $\frac{1}{5}$ fraction:

$$\begin{aligned}(5)(\frac{1}{5}x^2 - 10) &= \frac{1}{5}(x - n)(x + n)(5)\\x^2 - 50 &= (x - n)(x + n)\end{aligned}$$

Notice the Difference of Squares on both sides. Continuing forward, let's FOIL the right side of the equation:

$$\begin{aligned}x^2 - 50 &= (x - n)(x + n)\\x^2 - 50 &= x^2 - n^2\end{aligned}$$

Now use **Basic Algebra 1 to** remove the x^2 on both sides and keep cleaning up to find n:

$$\begin{aligned}x^2 - 50 &= x^2 - n^2\\-x^2 \quad &\quad -x^2\\-50 &= -n^2\\(-1) - 50 &= -n^2(-1)\\50 &= n^2\\\sqrt{50} &= \sqrt{n^2}\\\sqrt{50} &= n\end{aligned}$$

OK, so we actually have a workable value of n, which is $\sqrt{50}$. The only problem is, none of the answer choices look quite the same. We'll need to simplify the root, as we learned in the lesson on **Exponents & Roots**.

I'll use a factor pyramid to break down 50 into its smallest factors:

$$\begin{aligned}&50\\&=(5)(10)\\&=(5)(5)(2)\end{aligned}$$

Notice the *pair* of 5s, a set of two which we can pull out together from under the square root as a single number. That will leave the 2 trapped underneath the root (again, these techniques were covered in the lesson on **Exponents & Roots**):

$$\sqrt{50}=5\sqrt{2}$$

And now we're really done. The answer is Choice B.

12. **D.** This is a pretty tough question, although it doesn't use any new techniques. Don't get distracted by the wording about "graphs," "parabolas," and the "*xy*-plane." While it's true that all of these concepts can relate to Algebra 2, as we'll see soon, it's also true that none of them are essential for this question, which is entirely based on factoring and Difference of Squares.

Notice how all the answer choices are factored versions of... something. It's probably safe to assume they're factored versions of the original equation $y=4x^2-b$, otherwise none of this question would make much sense.

It sure would be easier if we had a simple x^2 instead of $4x^2$. Why don't we just factor out 4 and see what happens:

$$y=4(x^2-\frac{1}{4}b)$$

Notice that I've had to apply a fraction of $\frac{1}{4}$ to the b in order to accommodate factoring out a multiple of 4.

Now, do you see the Difference of Squares in $x^2-\frac{1}{4}b$? It can be factored like this:

$$x^2-\frac{1}{4}b=(x+\frac{1}{2}\sqrt{b})(x-\frac{1}{2}\sqrt{b})$$

Notice that $(\frac{1}{2})^2=\frac{1}{4}$. Also, we have to factor the $-b$ into $(+\sqrt{b})(-\sqrt{b})$, so that these factors multiply back to their original value of $-b$.

Yes, this factoring is a little tougher than most, perhaps. But if you FOIL back out, you can confirm for yourself that $(x+\frac{1}{2}\sqrt{b})(x-\frac{1}{2}\sqrt{b})=x^2-\frac{1}{4}b$.

Don't forget that we have to multiply the entire right side by 4, because we originally factored out a 4 back at the very beginning of the question:

$$4(x+\frac{1}{2}\sqrt{b})(x-\frac{1}{2}\sqrt{b})=4x^2-b$$

The correct answer, Choice D, will FOIL and distribute back to the original equation.

One alternate way to solve the question - especially if you're afraid of Factoring - would just be to work carefully through the four answer choices, FOILing them out and applying any distributing. Only Choice D will return you to the original equation of $y=4x^2-b$.

However, at this point you should be ready to show off your Factoring skills!

Lesson 17: The Quadratic Formula

Percentages

- 1.2% of Whole Test
- 2.5% of No-Calculator Section
- 0.5% of Calculator Section

Prerequisites

- Basic Algebra 2
- Exponents & Roots
- Basic Algebra 1

In this lesson, we'll learn about the Quadratic Formula: what it is, when to use it - and when not to. It's a new tool that we can add to our Algebra 2 toolkit.

We'll also cover the Discriminant, which is a minor but useful topic that many of my students have either never heard of or forgot about a long time ago.

Questions involving the Quadratic Formula appear reasonably often in the No-Calculator section, but they are very rare in the Calculator section.

Quadratic Formula Quick Reference

- The Quadratic Formula is used to solve Quadratic Equations of form $ax^2 + bx + c = 0$.
- We use the Quadratic Formula when we can't factor a Quadratic Equation by hand. Factoring (from **Basic Algebra 2**) is preferred when possible, because it is faster and less prone to **Careless Mistakes**.
- The Quadratic Formula is $X = \frac{-b \pm \sqrt{b^2 - 4ac}}{2a}$. We must have this formula committed to memory for the SAT Math test.
- The Discriminant is the portion of the Quadratic Formula located under the square root: $b^2 - 4ac$. It is used to quickly determine *how many* real solutions exist to a Quadratic Equation.
- Using these techniques on the right questions will save a lot of time and frustration. Using them on the wrong questions will waste time and encourage Careless Mistakes. Learn the signals that the SAT gives us when it's time to pull the Quadratic Formula or the Discriminant out of your pocket!

Introduction to the Quadratic Formula

First, let's understand what the Quadratic Formula is for and take a look at the formula itself.

The Quadratic Formula can be used to solve for the solutions of *any* Quadratic Equation $ax^2+bx+c=0$.

Here is the Quadratic Formula:

$$X=\frac{-b\pm\sqrt{b^2-4ac}}{2a}$$

When would we use this? Didn't the last lesson on **Basic Algebra 2** already teach us solve Quadratic Equations by factoring into the form $(x+a)(x+b)=0$?

Well, the problem is that not all Quadratic Equations can easily be factored by hand. For example, check out the equation below:

$$x^2-12x+17=0$$

Try to factor it first. I bet you can't!

This is a scenario that would be perfect for the Quadratic Formula. It's more work than factoring, and it's prone to **Careless Mistakes** - *but*, it can solve Quadratic Equations that we can't factor by hand, and that's why it's so valuable.

First, correctly identify your values for a, b, and c by comparing the equation $x^2-12x+17=0$ to the general Quadratic Equation $ax^2+bx+c=0$.

In our equation, a (the coefficient of x^2) is 1. The b (coefficient of x) is -12. The constant c is 17.

Now let's plug each of those into their places in the Quadratic Formula:

$$X=\frac{-b\pm\sqrt{b^2-4ac}}{2a}$$
$$X=\frac{-(-12)\pm\sqrt{(-12)^2-4(1)(17)}}{2(1)}$$

Can you see where everything (a, b, and c) got plugged into its proper place?

Now start cleaning up as much as possible:

$$X=\frac{12\pm\sqrt{144-68}}{2}$$
$$X=\frac{12\pm\sqrt{76}}{2}$$

At this stage, you might also need to simplify the root $\sqrt{76}$, depending on whether or not you have a calculator available. Review the lesson on **Exponents & Roots** for a deeper explanation. For now I'll just go ahead and simplify $\sqrt{76}$ to $2\sqrt{19}$ and keep cleaning up my Quadratic Formula:

$$X = \frac{12 \pm 2\sqrt{19}}{2}$$
$$X = 6 \pm \sqrt{19}$$

So, now we have our two solutions for x in our original equation $x^2 - 12x + 17 = 0$. The solutions are $x = 6 + \sqrt{19}$ and $x = 6 - \sqrt{19}$. Not the prettiest numbers, but valid solutions nonetheless.

And this is the value of the Quadratic Formula. There's *no way* we'd figured out the solutions $x = 6 \pm \sqrt{19}$ if we were factoring by hand. But the Quadratic Formula cuts through the problem like a hot knife through butter.

Now try a practice problem on your own. I suggest at least *trying* to factor it before you reach for the Quadratic Formula. You won't be able to, but at least you'll learn what it feels like to try factoring an "impossible" Quadratic Equation.

What are the two real solutions to the equation $x^2 + 8x + 4 = 0$?

OK, so hopefully you tried factoring first and realize it's basically impossible. There's no convenient pair of numbers that *multiply* to 4 and *add* to 8 (remember that we studied factoring in detail in the previous lesson on **Basic Algebra 2**).

Instead, it's time to reach for the Quadratic Equation.

$$X = \frac{-b \pm \sqrt{b^2 - 4ac}}{2a}$$

First we need to identify the values of a, b, and c in our given equation $x^2 + 8x + 4 = 0$. Remember that we get a, b, and c by comparing to the general Quadratic Equation $ax^2 + bx + c = 0$.

The a value (coefficient of x^2) is 1. The b value (coefficient of x) is 8. The value of the constant c is 4. Now let's plug them into their places in the Quadratic Equation:

$$X = \frac{-(8) \pm \sqrt{(8)^2 - 4(1)(4)}}{2(1)}$$

Do you understand where each of the numbers came from? If not, trace back each of the letters a, b, and c and where we got them. When you do understand, then go ahead and start cleaning up our equation:

$$X = \frac{-8 \pm \sqrt{64 - 16}}{2}$$
$$X = \frac{-8 \pm \sqrt{48}}{2}$$

I'll also clean up the $\sqrt{48}$ and simplify it to $4\sqrt{3}$ using the simplification techniques we used in the lesson on **Exponents & Roots** (review that lesson you don't understand how to do this).

$$X = \frac{-8 \pm 4\sqrt{3}}{2}$$
$$X = -4 \pm 2\sqrt{3}$$

And our final two solutions are $-4 + 2\sqrt{3}$ and $-4 - 2\sqrt{3}$.

As usual, when we use the Quadratic Formula, our answers aren't the prettiest numbers, but they *are* the two valid solutions to the original equation.

Getting the hang of it? Good! Now make sure you memorize the Quadratic Formula, because it's not going to be given to you on the SAT test.

Memorizing the Quadratic Formula

The Quadratic formula *must be memorized.* It will not be given to you on the SAT - you are expected to know it.

Note, that this formula can be perfectly sung to the melody of the classic children's song "Pop Goes the Weasel". This - and I'm 100% serious here - is how I've remembered this equation perfectly since 7th grade:

"X equals negative B

Plus or minus the square root

Of B squared minus four A C

All over two A"

I searched YouTube but I couldn't find a good video of someone singing it. Too bad I couldn't record my math teacher, Mrs. Butler, doing her rendition - because I've remembered it for nearly 20 years!

Pretest Question #1

Let's take a look at our first Pretest question on this topic. Try it yourself if you got it wrong the first time.

What are the solutions to $4x^2 + 16x - 8 = 0$?

(A) $x = -2 \pm \sqrt{6}$

(B) $x = -2 \pm \sqrt{3}$

(C) $x = 2 \pm \sqrt{6}$

(D) $x = 8 \pm 4\sqrt{6}$

This question is a perfect time to apply the Quadratic Formula. Even the answer choices offer powerful hints that we are meant to get "ugly" solutions to the equations involving roots. That's a dead giveaway that we aren't going to be able to factor the equation by hand.

Now, I'll recommend that you divide *everything* by 4 before you start plugging into the Quadratic Formula, just because it makes the numbers smaller and easier to work with, especially if you don't have a calculator available:

$$\frac{4x^2+16x-8}{4}=\frac{0}{4}$$
$$x^2+4x-2=0$$

Even now, we probably aren't able to factor this. I certainly can't think of a convenient pair of numbers that *multiply* to "-2" and *add* to "4".

By comparing to the basic Quadratic Equation form of $ax^2+bx+c=0$, we can identify that $a=1$, $b=4$, and $c=-2$. Go ahead and plug those into the Quadratic Formula:

$$X=\frac{-b\pm\sqrt{b^2-4ac}}{2a}$$
$$X=\frac{-(4)\pm\sqrt{(4)^2-4(1)(-2)}}{2(1)}$$

Do you understand where each number came from, and where it's supposed to plug in? Make sure you feel clear on this before continuing.

Now we can clean up and simplify our equation:

$$X=\frac{-4\pm\sqrt{16+8}}{2}$$
$$X=\frac{-4\pm\sqrt{24}}{2}$$

This looks a lot better, but it still doesn't quite fit the answer choices. Using what we learned in the lesson on **Exponents & Roots**, we can simplify the $\sqrt{24}$ into $2\sqrt{6}$:

$$X=\frac{-4\pm2\sqrt{6}}{2}$$

Now simplify the fraction:

$$X=-2\pm\sqrt{6}$$

This matches to **Choice A**, and we're done!

Review the steps: we recognized the opportunity to use the Quadratic Formula through the "ugly" answer choices and because we're not able to factor the given Quadratic Equation by hand. Then we divided everything by 4, because the opportunity presented itself and because it allows us to work with smaller numbers.

Next we compared our equation to the Quadratic Equation form $ax^2 + bx + c = 0$ and identified our a, b, and c values, which we then plugged straight into the Quadratic Formula.

From there we cleaned up and simplified - including simplifying a root - and after all was said and done, we ended up exactly on one of the given answer choices. Nice work!

The Discriminant

The "Discriminant" is a name for the portion of the Quadratic Formula that lives under the square root sign. I'll include the whole Quadratic Formula again for reference:

$$X = \frac{-b \pm \sqrt{b^2 - 4ac}}{2a}$$

Here is the Discriminant, which is found under the square root:

$$b^2 - 4ac$$

(Remember that a, b, and c come from the Quadratic Equation form $ax^2 + bx + c = 0$.)

The Discriminant is a quick way to find how *many* real solutions a Quadratic Equation will have. It doesn't tell you what the solutions *are*, just how *many* of them there will be.

- If the Discriminant is a positive number, there will be two real solutions to the equation.
- If the Discriminant is a negative number, there will be NO real solutions to the equation.
- If the Discriminant is zero, there will be one real solution to the equation.

Why is this true?

First of all, if the Discriminant is *negative*, then we can't take the square root of a negative number without using **Imaginary Numbers** (covered in an upcoming lesson). Typically, the SAT Math test is not interested in imaginary solutions to Quadratic Equations. Therefore, a negative-number Discriminant will return *no* real answers.

Now, if the Discriminant is *positive*, we can take the square root of it and get a real number. And remember the $\pm$ sign in the Quadratic Formula? We'll get *two* solutions to our Quadratic Equation because we'll *add* the square root to one solution and *subtract* it from the other solution, creating two distinct final solutions.

But if the Discriminant is *zero*, then we can certainly take the square root of it - but we will get 0 as a result (because $\sqrt{0} = 0$). And ± 0 only gives one solution, because there's no difference between adding "$+0$" and subtracting "-0". There *will* be a real solution - but only one.

So when could we use our knowledge of the Discriminant? We can use it on SAT questions that ask us for *how many* solutions there are to a Quadratic Equation, but not *what those solutions actually are*. These questions are not common, but they *do* occur on the SAT test.

Try using the Discriminant to answer the following practice question:

How many real solutions are there to the equation $5x^2 + 7x + 13 = 0$?

(A) 0

(B) 1

(C) 2

(D) 3

This question would be extremely difficult to factor by hand the way we did in the previous chapter on **Basic Algebra 2**. The numbers are simply too ugly. Furthermore, we're not even asked what the solutions *are*, just "how *many*" solutions the equation will have.

The equation is conveniently already in Quadratic Equation form of $0 = ax^2 + bx + c$. We can just read off our given equation that $a = 5$, $b = 7$, and $c = 13$. So, let's enter the correct values into the Discriminant:

$$\begin{aligned} &b^2 - 4ac \\ &= 7^2 - 4(5)(13) \\ &= 49 - 260 \\ &= -211 \end{aligned}$$

This Discriminant is a *negative* number, so we know the equation will have *no* real solutions. Our answer should be **Choice A**.

Pretest Question #2

Let's take a look at another Pretest question. Try it yourself before you look at my explanation below the question.

$$y = 3x^2 + 2x + 7$$
$$y = x^2 - 5x - 6$$

How many real solutions are there to the system of equations above?

(A) There are exactly 4 real solutions.

(B) There are exactly 2 real solutions.

(C) There is exactly 1 real solution.

(D) There are no real solutions.

This question is the perfect time to use the Discriminant. Note that the question gives us a pair of Quadratic Equations, but does *not* ask for the solution values - merely for the *number* of possible solutions. It's exactly what the Discriminant was made for!

First, though, we have to combine the two equations into one. Luckily, they're both set equal to y, so we can just set the two equations equal to each other:

$$3x^2+2x+7=x^2-5x-6$$

Now let's use **Basic Algebra 1** to move this into the Quadratic Equation form of $0=ax^2+bx+c$:

$$\begin{aligned}
&3x^2+2x+7=x^2-5x-6\\
&-x^2 \qquad\qquad\quad -x^2\\
&2x^2+2x+7=x-5x-6\\
&\qquad +5x+6 \qquad +5x+6\\
&2x^2+7x+13=0
\end{aligned}$$

Now, I'm pretty sure neither of us is going to be able to factor this equation by hand. It's time to use the Discriminant.

Remember, the Discriminant is calculated as b^2-4ac where a, b, and c come from the Quadratic Equation form $ax^2+bx+c=0$.

We can just read off from our equation that $a=2$, $b=7$, and $c=13$. Go ahead and plug those into the Discriminant formula:

$$\begin{aligned}
&b^2-4ac\\
&=(7)^2-4(2)(13)\\
&=49-104\\
&=-55
\end{aligned}$$

Our Discriminant is a *negative* number, which means there will be NO real solutions to this equation, which is **Choice D**. It's nice to save all that time we could have wasted, struggling to find the solutions, when they don't even exist!

Review & Encouragement

This lesson is relatively straightforward, since it's just teaching two additional techniques (the Quadratic Formula and the Discriminant) that can be used in a limited number of cases to solve certain Algebra 2 questions more effectively.

Make sure you commit the Quadratic Equation to memory, and that you understand *where* the Discriminant comes from and *why* it works - which is much easier than trying to memorize the Discriminant itself as an entirely separate concept.

Be on the lookout for the signals the SAT Math leaves to hint *when* we should use either the Quadratic Formula or the Discriminant. These techniques will save time if you use them at the right moment - but they will waste a lot of time if you use them on inappropriate questions that they weren't designed for.

Now complete the practice problems for this lesson to make sure you understand the Quadratic Formula and the Discriminant! You're leveling up your Algebra 2 skills with every completed question.

Quadratic Formula Practice Questions

DO NOT USE A CALCULATOR ON ANY OF THE FOLLOWING QUESTIONS UNLESS INDICATED.

1. Which of the following is a solution to the equation $8x = x^2 + 9$?

   (A) $4 - \sqrt{7}$

   (B) $4 + 2\sqrt{7}$

   (C) $8 + \sqrt{7}$

   (D) $8 - 2\sqrt{7}$

2. How many real solutions exist for the equation $x^2 - 4x + 12 = 0$?

   (A) There are infinite real solutions.

   (B) There are no real solutions.

   (C) There is exactly 1 real solution.

   (D) There are exactly 2 real solutions.

3. What is the sum of all values of x that satisfy the equation $2x^2 - 9x + 4 = 0$?

   (A) .5

   (B) 2.25

   (C) 3.5

   (D) 4.5

4. How many real solutions exist for the equation $2x^2 - 4x + 2 = 0$?

   (A) Exactly 4 real solutions.

   (B) Exactly 2 real solutions.

   (C) Exactly 1 real solution.

   (D) There are no real solutions.

$$h = -4.9t^2 + 46t$$

5. (CALCULATOR) FREE RESPONSE: The equation above gives the approximate height h, in meters of a ball t seconds after it is launched vertically upward from the ground with an initial velocity of 46 meters per second. After approximately how many seconds will the ball hit the ground? (Round to the nearest tenth of a second).

$$y = x^2 + 6x - 11$$
$$y = 4x + 2$$

6. How many real solutions exist for the system of equations above?

(A) There are no real solutions.

(B) Exactly 1 real solution.

(C) Exactly 2 real solutions.

(D) Exactly 4 real solutions.

$$y = 2x^2 - 7x + 1$$
$$y = x - 1$$

7. FREE RESPONSE: If a and b are solutions to the system of equations above, what is the value of $a + b$?

$$x^2 - \frac{n}{4}x = 3t$$

8. (CALCULATOR) In the quadratic equation above, n and t are constants. What are the solutions for x?

(A) $\frac{n}{8} \pm \frac{\sqrt{n^2 + 192t}}{8}$

(B) $\frac{n}{4} \pm \frac{\sqrt{n^2 + 192t}}{8}$

(C) $\frac{n}{8} \pm \frac{\sqrt{16n^2 + 12t}}{8}$

(D) $\frac{n}{4} \pm \frac{\sqrt{16n^2 + 12t}}{8}$

Quadratic Formula Answers

1. A
2. B
3. D
4. C
5. 9.4
6. C
7. 4
8. A

Quadratic Formula Explanations

1. **A.** We know by now that regardless of whether we factor or use the Quadratic Formula, we can't get anywhere with this problem until we first put it in the Quadratic Equation form $ax^2+bx+c=0$. So let's get that done first:

$$8x = x^2 + 9$$
$$-8x \quad -8x$$
$$0 = x^2 - 8x + 9$$

Now, can we factor this, using the techniques we learned in **Basic Algebra 2**? Well, *I* certainly can't find a pair of numbers that *multiply* to 9 and *add* to -8. It looks like we're going to need to bust out the Quadratic Formula.

$$X = \frac{-b \pm \sqrt{b^2-4ac}}{2a}$$

Our values are $a=1$, $b=-8$, and $c=9$. Go ahead and plug them in:

$$X = \frac{-(-8) \pm \sqrt{(-8)^2-4(1)(9)}}{2(1)}$$

Now clean up the equation:

$$X = \frac{8 \pm \sqrt{64-36}}{2}$$
$$= \frac{8 \pm \sqrt{28}}{2}$$

Now simplify $\sqrt{28}$ using the techniques we learned for simplifying roots in the lesson on **Exponents & Roots**:

$$\frac{8 \pm 2\sqrt{7}}{2}$$

And then simplify the fraction:

$$4 \pm \sqrt{7}$$

The question only asks for *one* of the two solutions, so skim through the options and find the choice that matches either $4+\sqrt{7}$ or $4-\sqrt{7}$. Choice A matches $4-\sqrt{7}$, so that's the one we want.

2. **B.** If you studied the section on Discriminants, you know that this question is the perfect time to use them. We aren't asked for the *actual* solutions to the equation - just *how many* solutions there are.

Remember, the Discriminant is the part of the Quadratic Formula underneath the square root sign, which is:

$$b^2 - 4ac$$

By comparing to the basic Quadratic Equation form of $ax^2+bx+c=0$, we can identify that $a=1$, $b=-4$, and $c=12$. Go ahead and plug those into the Discriminant formula:

$$(-4)^2 - 4(1)(12)$$
$$= 16 - 48$$
$$= -32$$

Our Discriminant is -32, a *negative* number, which means there are no real solutions to this equation, or Choice B.

3. **D.** So, this equation is already in the Quadratic Equation form $ax^2+bx+c=0$. You should typically try factoring first, because it saves time compared to using the whole Quadratic Formula workflow, but in this case, the Answer Choices themselves demonstrate that it's unlikely that factoring will work for us.

How can I tell? Well, all of the Answer Choices have decimals in them. But when we factor successfully, we're typically working only with whole numbers. So, you can try to factor if you want, but odds are we'll just have to use the Quadratic Formula.

By comparing to the basic Quadratic Equation form of $ax^2+bx+c=0$, we can identify that $a=2$, $b=-9$, and $c=4$. Go ahead and plug those into the Quadratic Formula:

$$X = \frac{-b \pm \sqrt{b^2-4ac}}{2a}$$
$$X = \frac{-(-9) \pm \sqrt{(-9)^2-4(2)(4)}}{2(2)}$$

Now that everything is plugged in its place, let's start cleaning up and simplifying:

$$X = \frac{9 \pm \sqrt{81-32}}{4}$$
$$X = \frac{9 \pm \sqrt{49}}{4}$$
$$X = \frac{9 \pm 7}{4}$$
$$X = \frac{9}{4} \pm \frac{7}{4}$$

So, we have our two solutions for x now:

$$X = \frac{9}{4} + \frac{7}{4} = \frac{16}{4} = 4$$
$$X = \frac{9}{4} - \frac{7}{4} = \frac{2}{4} = .5$$

The two solutions are "4" and "$.5$". Now let's finish the question, which asks for the *sum* of these two solutions. Of course, $4 + .5 = 4.5$, so we're finished with **Choice D**.

4. **C.** This is another perfect time to use the Discriminant, just as in Question 2. We aren't asked for any *actual* solutions, just *how many* solutions there are.

Remember, the Discriminant is the part of the Quadratic Formula underneath the square root sign, which is:

$$b^2 - 4ac$$

By comparing to the basic Quadratic Equation form of $ax^2 + bx + c = 0$, we can identify that $a = 2$, $b = -4$, and $c = 2$. Go ahead and plug those into the Discriminant formula:

$$(-4)^2 - 4(2)(2)$$
$$= 16 - 16$$
$$= 0$$

Our Discriminant is *zero*, which means there is exactly one real solution to this equation, or Choice C.

5. **9.4.** In this question we're asked when the ball will hit the ground. When a ball hits the ground, it will be at a height of 0 meters. So, plug in "0" for the height:

$$0 = -4.9t^2 + 46t$$

Now, if you reach straight for the Quadratic Formula just out of habit, you're making a big mistake! I threw this question into the chapter as a trick to see if you're awake. This equation can be factored by pulling out a t from the equation:

$$0 = t(-4.9t + 46)$$

Now we can find the two possible values for t that make the equation true, as we did in the lesson on **Basic Algebra 2**. If $t = 0$, the equation is true. But, at a time of 0 seconds, the ball hasn't been launched yet - it can't hit the ground before it's launched!

Instead, consider that $0 = -4.9t + 46$. This can be solved for a second possible value of t:

$$0 = -4.9t + 46$$
$$-46 \qquad\qquad -46$$
$$-46 = -4.9t$$
$$\frac{-46}{-4.9} = \frac{-4.9t}{-4.9}$$
$$9.3877... = t$$

Rounding to the nearest tenth of a second, as the question asks, we can give the time the ball hits the ground as **9.4 seconds**.

6. **C.** This is another perfect time to use the Discriminant, just as in Questions 2 and 4. We aren't asked for any *actual* solutions, just *how many* solutions there are.

Remember, the Discriminant is the part of the Quadratic Formula underneath the square root sign, which is:

$$b^2 - 4ac$$

But before we can compare our equations to the basic Quadratic Equation form of $0 = ax^2 + bx + c$,we have to combine the two equations (they're both equal to y, so they can be set equal to each other):

$$4x + 2 = x^2 + 6x - 11$$

Now arrange this into Quadratic Equation form:

$$4x+2=x^2+6x-11$$
$$-4x-2 \qquad -4x \ -2$$
$$0=x^2+2x-13$$

Remember, we're not *solving* this equation. We're just testing the number of real solutions using the Discriminant. We can identify that $a=1$, $b=2$, and $c=-11$. Go ahead and plug those into the Discriminant formula:

$$(2)^2-4(1)(-11)$$
$$=4-(-44)$$
$$=48$$

Our Discriminant is a *positive* number, which means there are exactly two real solutions to this equation, or Choice C.

7. **4.** This question has the apparent form and signs of a Quadratic Equation (for example, "two solutions" and an equation resembling ax^2+bx+c), but we also have a **System of Equations** that must be combined before we can apply either factoring or the Quadratic Formula.

Luckily, both equations are already equal to "y", so we can set the two equations equal to each other without any difficulty:

$$2x^2-7x+1=x-1$$

Now maneuver everything into Quadratic Equation form and get it equal to 0 on one side:

$$2x^2-7x+1=x-1$$
$$-x+1 \ -x+1$$
$$2x^2-8x+2=0$$

If you like, you can also divide both sides by "2" just to work with lower numbers, which would be advantageous if you don't have a calculator:

$$\frac{2x^2-8x+2}{2}=\frac{0}{2}$$
$$x^2-4x+1=0$$

Can you factor this by hand? I sure can't. There isn't a convenient pair of numbers that *multiplies* to "1" and *adds* to "4".

Instead, let's use the Quadratic Formula. By comparing to the basic Quadratic Equation form of $ax^2+bx+c=0$, we can identify that $a=1$, $b=-4$, and $c=1$:

$$X=\frac{-b\pm\sqrt{b^2-4ac}}{2a}$$
$$X=\frac{-(-4)\pm\sqrt{(-4)^2-4(1)(1)}}{2(1)}$$

Now simplify and clean up as much as possible:

$$X=\frac{4\pm\sqrt{16-4}}{2}$$
$$X=\frac{4\pm\sqrt{12}}{2}$$

The $\sqrt{12}$ can be simplified to $2\sqrt{3}$, as we learned in the lesson on **Exponents & Roots**:

$$X=\frac{4\pm2\sqrt{3}}{2}$$

Now simplify the fraction:

$$X=2\pm\sqrt{3}$$

Since a and b are the two solutions to the equation, we can surmise that $a=2+\sqrt{3}$ and $b=2-\sqrt{3}$. It doesn't really matter which is which, since we're going to add them together anyway.

To finish the question, give the sum of $a+b$:

$$(2+\sqrt{3})+(2-\sqrt{3})$$
$$=2+2+\sqrt{3}-\sqrt{3}$$
$$=4$$

Our final answer is nice and clean - just **4.**

8. **A.** This is by far the hardest question in this lesson, and probably the hardest Quadratic Formula-based question you'll ever see on the SAT. But it's not impossible with the right moves and mindset.

One of the most obnoxious thing is how they've worked in the unknown constants n and t in the place of the nice, simple numbers we've grown accustomed to. That's OK, we'll just treat these constants as part of the same methods we've been using up until now.

It's still a Quadratic Equation, so everything begins by putting the equation into $ax^2 + bx + c = 0$ form:

$$x^2 - \frac{n}{4}x = 3t$$
$$-3t \quad -3t$$
$$x^2 - \frac{n}{4}x - 3t = 0$$

Now, I guarantee we won't be able to factor this, so you might reach immediately for the Quadratic Formula. But, there's one thing we can do right now to make this whole thing *much* easier to deal with, and that's to multiply everything by "4" to get rid of that disgusting fraction we have in the middle of our equation.

Trust me on this: the Quadratic Formula is annoying enough *without* extra fractions. So let's get rid of any fractions early on:

$$(4)(x^2 - \frac{n}{4}x - 3t) = 0(4)$$
$$4x^2 - nx - 12t = 0$$

OK, *now* let's use the Quadratic Formula. By comparing to the basic Quadratic Equation form of $ax^2 + bx + c = 0$, we can identify that $a = 4$, $b = -n$, and $c = -12t$.

Note well: our b and c values have the unknown constants n and t in them. Don't freak out! Remember, "constants" are still *numbers* - they're just numbers we don't happen to know yet.

Set up the Quadratic Formula using the a, b, and c values that we've identified:

$$X = \frac{-b \pm \sqrt{b^2 - 4ac}}{2a}$$
$$X = \frac{-(-n) \pm \sqrt{(-n)^2 - 4(4)(-12t)}}{2(4)}$$

Pretty weird looking if you ask me, but let's keep our heads down and just keep working through it. Clean up and simplify as much as you can:

$$X = \frac{-(-n) \pm \sqrt{(-n)^2 - 4(4)(-12t)}}{2(4)}$$
$$X = \frac{n \pm \sqrt{n^2 + 192t}}{8}$$

Be sure to double-check your negative signs, which can easily get sloppy in work like this. Write neatly and show every step!

OK, we're almost done. The only difference between our answer and their choices is that they've separated the big fraction into two smaller fractions. That's easy enough - **Choice A** is the same as ours, just separated into two fractions.

This question is a great example of how a confident Quadratic Formula setup can cut through even the ugliest Quadratic Equations - regardless of having ugly fractions or unknown constants like n and t in our equation.

Hopefully by now you can also recognize the signals of the Answer Choices, which are major hints that we should use the Quadratic Formula to set this question up - even with unknown constants, it will still work just fine.

Again, I highly recommend getting rid of fractions *before* entering values into the Quadratic Formula, if possible!

If you understand this question and solution, you're probably more than ready to take on anything the SAT Math test throws at you on the topic of Quadratic Formula and Discriminants. Congratulations!

Lesson 18: Advanced Algebra 2

Percentages

- 2.2% of Whole Test
- 4.5% of No-Calculator Section
- 1.1% of Calculator Section

Prerequisites

- Basic Algebra 2
- The Quadratic Equation
- Basic Algebra 1

In this lesson we'll be looking at some harder Algebra 2 questions. This topic demonstrates some "unusual" variations on the content we saw in the previous lesson on **Basic Algebra 2**.

Many questions are still based around FOILing, Factoring, and the Quadratic Equation form $ax^2 + bx + c = 0$. But, some of these questions feature uniquely "SAT-style" twists on the basics.

Occasionally we'll encounter more difficult or unintuitive factoring situations. Some questions give us "higher-order" polynomials based on x^3 or x^4, instead of the x^2 we've grown accustomed to.

Other questions incorporate more words into the problem or involve confusing setups that may make us feel like we "just don't know where to start."

There aren't any quantum leaps or major new math techniques in this lesson; just some new angles on familiar concepts that lead a lot of my SAT Math students to feel completely stumped.

Advanced Algebra 2 Quick Reference

- This lesson doesn't add any brand-new skills. It just pushes our existing Algebra 2 abilities a bit further.
- Before studying this lesson, be sure you're confident on everything from **Basic Algebra 2** and **The Quadratic Equation**.
- When you're not sure what else to do with an Algebra 2 question, look aggressively for opportunities to factor.
- Don't give up too soon on these questions. One key insight may be all you need to crack a question wide open.

As I mentioned in the introduction to this lesson, we're not really learning any new skills in this lesson. We'll just use what we already know and learn about some interesting variations on the concepts we studied in **Basic Algebra 2**.

We won't use this lesson to learn any specific new "techniques" - it's more focused on gaining experience with some weirder Algebra 2 questions and familiarizing ourselves with certain tougher types of Algebra 2 questions the SAT will give us.

Pretest Question #1

Let's take a look at our first Pretest question on this topic. Try it yourself if you got it wrong the first time.

> FREE RESPONSE: If $x > 0$, what is one possible solution to the equation $x^5 - 5x^3 + 4x = 0$?

This question is all about factoring, which we studied in the lesson on **Basic Algebra 2**. Many harder Algebra 2 questions require some insightful factoring to make progress on them.

We can factor out an x from everything on the left-hand side:

$$x(x^4 - 5x^2 + 4) = 0$$

Then we can factor the portion in parentheses:

$$x(x^2 - 4)(x^2 - 1) = 0$$

Each of the parentheses are a Difference of Squares and can be factored further:

$$x(x+2)(x-2)(x+1)(x-1) = 0$$

So, what values of x would make this equation true? There are five possible values for x that would cause everything to multiply to 0:

$$x = 0, -2, 2, -1, 1$$

The question only wants solutions that are greater than 0, which limits us to only $x = 1$ or $x = 2$. Either of these answers is acceptable.

We haven't used anything new for this question. It's just that the factoring we studied in **Basic Algebra 2** has gotten a little tougher.

Pretest Question #2

Let's take a look at another Pretest question. Try it yourself before you look at my explanation.

$$(ax+1)(2x^2+bx-3)=4x^3+4x^2-5x-3$$

FREE RESPONSE: The equation above is true for all x, where a and b are constants. What is the value of $a+b$?

The first step of solving this problem should be to distribute out all of the multiplication on the left-hand side of the equation. Be sure not to miss anything!

$$\begin{aligned}&(ax+1)(2x^2+bx-3)\\&=2ax^3+abx^2-3ax+2x^2+bx-3\\&=2ax^3+abx^2+2x^2-3ax+bx-3\end{aligned}$$

After distributing, I've also rearranged to group my terms in a neat, orderly way - so that x^2s are next to each other, x's are next to each other, and so forth.

Since the equation is already balanced, we know the left side and the right side are equal.

$$2ax^3+abx^2+2x^2-3ax+bx-3=4x^3+4x^2-5x-3$$

If we add $+3$ to both sides, we can simplify slightly by removing the "-3" on both sides:

$$2ax^3+abx^2+2x^2-3ax+bx=4x^3+4x^2-5x$$

Now, compare our highest-order terms - the x^3 terms on the left and right sides. Since the equations are equal, it's reasonable to assume that the x^3 terms on the left and right sides are equal to each other:

$$2ax^3=4x^3$$

And then we can solve for a:

$$\begin{aligned}\frac{2ax^3}{2}&=\frac{4x^3}{2}\\ax^3&=2x^3\\\frac{ax^3}{x^3}&=\frac{2x^3}{x^3}\\a&=2\end{aligned}$$

Now that we know the value of a is 2, we can plug it back in to our simplified version of the original equation:

$$\begin{aligned}2ax^3+abx^2+2x^2-3ax+bx&=4x^3+4x^2-5x\\2(2)x^3+(2)bx^2+2x^2-3(2)x+bx&=4x^3+4x^2-5x\end{aligned}$$

And clean that up as much as we can:

$$4x^3+2bx^2+2x^2-6x+bx=4x^3+4x^2-5x$$

We can also subtract the $4x^3$ term from both sides, just to keep simplifying:

$$2bx^2 + 2x^2 - 6x + bx = 4x^2 - 5x$$

OK, now we could compare the x^2 terms on both sides. Again, since the original equation is balanced, we can assume that the x^2 terms on the left and right sides of the equation will be equal to each other:

$$2bx^2 + 2x^2 = 4x^2$$

Again, since the equations are equal, it's reasonable to assume that the x^2 terms are equal on the left and right, just as we assumed the x^3 terms were balanced. We can now proceed to solve for b:

$$\begin{gathered} 2bx^2 + 2x^2 = 4x^2 \\ -2x^2 \quad -2x^2 \\ 2bx^2 = 2x^2 \\ \frac{2bx^2}{2} = \frac{2x^2}{2} \\ bx^2 = x^2 \\ \frac{bx^2}{x^2} = \frac{x^2}{x^2} \\ b = 1 \end{gathered}$$

And now we have a value for b. Finish the question with the value of $a+b$:

$$a+b = (2)+(1) = 3$$

Our final answer should be **3.**

In this question, the toughest breakthrough is the idea of comparing the *terms* of both sides - for example, the x^3 terms on the left to the x^3 terms on the right, then the x^2 terms on the left to the x^2 terms on the right. We have to think a bit more creatively than "just set equal to 0 and factor" or "just use the **Quadratic Formula**."

Review & Encouragement

This lesson doesn't add anything particularly new to our math toolkit - it just stretches our previous skills a little further.

Keep looking for opportunities to factor. Above all else, strong factoring skills seem to be the most important technique when it comes to more advanced Algebra 2 questions.

Also look for opportunities to compare similar terms across an equal sign - for example, x^3 terms on the left to x^3 terms on the right, as we did in Pretest Question 2.

Now, be sure to work through all of the following practice problems. Each one has something interesting to teach you about Advanced Algebra 2 questions on the SAT.

Advanced Algebra 2 Practice Questions

DO NOT USE A CALCULATOR ON ANY OF THE FOLLOWING QUESTIONS.

$$x-2=\frac{8}{x-2}$$

1. In the equation above, which of the following is a possible value of $x-2$?

(A) $2\sqrt{2}-2$

(B) 2

(C) $2\sqrt{2}$

(D) $2+2\sqrt{2}$

2. In the xy-plane, the graph of the function $f(x)=x^2+2x-35$ has two x-intercepts. What is the distance between the x-intercepts?

(A) 2

(B) 9

(C) 12

(D) 35

3. In the xy-plane, the graph of function f has x-intercepts at -6, -2, and 2. Which of the following could define f?

(A) $f(x)=(x-6)(x-2)(x+2)$

(B) $f(x)=(x+6)(x-2)(x+2)$

(C) $f(x)=(x-2)^2(x+6)$

(D) $f(x)=(x+2)^2(x-6)$

$$(ax-3)(6x+3)+9-2x^2$$

4. In the expression above, a is a constant. If the expression is equivalent to bx, where b is a constant, what is the value of b?

(A) -17

(B) -12

(C) -9

(D) -6

5. FREE RESPONSE: In the xy-plane, the graph of $y=4x^2-17x$ intersects the graph of $y=-x$ at the points $(0,0)$ and (b,b). What is the value of b?

$$x^6 - 13x^4 = -36x^2$$

6. FREE RESPONSE: If $x > 0$, what is one possible solution to the equation above?

$$3x^2 - 6x = n$$

7. In the equation above, n is a constant. If the equation has no real solutions, which of the following could be the value of n?

(A) -4

(B) -3

(C) 1

(D) 3

$$f(x) = \frac{1}{(x-3)^2 + 5(2x-4) + 6}$$

8. FREE RESPONSE: If $x > 0$ in the function f above, for what value of x is the function undefined?

$$x^3 - 4x^2 - 2x + 8 = 0$$

9. FREE RESPONSE: For what integer value of x is the equation above true?

$$(ax+2)(6x^2 - bx + 5) = 18x^3 - 3x^2 + 5x + 10$$

10. FREE RESPONSE: The equation above is true for all x, where a and b are constants. What is the value of ab?

Advanced Algebra 2 Answers

1. C
2. C
3. B
4. A
5. 4
6. 2 or 3
7. A
8. 1
9. 4
10. 15

Advanced Algebra 2 Explanations

1. **C.** The first thing we should do is multiply both sides by $x-2$, which gets rid of the fraction and puts all our x terms on the same side:

$$(x-2)(x-2)=\frac{8}{x-2}(x-2)$$
$$(x-2)(x-2)=8$$

Here's where the biggest trick comes in. Notice the left side of the equation and remember the question is asking, not for x, but for the value of $x-2$. Instead of FOILing, it's more efficient to reduce $(x-2)(x-2)$ into $(x-2)^2$:

$$(x-2)^2=8$$

Now we can square root both sides:

$$\sqrt{(x-2)^2}=\sqrt{8}$$
$$(x-2)=\sqrt{8}$$

The $\sqrt{8}$ can be simplified to $2\sqrt{2}$ using what we learned in the lesson on **Exponents & Roots.**

$$(x-2)=2\sqrt{2}$$

And we're done! The answer must be Choice C.

Note that we *can* solve this question using a combination of FOILing, setting equal to the Quadratic Equation form $ax^2+bx+c=0$, then using the **Quadratic Formula**, but it's *much* more efficient to use the method above. The question was *designed* to be solved this way. You can go the slow & difficult way, or you can focus on finding $x-2$ rather than going on autopilot and solving for x like we usually would.

2. **C.** Remember from the previous lesson on **Basic Algebra 2** that "x-intercepts" is just another word for "solutions," "roots," or "zeros." In other words, we just have to set this equation equal to "0" and solve:

$$0=x^2+2x-35$$

We should be able to solve this by factoring:

$$0=(x+7)(x-5)$$

And we see that our solutions are $x=-7$ and $x=5$.

The question asks for the *distance* between these two coordinates. As we learned in **Absolute Value**, *distance* is a subtraction problem with an absolute value applied to the result (since distance can't be negative):

$$|(-7)-(5)|$$
$$=|-7-5|$$
$$=|-12|$$
$$=12$$

Our final answer is 12, or Choice C.

3. **B.** As in Question 2 and the lesson on **Basic Algebra 2**, remember that "x-intercepts" is just a different name for "solutions," "roots," or "zeros." We should be able to plug the given *x*-intercepts from the question into x for one of the equations in the answer choices and find that *each* of the given intercepts provides a final value of "0".

If we start by testing the given value "-6" and plugging it in for x into the four answer choices, we find that Choices A and D do *not* provide "0" as a result. That eliminates those two choices.

Now let's test plugging in the next given *x*-intercept, "-2", for x. It gives "0" as a result for Choice B, but not for Choice C. So, we should eliminate Choice C.

Just to be sure the correct final answer is actually Choice B, let's test the last *x*-intercept of "2" given in the question. When plugged into x for Choice B, we get "0" again, just as we hoped. The correct answer must be Choice B.

In other words, Choice B is correct because it's the only answer where each of the three given *x*-intercepts of -6, -2, and 2 can be plugged into the function to get a "zero" or *x*-intercept.

If you're getting good at finding solutions from factors, this question can be even easier, without the need to even test the answers. You know that a zero of $x=-6$, for example, means that one of the factors must be $(x+6)$, and that a zero of $x=2$ means another factor must be $(x-2)$, and so forth. So you can just skim through the answers and quickly eliminate any that don't have the correct factors.

4. **A.** The first step in this question is to FOIL out the parentheses on the left side, as we learned in the previous lesson on **Basic Algebra 2**. Then clean up:

$$(ax-3)(6x+3)+9-2x^2$$
$$=6ax^2+3ax-18x-9+9-2x^2$$
$$=6ax^2-2x^2+3ax-18x$$

Now set this equal to bx, since the question gives it as "equivalent" to our original expression:

$$6ax^2-2x^2+3ax-18x=bx$$

OK, so here's the interesting insight for this question. The right side of the equation doesn't have *any* x^2 terms. But, we know the two sides are *equal*. Therefore, it's reasonable to assume that all of the x^2 terms on the left side of the equation cancel each other out to "0". Otherwise, how could there be no x^2s on the right side?

If the x^2 terms on the left side cancel each other out, then $6ax^2-2x^2$ must equal 0, and we can solve for a:

$$6ax^2-2x^2=0$$
$$\frac{6ax^2-2x^2}{x^2}=\frac{0}{x^2}$$
$$6a-2=0$$
$$+2\ +2$$
$$6a=2$$
$$\frac{6a}{6}=\frac{2}{6}$$
$$a=\frac{1}{3}$$

OK, we've found the value of $a=\frac{1}{3}$ and that's progress. But the question asked for the value of b, not a. We need to plug back in $a=\frac{1}{3}$ and see what happens. I'll return to the cleaned-up version of the equation we were using earlier:

$$6ax^2-2x^2+3ax-18x=bx$$
$$6(\frac{1}{3})x^2-2x^2+3(\frac{1}{3})x-18x=bx$$
$$2x^2-2x^2+x-18x=bx$$

We can keep cleaning this up by combining like terms, which gives an equation we can solve for b:

$$-17x=bx$$
$$\frac{-17x}{x}=\frac{bx}{x}$$
$$-17=b$$

And there we go: $b=-17$, which is Choice A.

5. **4.** We learned in **Basic Algebra 1** that the *intersection point* of two graphs is the point where both equations are *equal*. In other words, we can set equations equal to each other, then solve the Algebra, to find the point where they intersect.

So, let's set the two equations equal to each other.

$$-x=4x^2-17x$$

Now set equal to 0, as we usually do with Quadratic-based questions:

$$-x=4x^2-17x$$
$$+x\qquad\quad +x$$
$$0=4x^2-16x$$

And factor:

$$0=4x(x-4)$$

Our two solutions to this factored equation would be $x=0$ (which the question already gave us) and $x=4$, which is more interesting.

It *seems* that our solution is $x=4$ (and therefore $b=4$ because b is an *x*-coordinate in the question), but it would be good to check. Let's plug $x=4$ back into the intersection equation to make sure everything works out.

$$-x=4x^2-17x$$
$$-(4)=4(4)^2-17(4)$$
$$-4=4(16)-68$$
$$-4=64-68$$
$$-4=-4$$

OK, looking good! Note that this final stage is simply checking that our answer was correct. It's not strictly necessary, but if you have an extra minute on SAT test day, it's always a good idea to test your answers when possible.

6. **2** or **3.** The trick to this question is just to treat it like the $ax^2 + bx + c = 0$ we're familiar with - mostly ignoring the fact that it has higher-order exponents - and then just keep factoring.

First, we have to set everything equal to 0:

$$x^6 - 13x^4 = -36x^2$$
$$+36x^2 \quad +36x^2$$
$$x^6 - 13x^4 + 36x^2 = 0$$

Now factor out an x^2 from everything:

$$x^2(x^4 - 13x^2 + 36) = 0$$

This equation above can be factored further. We'll just have to use x^2 in our factors, instead of the x we're more accustomed to:

$$x^2(x^4 - 13x^2 + 36) = 0$$
$$x^2(x^2 - 4)(x^2 - 9) = 0$$

You can check for yourself that this works, if you want, by FOILing out the factored version. You'll return to the original equation.

OK, now notice that both of the new factors are also a Difference of Squares (as we studied in **Basic Algebra 2**). They can each be factored even further:

$$x^2(x^2 - 4)(x^2 - 9) = 0$$
$$x^2(x-2)(x+2)(x-3)(x+3) = 0$$

And now we can read off our solutions for x: it could be $x = 0$, $x = 2$, $x = -2$, $x = 3$, or $x = -3$. The question requires that $x > 0$, so only our positive solutions are included.

However, we still have two options for x, and either $x = 2$ or $x = 3$ are valid answers.

7. **A.** This is the perfect time to use the Discriminant, which we learned about in the previous lesson on **The Quadratic Formula.**

Remember that the Discriminant is the portion of the Quadratic Formula that's underneath the square root, $b^2 - 4ac$, where the values for a, b, and c come from the Quadratic Equation $ax^2 + bx + c = 0$. The Discriminant can tell us if a Quadratic Equation will have *two* solutions (if the Discriminant is a positive number), *one* solution (if the Discriminant is zero), and *no* solutions (if the Discriminant is a negative number).

For this question, we're looking for a situation in which the equation has *no* real solutions, which means the Discriminant must be a *negative* number, or "less than 0".

First, we need to put our starting equation in $ax^2 + bx + c = 0$ form:

$$3x^2 - 6x = n$$
$$-n \quad -n$$
$$3x^2 - 6x - n = 0$$

Now, set up the Discriminant with the values $a = 3$, $b = -6$, and $c = -n$. Plug these into the Discriminant formula and clean it up:

$$b^2 - 4ac$$
$$= (-6)^2 - 4(3)(-n)$$
$$= 36 + 12n$$

Remember that to have *no* real solutions, this Discriminant must be a *negative* number, or *less than 0*:

$$36 + 12n < 0$$

Now use **Basic Algebra 1** to solve the inequality:

$$36 + 12n < 0$$
$$-36 \qquad -36$$
$$12n < -36$$
$$\frac{12n}{12} < \frac{-36}{12}$$
$$n < -3$$

This shows that if we want a *negative* Discriminant, meaning that the original equation has *no solutions*, then n must be less than "-3." Our only possible answer is Choice A, since -4 is less than -3.

8. **1.** We know that any fraction that "divides by 0" will give an "undefined" result. Therefore, the bottom of our fraction should be set equal to 0:

$$(x-3)^2+5(2x-4)+6=0$$

Now FOIL, distribute, and just clean up in general:

$$\begin{aligned}(x-3)^2+5(2x-4)+6&=0\\ x^2-6x+9+10x-20+6&=0\\ x^2+4x-5&=0\end{aligned}$$

This equation is already in $ax^2+bx+c=0$ Quadratic Equation form, and can also be easily factored. Let's do that:

$$\begin{aligned}x^2+4x-5&=0\\ (x+5)(x-1)&=0\end{aligned}$$

And our solutions for x would be $x=-5$ and $x=1$. The question says $x>0$, so our answer should be $x=1$.

9. **4.** I'll be totally honest and admit that even *I* get stuck on this type of question when I see it on the SAT. I've also seen a lot of smart students play around with it for a while and still have no clue what to do with it.

The key lies in factoring - but it's not the sort of "obvious" factoring that we're probably good at by now.

Look at my steps below. Pay attention to how I find a factor of $(x-4)$ in two parts of the equation.

$$\begin{aligned}x^3-4x^2-2x+8&=0\\ x^2(x-4)-2x+8&=0\\ x^2(x-4)-2(x-4)&=0\end{aligned}$$

Notice that we've been able to pull out a factor of $(x-4)$ in two parts of the equation. If you didn't see this, don't feel alone. I wrote the dang question and I *still* have a hard time figuring this one out.

Now we can rewrite the equation by factoring out the $(x-4)$:

$$\begin{aligned}x^2(x-4)-2(x-4)&=0\\ (x-4)(x^2-2)&=0\end{aligned}$$

If you check this by FOILing this back out, you'll return to the original equation.

Now we can find the zeros or "solutions" for this equation, one factor at a time:

$$\begin{aligned}x-4&=0\\ +4\ &+4\\ x&=4\end{aligned}$$

$$\begin{aligned}x^2-2&=0\\ +2\ &+2\\ x^2&=2\\ \sqrt{x^2}&=\sqrt{2}\\ x&=\pm\sqrt{2}\end{aligned}$$

The question asked us for an "integer value" of x, which means only *whole* numbers. Our only integer solution to x is $x=4$, which is the correct answer.

10. **15.** This question is like an advanced version of Question 4.

First, let's distribute out the parentheses on the left side of the equation. First, multiply everything by ax and then multiply everything by 2. Don't miss anything!

$$(ax+2)(6x^2-bx+5)=18x^3-3x^2+5x+10$$
$$6ax^3-abx^2+5ax+12x^2-2bx+10=18x^3-3x^2+5x+10$$

Now clean it up and organize it by powers of x. Notice that we can also subtract a 10 from both sides.

$$6ax^3+12x^2-abx^2+5ax-2bx=18x^3-3x^2+5x$$

Now, like we did in Question 4, compare the coefficients of the x^3 terms on both sides. The equations are balanced, so these two x^3 terms must be equal to each other:

$$6ax^3=18x^3$$

We can solve this for the value of a:

$$\frac{6ax^3}{6}=\frac{18x^3}{6}$$
$$ax^3=3x^3$$
$$\frac{ax^3}{x^3}=\frac{3x^3}{x^3}$$
$$a=3$$

OK, that's a good sign. We've found that the value of *a* is "3."

We can use $a=3$ by plugging it back into our original equation:

$$6ax^3+12x^2-abx^2+5ax-2bx=18x^3-3x^2+5x$$
$$6(3)x^3+12x^2-(3)bx^2+5(3)x-2bx=18x^3-3x^2+5x$$

Now clean up as much as possible:

$$6(3)x^3+12x^2-(3)bx^2+5(3)x-2bx=18x^3-3x^2+5x$$
$$18x^3+12x^2-3bx^2+15x-2bx=18x^3-3x^2+5x$$

OK, just like we looked at the coefficients of the x^3 terms on both sides before, now we're going to look at the coefficients of the x^2 terms on both sides, and set them equal to each other:

$$12x^2-3bx^2=-3x^2$$

Notice I've removed everything from the equation that's not an x^2 term. I should be able to solve for b from here:

$$12x^2 - 3bx^2 = -3x^2$$
$$-12x^2 \qquad\qquad -12x^2$$
$$-3bx^2 = -15x^2$$
$$\frac{-3bx^2}{-3} = \frac{-15x^2}{-3}$$
$$bx^2 = 5x^2$$
$$\frac{bx^2}{x^2} = \frac{5x^2}{x^2}$$
$$b = 5$$

Congratulations! We've found the value of a and b. Now finalize the question with the value of ab:

$$ab = (3)(5) = 15$$

And, our final answer is 15 - done.

This lesson really tested the limits of our factoring and FOILing skills. Be ready for more on the day of the SAT test!

Lesson 19: Algebra 2 Parabolas

Percentages

- 3.3% of Whole Test
- 3% of No-Calculator Section
- 3.4% of Calculator Section

Prerequisites

- Basic Algebra 2
- The Quadratic Equation
- Basic Algebra 1
- Linear Equations (Recommended)
- Absolute Value (Recommended)

In this lesson we'll cover the concept of Parabolas - the essential graphs of Algebra 2. Just as a graph of a straight line - or a **Linear Equation** - was the "defining" graph of Algebra 1, so we can think of the parabola as the defining graph of Algebra 2.

A parabola is a graph of a Quadratic Equation $y = ax^2 + bx + c$, just as a straight line was a graph of a Linear Equation $y = mx + b$. Instead of the straight lines we've studied before, in this lesson we'll be dealing with curves - specifically, the "U-shaped" curves of Parabolas.

This lesson builds heavily upon the concepts we explored in the previous lesson on **Basic Algebra 2**. Make sure you understand everything in that lesson before you approach this one.

Algebra 2 Parabolas Quick Reference

- A parabola is a graph of a Quadratic Equation $y = ax^2 + bx + c$. Parabolas are the defining graph of Algebra 2, just as Linear Equations are the defining graph of Algebra 1.
- Parabolas have a characteristic "U-shaped" visual appearance on a graph.
- The coefficient of x^2 in $y = ax^2 + bx + c$ determines the steepness of the parabola and whether it opens upwards (positive) or downwards (negative).
- The *x*-intercepts of the parabola are the "zeros" or solutions to the equation $0 = ax^2 + bx + c$.
- The vertex is the highest or lowest point of the parabola. It is always at the center of the parabola.
- The *x*-midpoint at the center of the parabola is also be the *x*-coordinate of the vertex point - find this x value and plug it into the parabola's equation to get the *y*-coordinate of the vertex.
- "Vertex Form" and "Parabola Transformations" have limited utility on the SAT, and I consider them unworthy of in-depth study for this test - but we will take a quick look at them, just in case.
- Watch out for "Identified as constants or coefficients" questions and recognize their traps.

Introduction to Parabolas

So, what is a Parabola? A parabola is a graph of a Quadratic Equation $y = ax^2 + bx + c$. This is an equation we studied extensively in the lesson on **Basic Algebra 2**

The distinctive symmetrical U-shape of a parabola, seen below, is the classic graph of Algebra 2 and of the Quadratic Equation $y = ax^2 + bx + c$.

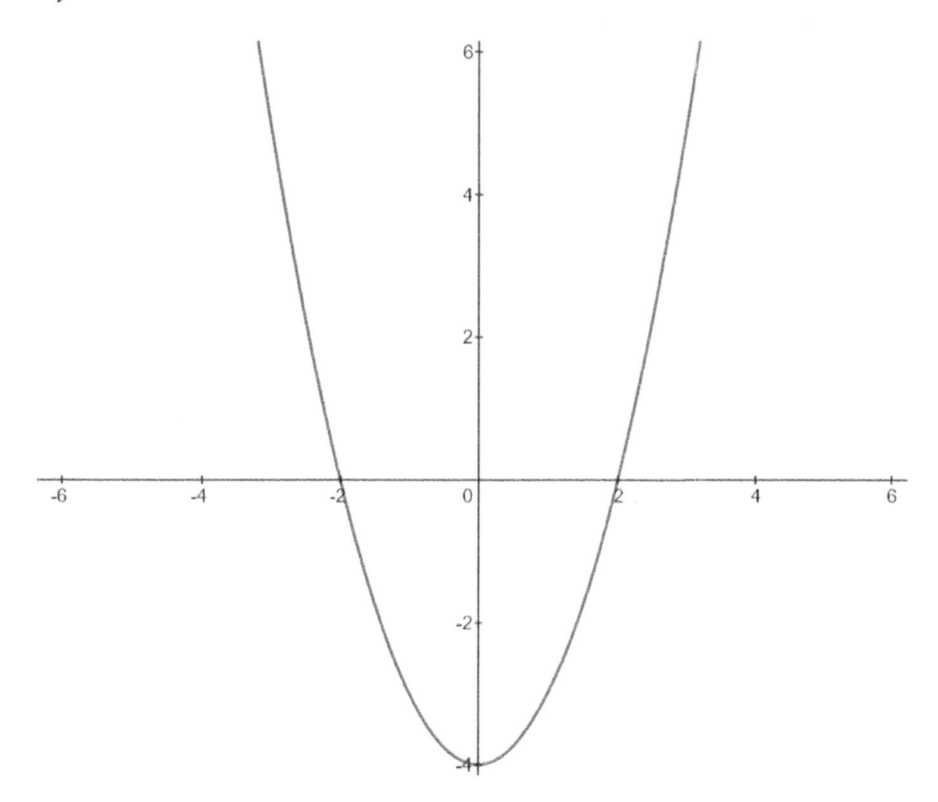

As we've encountered in **Basic Algebra 2**, the most distinguishing feature of the Quadratic Equation is its x^2 term.

Now, I want you to think back to the basic **Linear Equation** form of $y = mx + b$. In that equation, the value of m (the coefficient of x) told us a lot about the graph of the line. It gave us the slope, and whether the line rose (with a positive slope) or fell (with a negative slope).

In a parabola, the coefficient of x^2 is equally important. It tells us whether the shape of the parabola opens *upwards* (with a positive coefficient) or *downwards* (with a negative coefficient).

In the graph above, do you think the coefficient of x^2 is positive or negative?

That's right - it's positive. We know because the parabola opens upwards. It makes a weird sort of sense - a positive x^2 parabola will open upwards.

In the graph below, the solid-line graph of $y = x^2$ has a positive coefficient, opening upwards. The dotted line graph of $y = -x^2$ has a negative coefficient, opening downwards.

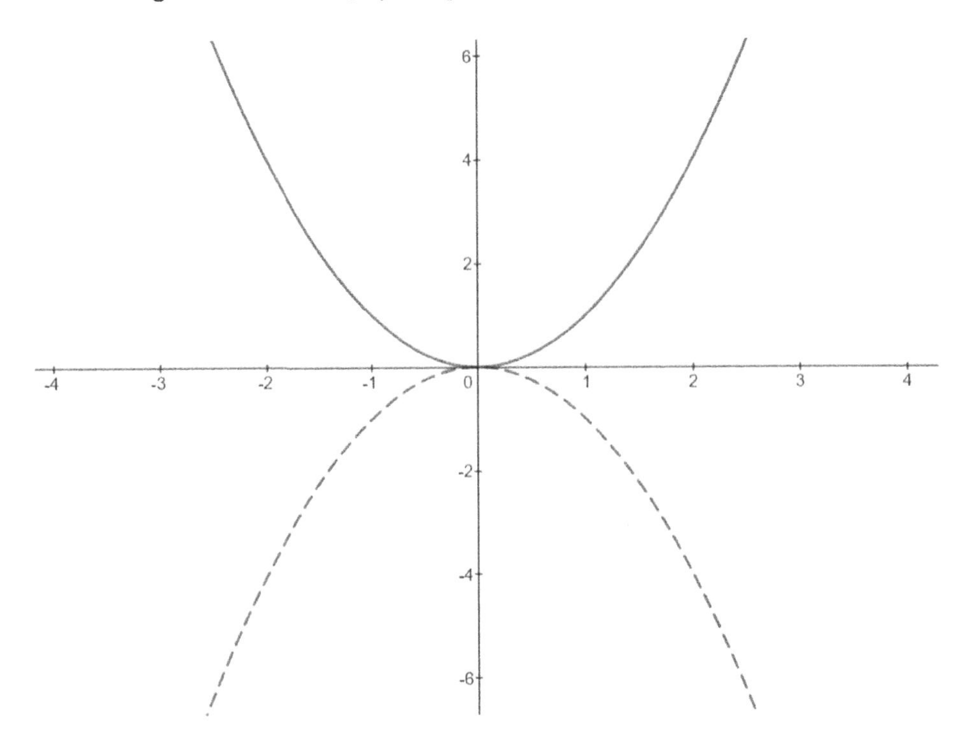

Get it? A positive coefficient parabola (x^2) opens upwards. A negative coefficient ($-x^2$) opens downwards.

The coefficient of the x^2 also tells us how wide or narrow the parabola will be. Higher coefficients create steeper and narrower parabolas. Lower coefficients are more shallow and wider. In the graph below, the solid line shows $y = 4x^2$ and the dotted line shows $y = \frac{1}{4}x^2$. Notice how the graph of the higher coefficient is steeper and narrower, and the graph of the lower coefficient is wider and shallower.

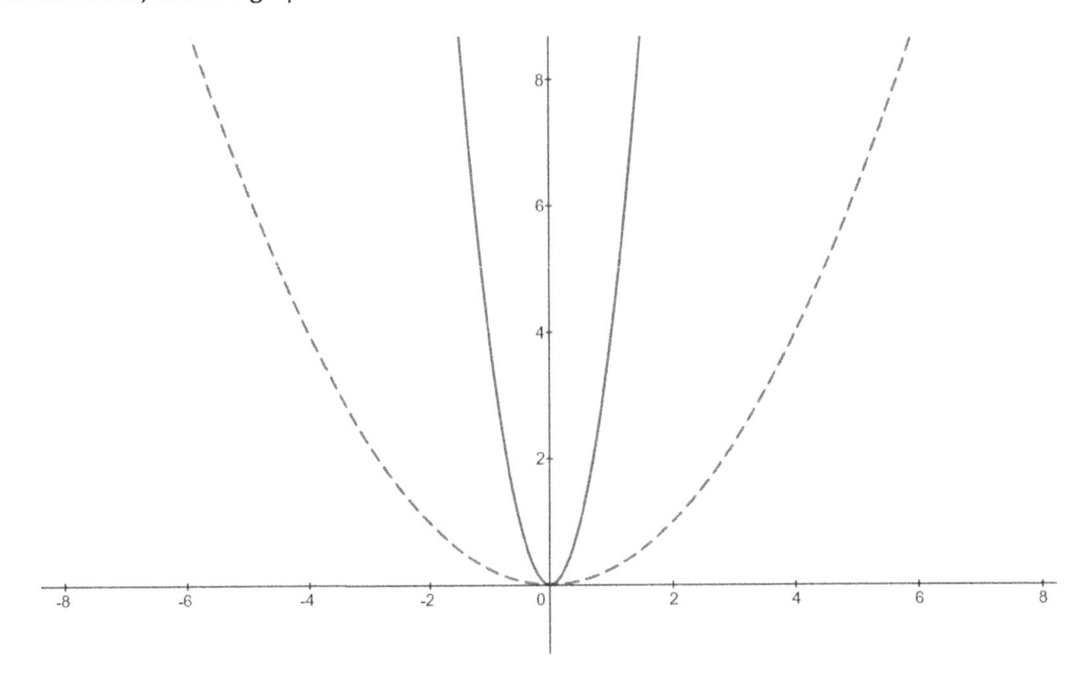

In some ways, we can imagine the coefficient of x^2 holding a similar relationship to a parabola as the slope coefficient (m) of x held to a linear equation in $y = mx + b$. Both coefficients relate to the steepness and to the overall direction of their graphs. This is an imperfect comparison, but a useful one nevertheless.

Finding the Intercepts

When we studied **Linear Equations**, we put a lot of emphasis on the *y*-intercept. We often used the *y*-intercept as a reference point or starting point for sketching the line.

Just like a Linear Equation, a Parabola has some key reference points. However, instead of using the *y*-intercept, we usually use the *x*-intercepts.

Think about the Quadratic Equation $y = ax^2 + bx + c$. In previous lessons, especially **Basic Algebra 2**, we frequently plugged in "0" for y and solved for the "zeros", which can also be called "roots," "solutions," or - get this! - the "*x*-intercepts."

In other words, you already *know* how to find these valuable *x*-intercepts of a parabola. Just take the Quadratic Equation, set it equal to 0, and factor it to find the two solutions to x. If factoring the equation by hand is impossible, you can always fall back on the **Quadratic Equation**.

Notice how easy it is to apply your previous knowledge. Parabolas are *not* new to us; they're just a *picture* of the equations we've been working with in Algebra 2. Parabolas are simply a visual representation - a *graph* - of the equation form we've already become familiar with.

For example, below is the graph of $y = x^2 - 4$. We could factor this equation as $y = (x+2)(x-2)$.

By setting the y value to 0, we would get the equation $0 = (x+2)(x-2)$. The two solutions to this equation are $x = -2$ and $x = 2$.

Notice that the graph also has its *x*-intercepts at $(-2,0)$ and $(2,0)$.

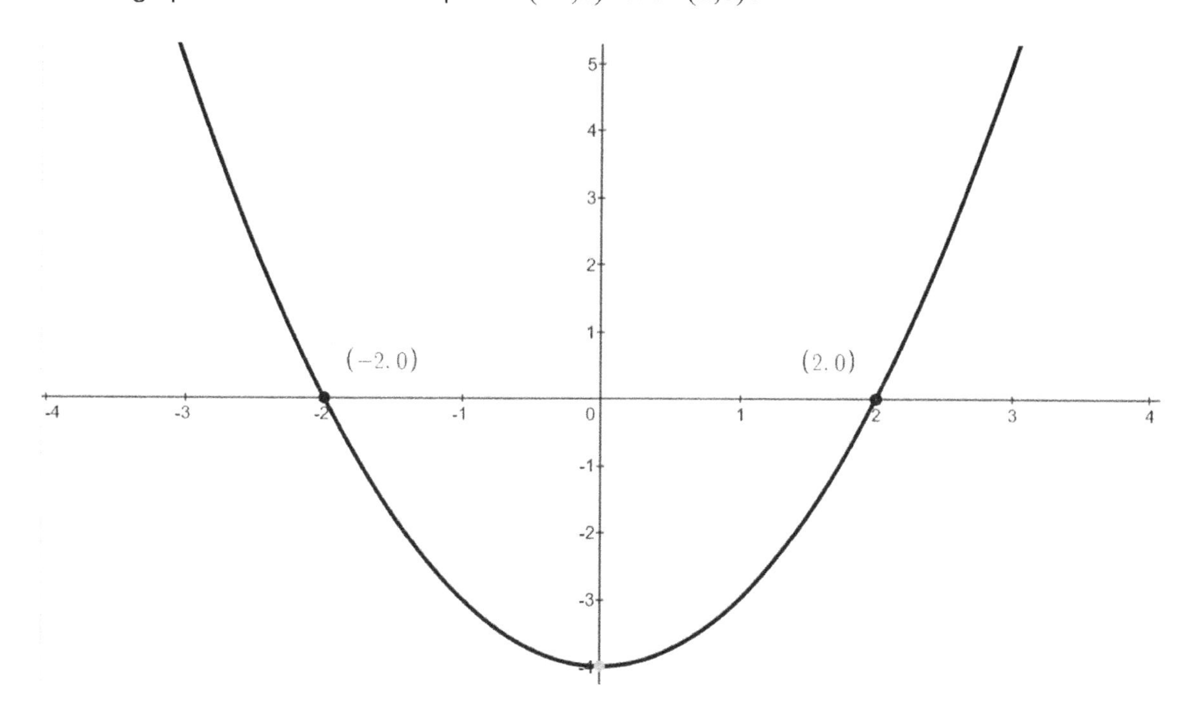

Calculating *x*-intercepts this way is an important and common technique when working with parabolas.

Although we're usually more interested in the *x*-intercepts of a parabola, it's also easy to solve for the *y*-intercept of a parabola. Just plug in "0" for x into your Quadratic Equation, and you'll find the *y*-intercept. This makes sense, because *y*-intercepts are always found where the *x*-coordinate is 0.

For example, we could easily find the *y*-intercept of the parabola defined by $y = x^2 - 4x + 3$ by simply plugging in "0" for x.

If we do so, we'd get:

$$y = (0)^2 - 4(0) + 3$$
$$y = 3$$

And indeed, as we look at the graph below, the parabola's *y*-intercept is at $(0,3)$.

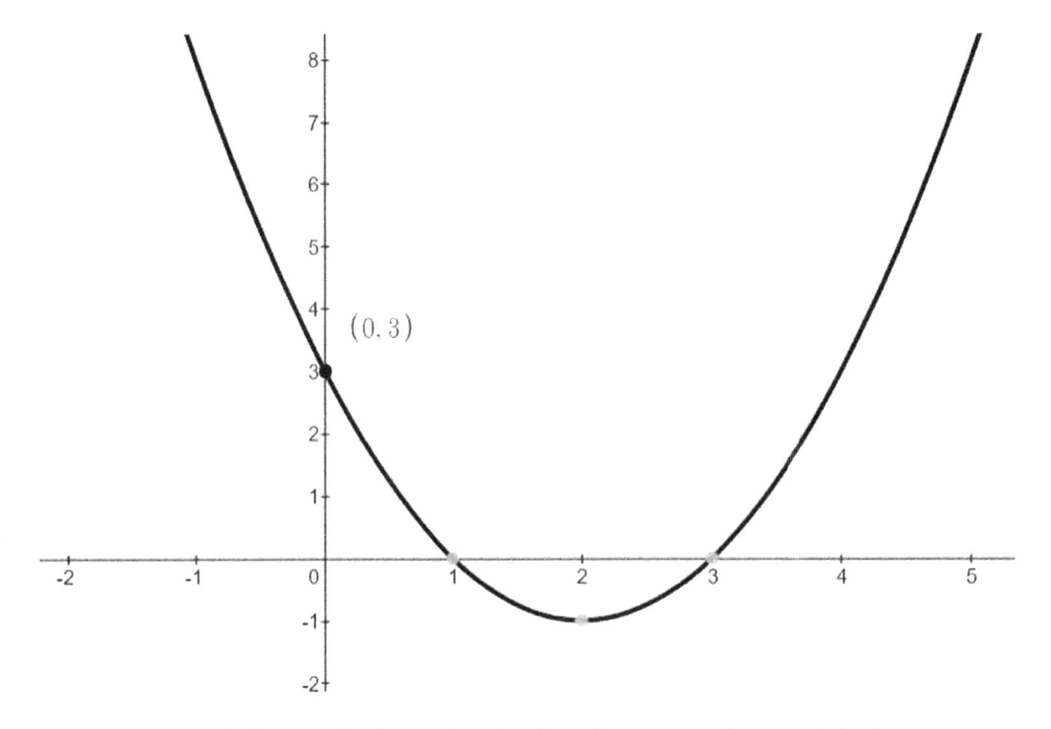

Although there *are* times we will be required to solve for the *y*-intercept of a parabola, it is more common to solve for the *x*-intercepts. That's just in the nature of this topic.

The Vertex

This brings us to our next major feature of parabolas - the *Vertex*.

In simplest terms, the *vertex* of a parabola is its highest or lowest point.

If the parabola is positive (opens upwards,) it will have a *lowest* point or *minimum*. If the parabola is negative (opens downwards), it will have a *highest* point or *maximum*. Either way, this minimum or maximum point is the "vertex" of the parabola.

In the graph below, I've marked the vertex at coordinate $(0,2)$. Notice that it is the very *lowest* point on this parabola.

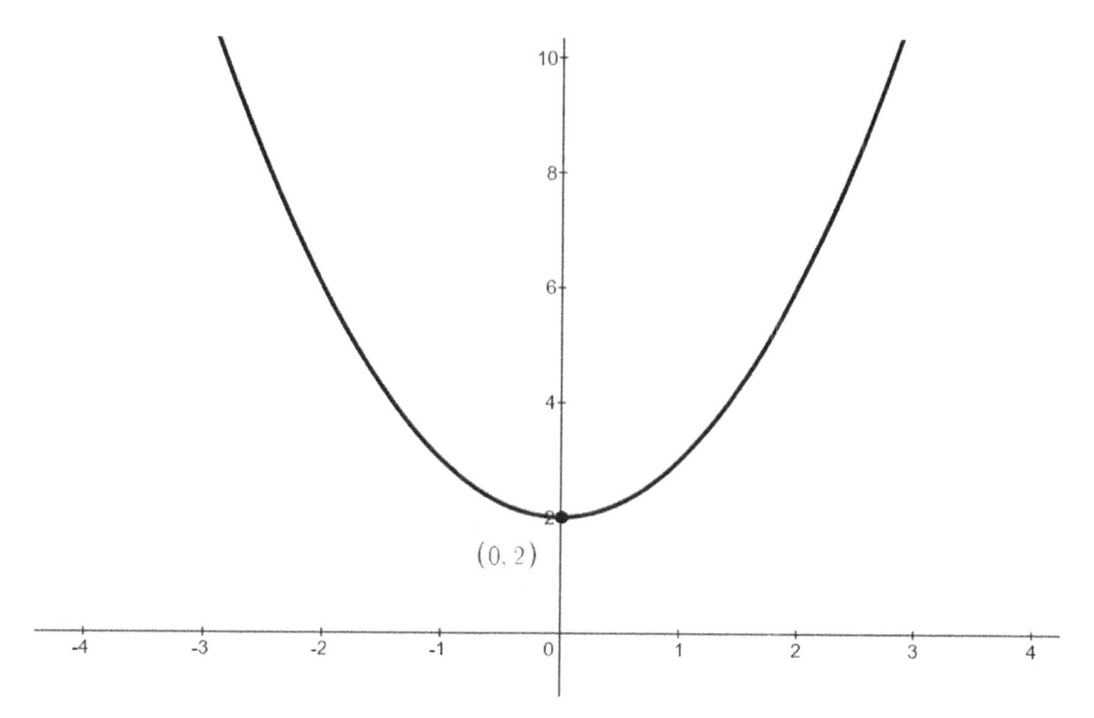

The vertex is not only the minimum or maximum of the parabola; it's also the exact point where the graph changes direction from down to up (or vice-versa if the parabola is negative).

Points where a curve changes direction from up to down (or from down to up) are known as "inflection points." More technically, inflection points are where the slope of a curve goes to 0, but this is outside the scope of what we need to know for the SAT.

There are also graphs of "higher-order" equations that have more than one inflection point, but they are extremely rare on the SAT.

For the moment, I'm going to keep it simple. I'll leave further details about inflection points for another time, since they aren't relevant to the huge majority of SAT Math questions.

Calculating the Vertex of a Parabola

Calculating the coordinates of a parabola's vertex is easier than many students think. You probably know by now that I'm not big on *memorizing* facts; I'd rather *understand* the underlying concepts. So let's think this through together.

First of all, parabolas are always *symmetrical*. The left and right sides are mirror images of each other. That means that all parabolas have a *middle*. And that exact middle is where the vertex lives.

Furthermore, we've already learned to calculate the *x*-intercepts of a parabola by setting a Quadratic Equation equal to zero and solving for the two values of x.

So, if we go to the *middle* of the two *x*-intercepts, then we will be at the center of the parabola.

Once you've found the x in the middle, you can plug this *x*-midpoint into the equation of the parabola to find the corresponding *y*-value of the vertex.

For example, we'll again consider the equation $y = x^2 - 4$ and its graph.

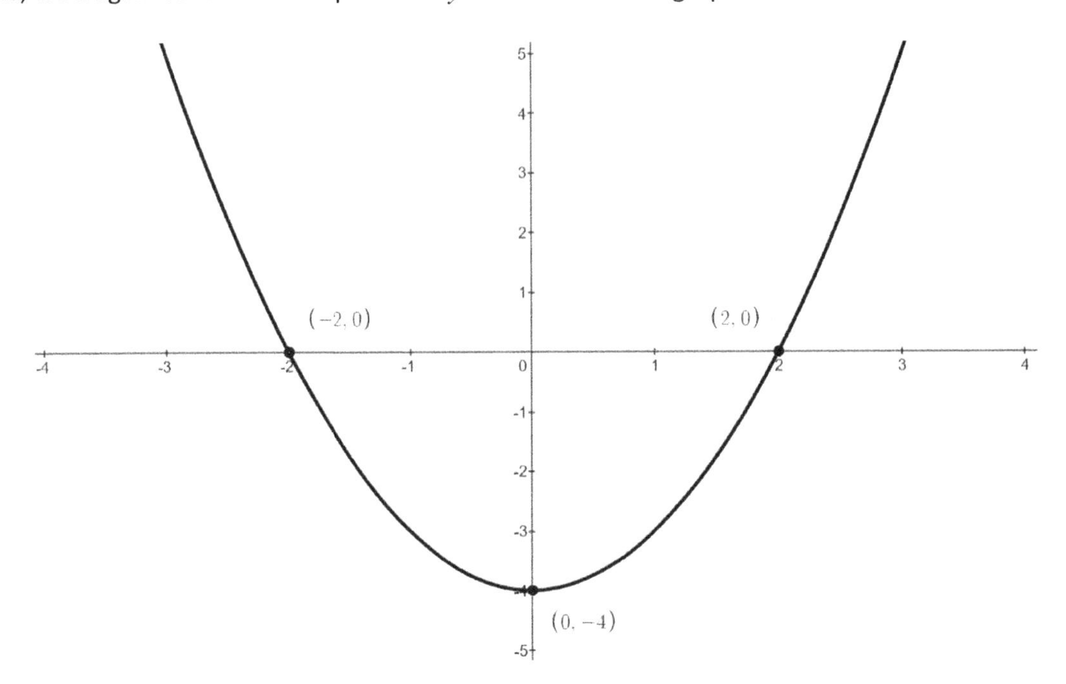

We already found that the *x*-intercepts are at $(-2,0)$ and $(2,0)$.

Now, find the midpoint of $x = -2$ and $x = 2$. You can do this by adding the two values of x and dividing by two, or even simply by counting on your fingers. Either way, the *x*-midpoint is $x = 0$. This will be the *x*-coordinate of the vertex.

Now plug the *x*-coordinate of the vertex ($x = 0$) into the equation of the parabola:

$$
\begin{aligned}
y &= x^2 - 4 \\
y &= (0)^2 - 4 \\
y &= 0 - 4 \\
y &= -4
\end{aligned}
$$

This tells us the *y*-coordinate of the vertex. And indeed, the graph above shows that our vertex point is at $(0,-4)$.

Try calculating the coordinates of a vertex for yourself:

What are the coordinates of the vertex of the parabola defined by the equation $y = x^2 - 9$?

First, we need to find the *x*-intercepts. Set y equal to 0.

$$0 = x^2 - 9$$

Then, this equation can be easily factored, as we learned in **Basic Algebra 2**. It is a Difference of Squares:

$$0 = (x-3)(x+3)$$

The solutions are $x = 3$ and $x = -3$. These are our *x*-intercepts. Now find the *x*-midpoint between them.

$$x = \frac{(3)+(-3)}{2}$$
$$x = \frac{0}{2}$$
$$x = 0$$

Our *x*-midpoint - the middle of the parabola, and therefore the *vertex* - is at $x = 0$. Plug it back into the equation to find the corresponding *y*-coordinate:

$$y = x^2 - 9$$
$$y = (0)^2 - 9$$
$$y = 0 - 9$$
$$y = -9$$

The *y*-coordinate is "-9", and the coordinates of our vertex must be at $(0,-9)$. Below, I've included a graph of the parabola defined by $y = x^2 - 9$. Notice the coordinates of the vertex and the *x*-intercepts are exactly where we predicted they would be.

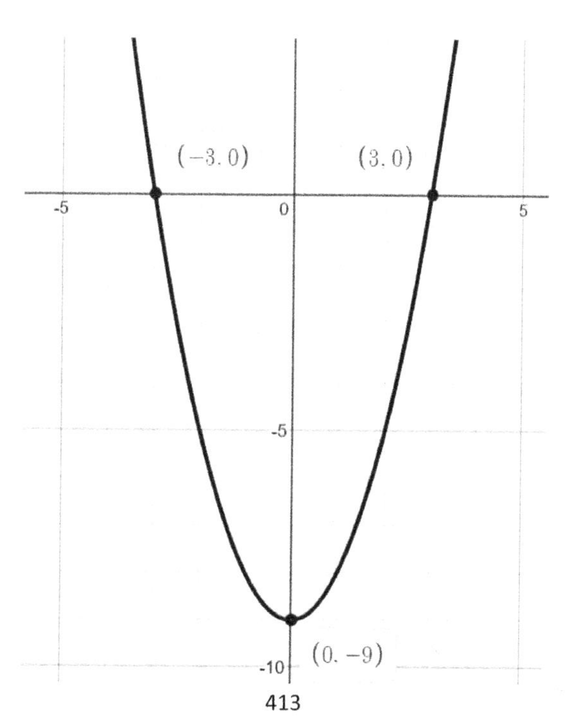

Putting It All Together

So, we come to a topic of central importance: how do we sketch a parabola from its given equation? By now, we have the tools and understanding required.

There's more than one way to accomplish this - but my favorite way is to put the equation into $ax^2 + bx + c = 0$ form, then use either Factoring or the Quadratic Equation to find the solutions for *x*. These will be our two *x*-intercepts.

Then you can find the midpoint of the *x*-intercepts. This will be the *x*-coordinate of the vertex. Then plug this *x*-value into the original equation to find the *y*-coordinate of the vertex. Now you can plot your vertex point.

With two *x*-intercepts and the vertex, you should be able to draw a reasonably accurate sketch of the parabola. But, if you want one more reference point, you could also plug "$x = 0$" into the equation to calculate the *y*-intercept.

Let's try graphing the parabola defined by the equation $y = x^2 + 2x - 8$.

First, let's set y equal to 0, and factor to find the *x*-intercepts:

$$0 = x^2 + 2x - 8$$
$$0 = (x-2)(x+4)$$

Our two solutions are $x = 2$ and $x = -4$, which are the *x*-intercepts.

Now let's calculate the vertex. It will be at the center of the parabola - right in the middle of the *x*-intercepts. We can calculate this midpoint by adding the two *x*-intercepts and dividing by 2:

$$x = \frac{(2)+(-4)}{2}$$
$$x = \frac{-2}{2}$$
$$x = -1$$

The *x*-coordinate of our vertex must be "-1". Now plug that back into the equation to find the corresponding *y*-coordinate of the vertex:

$$y = x^2 + 2x - 8$$
$$y = (-1)^2 + 2(-1) - 8$$
$$y = 1 - 2 - 8$$
$$y = -9$$

The *y*-coordinate of our vertex must be "-9", and the complete vertex coordinate is $(-1,-9)$.

If we also wanted to find the *y*-intercept of our parabola, we could just plug $x = 0$ into the equation:

$$y = x^2 + 2x - 8$$
$$y = (0)^2 + 2(0) - 8$$
$$y = 0 + 0 - 8$$
$$y = -8$$

The *y*-intercept must be at $(0, -8)$.

Now, with four key points clearly defined, we can graph our parabola quite easily.

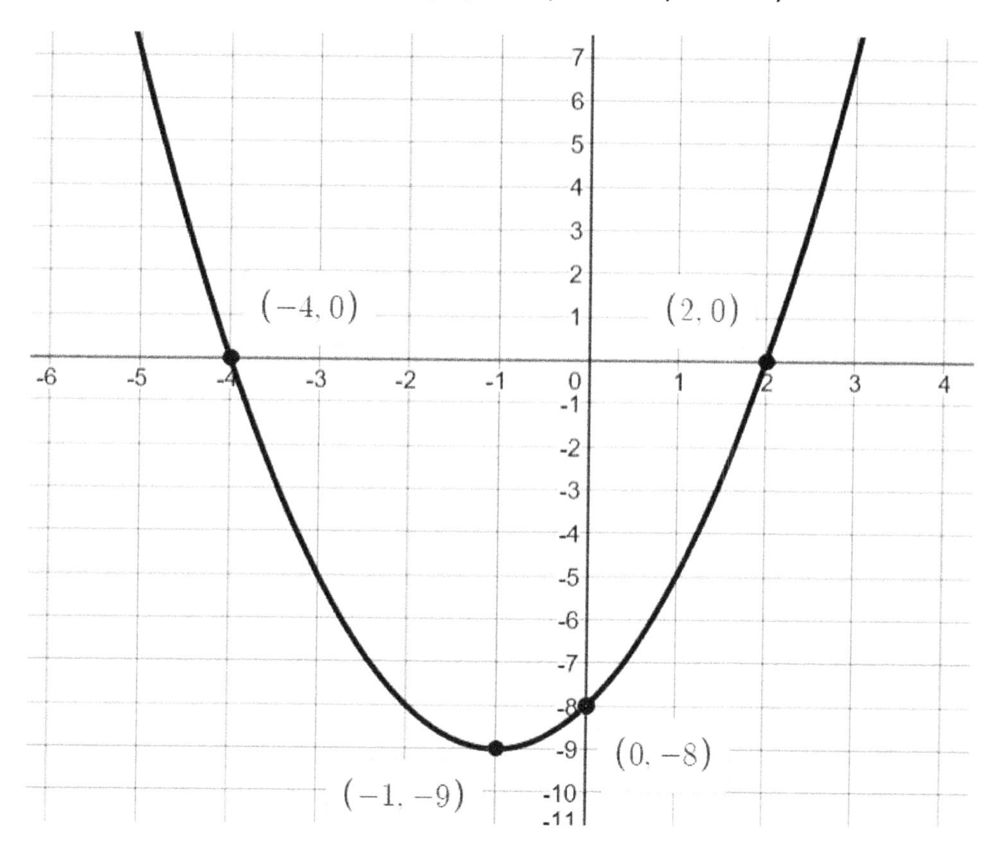

Now try sketching a parabola from an equation on your own on the blank axis below.

Sketch the parabola defined by $y = x^2 - 8x + 12$ on the xy-plane below.

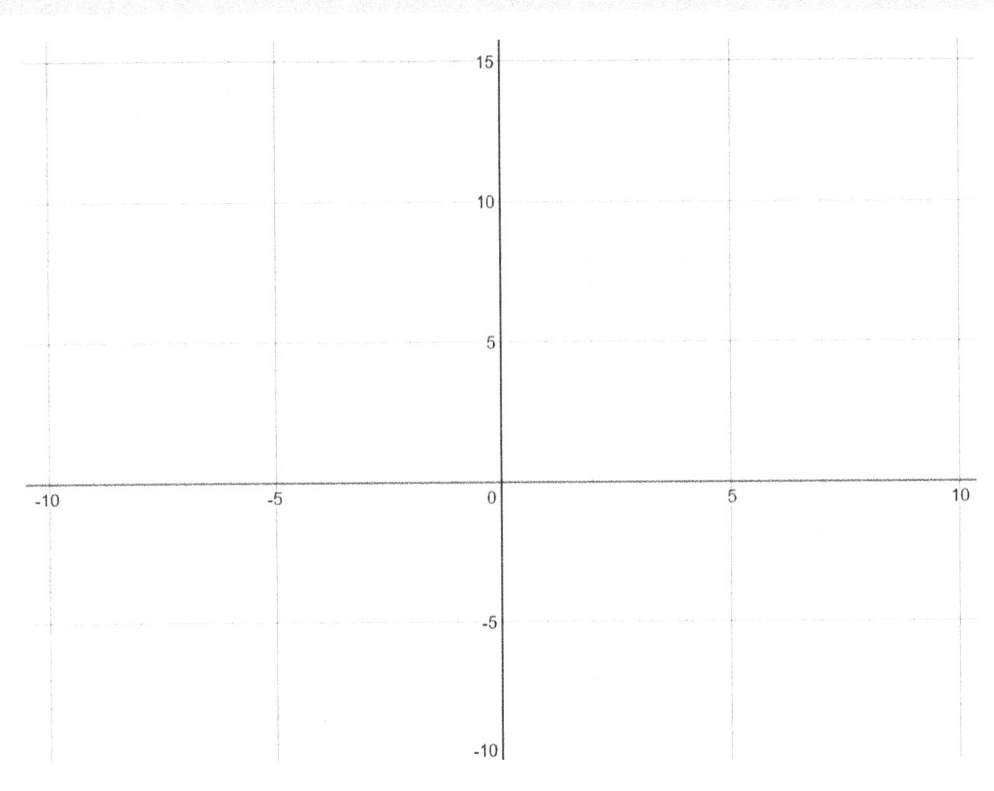

Remember the steps we've used before. First, set the y in the Quadratic Equation equal to 0 and factor it to find the solutions or *x*-intercepts:

$$0 = x^2 - 8x + 12$$
$$0 = (x-2)(x-6)$$
$$x = 2,6$$

Our *x*-intercepts must be at $x = 2$ and $x = 6$.

Now find the *x*-midpoint, which will help define the vertex. Add the two *x*-intercepts and divide by 2:

$$x = \frac{(2)+(6)}{2}$$
$$x = \frac{8}{2}$$
$$x = 4$$

The *x*-coordinate of our vertex must be "4." Now plug $x = 4$ back into the equation to find the corresponding *y*-coordinate of the vertex:

$$y = x^2 - 8x + 12$$
$$y = (4)^2 - 8(4) + 12$$
$$y = 16 - 32 + 12$$
$$y = -4$$

The *y*-coordinate of our vertex is "-4", and the complete coordinates of the vertex are $(4,-4)$.

Finally, find the *y*-intercept just to make sure your parabola is looking sharp and accurate. Plug in "0" for the value of x in the original equation to calculate the *y*-intercept:

$$y = x^2 - 8x + 12$$
$$y = (0)^2 - 8(0) + 12$$
$$y = 0 - 0 + 12$$
$$y = 12$$

Now we know the *y*-intercept is at $(0,12)$. With these four points marked on the coordinate plane, we should be able to draw a pretty smooth sketch of the parabola.

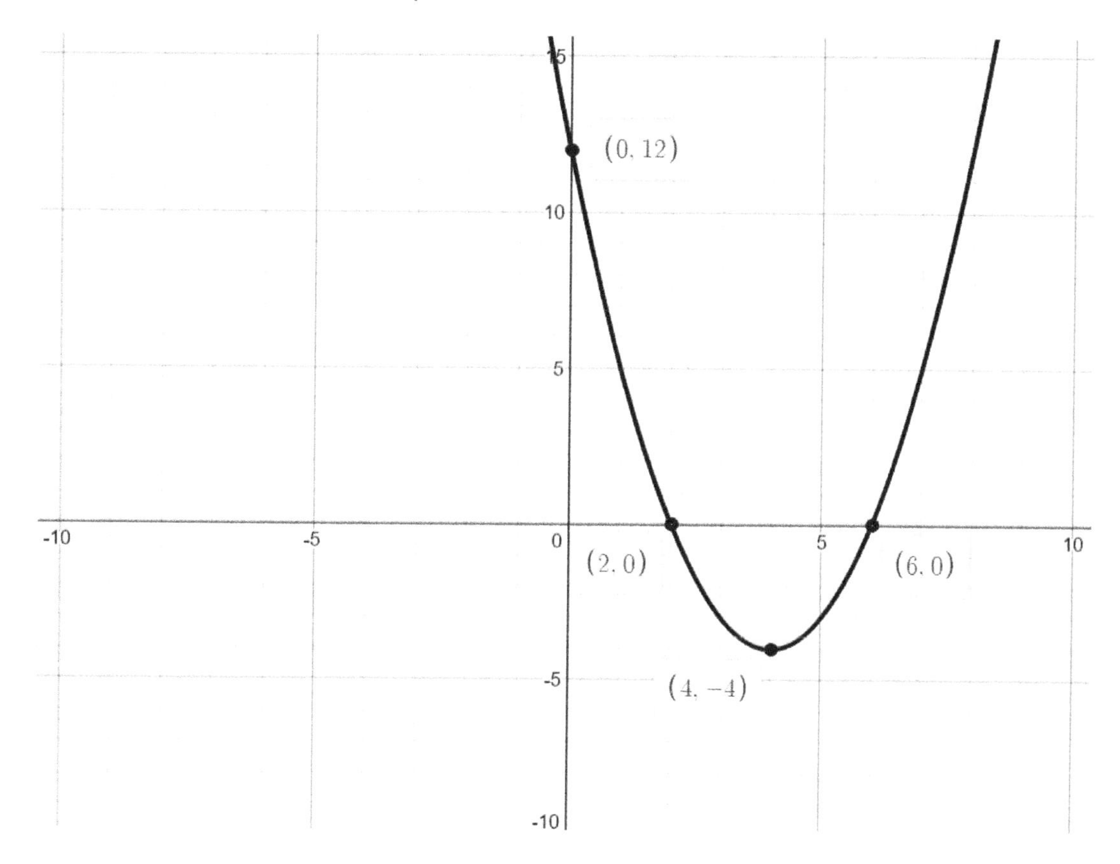

Notice that all four key points are exactly where we predicted they would be.

Pretest Question #1

Let's take a look at our first Pretest question on this topic. Try it yourself if you got it wrong the first time.

(CALCULATOR) The parabola represented by the equation $y = x^2 + x + b$ is graphed in the *xy*-plane, where b is a constant. The parabola passes through the point $(0,-20)$ and the vertex of this parabola is at the coordinate point (m, n). What is the value of $m + n$?

(A) -20

(B) -20.25

(C) -20.75

(D) -21.25

OK, there's a bit of work to do on this question. Before we can set the equation equal to 0 and factor, we have to figure out the value of the constant b.

First, plug the given coordinate $(0,-20)$ into the equation for x and y, then solve for b:

$$y = x^2 + x + b$$
$$-20 = (0)^2 + (0) + b$$
$$-20 = 0 + 0 + b$$
$$-20 = b$$

OK, now we know the value of the constant $b = 20$, and we can plug it back into the equation:

$$y = x^2 + x - 20$$

Now let's set y equal to 0 and factor, as we learned in **Basic Algebra 2**:

$$0 = x^2 + x - 20$$
$$0 = (x+5)(x-4)$$

The solutions to this equation are $x = -5$ and $x = 4$, which are the *x*-intercepts of the parabola.

Now add them up and divide by 2 to find the *x*-midpoint of the parabola, which will also be the *x*-coordinate of the vertex:

$$x = \frac{(-5)+(4)}{2}$$
$$x = \frac{-1}{2}$$
$$x = -.5$$

So, the *x*-coordinate of the vertex must be "$-.5$". Plug $x = -.5$ back into the equation to find the corresponding *y*-coordinate of the vertex.

$$y = x^2 + x - 20$$
$$y = (-.5)^2 + (-.5) - 20$$
$$y = .25 - .5 - 20$$
$$y = -20.25$$

The exact coordinate of the vertex must be $(-.5,-20.25)$. The question gives this as the point (m,n). So, $m = -.5$ and $n = -20.25$.

The question asks for the value of $m+n$, so finish by adding them together:

$$(-.5) + (-20.25) = -20.75$$

Our final answer is **Choice C**.

Parabola Transformations

"Transformations" are changes to the original parabola equation that create corresponding changes in the shape and position of the graph.

This is a topic that tends to be emphasized in high school math classes, *without* having a corresponding level of importance on the SAT Math test.

I struggled to decide if I should include this section on Parabola Transformations in the lesson - mainly because I barely (or never) find myself using them on the SAT. I decided to compromise and include a small amount of basic info on Parabola Transformations without going overly deep into the details.

At a minimum, we understand from earlier in the lesson that the coefficient of the x^2 term can either make the parabola narrower and steeper (with higher coefficients) or wider and more shallow (with lower coefficients).

Also as we saw earlier, if you change the coefficient of the x^2 term from positive to negative (or vice-versa), the direction of the parabola will flip upside-down. This could also be called a "reflection across the *x*-axis."

Furthermore, you may have realized by now that in the Quadratic Equation $y = ax^2 + bx + c$, the value of c determines the *y*-intercept of the parabola. Therefore, you can vertically shift a parabola up or down by increasing or decreasing the value of *c*.

Total honesty here - I really don't know much more about Parabola Transformations, yet I'm still able to easily get a perfect score on the SAT Math test. My careful research also shows that it's a rare topic on the SAT and you can always easily work around the problem using all the essential information we've already covered.

The reality is, I don't ever find myself wishing I knew more about Parabola Transformations when taking the SAT. As always, I prefer to *remember* less, and *understand* more, to keep my brain from getting too cluttered up with excessive information.

Therefore, I'll leave this topic here, and you can do further explorations yourself if you're interested.

Other Parabola Equation Forms

There are two main ways to represent the equation of a parabola. One of those forms is the Quadratic Equation $y = ax^2 + bx + c$, which we've already grown accustomed to.

The other is called "Vertex Form" and is written as $y = a(x-h)^2 + k$, where (h,k) are the coordinates of the vertex.

True mastery of the Vertex Form requires an additional skill, called "Completing the Square," which we will cover in a future lesson on the **Equation of a Circle**.

Furthermore, *I do not ever use Vertex Form* in my own work on the SAT test. That's right - I truly do *not* remember or use Vertex Form in any of my calculations.

Is this a bit of a blind spot on my part? Perhaps. I have no doubt that occasionally, I could save between 10-30 seconds on certain SAT Math questions if I had Vertex Form memorized.

However, I simply don't want to clutter my brain with more information when everything we've covered so far allows me to solve every Parabola question the SAT gives me.

I've seen a handful of students who are very confident in their Vertex Form who can answer certain Parabola questions a few moments before I do. This only represents about 1 in 100 parabola questions, and I usually end up getting the right answer only a few seconds later. I'm always impressed by my students when this happens.

But just as with Parabola Transformations, I'd rather memorize the least-possible amount of information necessary. Instead of memorizing more forms and equations, I prefer getting the greatest mileage out of the techniques I already know, instead of memorizing more.

It's kind of like a toolkit: do you *really* need five different sizes of hammer for every imaginable situation? Those hammers - like extra equations - are awfully heavy to carry around, plus you have to keep them clean and rust-free. What if I can get along just fine with only *two* hammers? Isn't that more efficient and lighter to carry?

So again, I will leave this topic here, and if you'd like to do further explorations on your own, feel free. I do think, thought, that your SAT prep time would be better spent elsewhere.

"Displayed as Constants or Coefficients"

There is one final type of Parabola-based question to be aware of on the SAT test. These questions present us with the graph of a parabola and ask us to choose an equation where some key point is given as "constants or coefficients" (the SAT's words) in the equation.

Keep in mind that "constants or coefficients" is just a fancy phrase meaning "numbers." A "constant" is just a number like the 5 in $x+5$, and a "coefficient" is just a number that's in front of a variable like the 3 in $3x$.

These types of questions may ask us for the *x*-intercepts, the *y*-intercept, or the coordinates of the vertex to be clearly visible within the equation as constants or coefficients (as "numbers").

Most of the time, these questions give us a variety of equation forms, and *most* of them are "equivalent" to the original equation. However, only *one* of the equivalent forms will also display the desired point(s) as numbers within the equation.

This might sound pretty abstract to you, so let's take a look at one of these questions now.

Pretest Question #2

Here's a "displayed as constants or coefficients" question that comes straight from the Pretest. Try it yourself before you look at my explanation below the question.

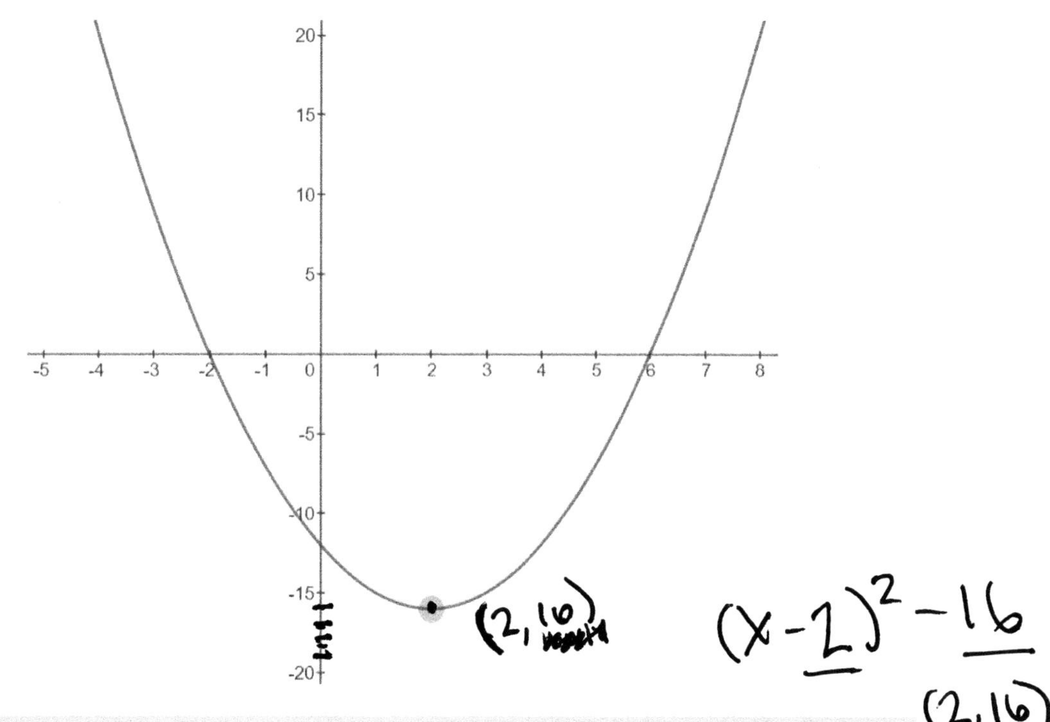

The equation of the parabola in the *xy*-plane above is $y = x^2 - 4x - 12$. Which of the following is an equivalent form of the equation from which the coordinates of the vertex can be identified as constants or coefficients in the equation?

(A) $y = (x - 6)(x + 2)$

(B) $y = (x + 6)(x - 2)$

(C) $y = x(x - 4) - 12$

(D) $y = (x - 2)^2 - 16$

This question is a variation of one of my favorites in the SAT Math test. It always seems to catch students unaware!

Notice from the wording that this is an "displayed as constants or coefficients"-type question.

Here's the first thing students do wrong. They completely ignore that key part of the word problem and fixate on the words "which of the following is an *equivalent form* of the equation."

Then they dive right into Answer Choice A, which, when we FOIL it out (as in **Basic Algebra 2**), *does* give an equivalent to the original equation $y = x^2 - 4x - 12$. These students pick Choice A, and look up confusedly, like "shouldn't that have been harder?"

And yes, it is harder. Or at least it's "trickier." It's true that Choice A *is* equivalent to the original equation - but just as importantly, we were asked for an equation where "the coordinates of the vertex can be identified"!

In fact, if you distribute out Answer Choices A, C and D, you will find that they are *all* equivalent forms of the original equation.

So, let's turn our attention to the vertex. It's pretty clearly at an *x*-coordinate of about "2", even just by eyeballing the graph. Furthermore, the *y*-coordinate of the vertex seems to be just a *bit* below $y = -15$.

There's really only choice that's even close to having values in it that are close to the coordinate $x = 2$ and $y =$"a bit less than -15". And that's Choice D, with the constants "2" and "-16".

Sharp-eyed students will notice that Choice D is also the equation put correctly into Vertex Form. But again, that's not a math topic I pay attention to, personally, so it doesn't really help me.

If you want to be more thorough, you could even go through the whole process for finding the vertex: set the equation to $y = 0$, factor it, and find the solutions for x. Then find their *x*-midpoint and plug it back into the equation for x to get the *y*-coordinate of the vertex. You'll end up with the point $(2,-16)$.

But frankly I think it's overkill to do so, since we can gather almost all of that information just by looking at the graph and estimating the coordinates of the vertex. More important than the *exact* coordinates of the vertex is to *not* misread the question. If we read correctly and know what we're looking for, the Answer Choices and graph help us do the rest.

Of course, our final answer should be **Choice D**, which is an equivalent form of the original equation that, most importantly, *also* displays the coordinates of the vertex at $(2,-16)$.

Review & Encouragement

Parabolas are the definitive graph of Algebra 2 - immediately recognizable with their *U*-shaped curves and x^2 terms in their equations.

These graphs are nothing more than visual representations of the Quadratic Equations that we've been studying for most of Algebra 2.

The key skill for Parabola mastery is understanding how to find their *x*-intercepts and vertex points. From there we can easily sketch a graph of the parabola, if necessary.

Parabolas do not have *slopes* in the traditional sense that a $y = mx + b$ **Linear Equation** does. Don't make the common error of worrying about the "slope" of a parabola - this is a concept we don't explore until Calculus (which isn't on the SAT Math test) and for the purposes of the SAT Math test, we can completely forget about any concept of "slope" in regards to parabolas. It just doesn't apply.

We've also raised the topics of Parabola Transformations and Vertex Form as concepts of questionable usefulness on the SAT test. Again, I barely know anything about these two topics, and I still find it easy to get 100% of the Parabola questions on the SAT test correct.

Watch out for SAT questions that ask for a certain point of the parabola to be "displayed as a constant or coefficient." These questions are more bark than bite - no harder than any other parabola question, if you understand the lessons in this chapter. But if you misread or ignore the directions, you'll get bitten.

There's a lot to practice here, so work through the following question set until you feel confident in your understanding and mastery of Parabola questions on the SAT!

Algebra 2 Parabolas Practice Questions

DO NOT USE A CALCULATOR ON ANY OF THE FOLLOWING QUESTIONS UNLESS INDICATED.

1. The equation $y = 3x^2 + 12x - 15$ is graphed in the xy-plane. If the graph crosses the y-axis at the point $(0, t)$, what is the value of t?

(A) -15

(B) -12

(C) 12

(D) 15

2. The function f is defined by $f(x) = (x+6)(x+2)$. The graph of f in the xy-plane is a parabola. Which of the following intervals contains the x-coordinate of the vertex of the graph of f?

(A) $-6 < x < -4$

(B) $-6 < x < 2$

(C) $-4 < x < 2$

(D) $2 < x < 6$

$$h(t) = -22t^2 + 75t + 14$$

3. The quadratic function above models the height above the ground h, in feet, of a tennis ball t seconds after it was launched vertically from an air cannon resting on an elevated platform. If $y = h(t)$ is graphed in the xy-plane, which of the following represents the real-life meaning of the positive x-intercept of the graph?

(A) The height, in feet, of the elevated platform.

(B) The time, in seconds, at which the tennis ball hits the ground.

(C) The maximum height, in feet, of the tennis ball.

(D) The total distance, in feet, that the tennis ball travels.

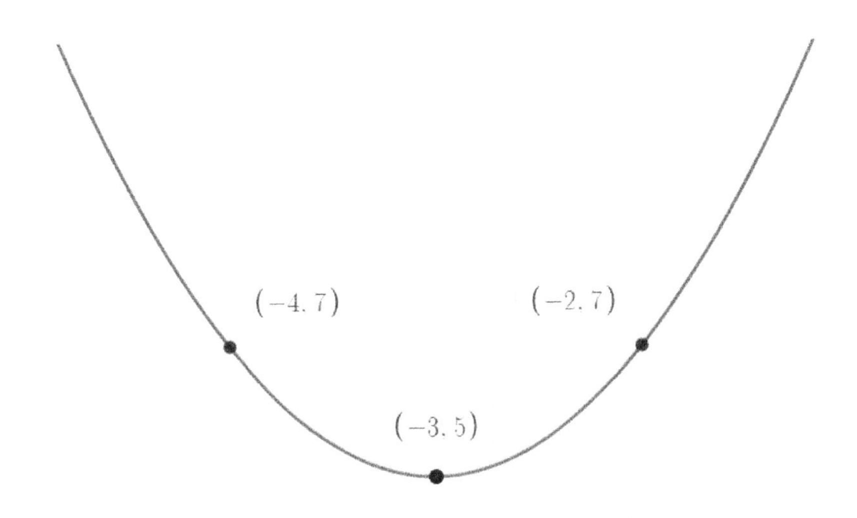

4. The graph of the function f in the xy-plane above (axes not shown) is a parabola. Which of the following defines f?

(A) $y = 2(x-3)^2 + 5$

(B) $y = 2(x+3)^2 - 5$

(C) $y = (x+3)^2 + 5$

(D) $y = 2(x+3)^2 + 5$

$$y = -(x-4)^2 - a$$

5. In the equation above, a is a constant. The graph of the equation in the xy-plane is a parabola. Which of the following is true about the parabola?

(A) Its minimum occurs at $(-4,-a)$.

(B) Its minimum occurs at $(4,a)$.

(C) Its maximum occurs at $(4,-a)$.

(D) Its maximum occurs at $(-4,a)$.

6. In the xy-plane, the parabola with equation $y = (x-7)^2$ intersects the line with equation $y = 36$ at two points, A and B. What is the length of $\overline{AB}$?

(A) 14.5

(B) 14

(C) 12

(D) 7

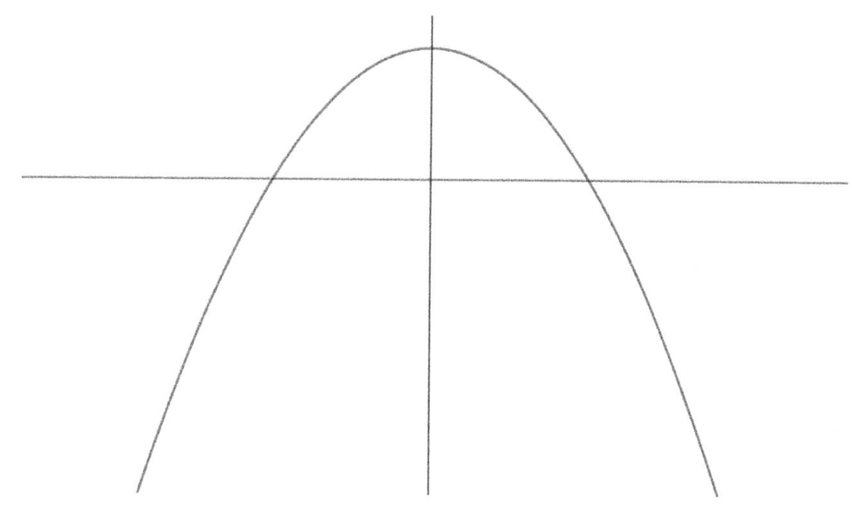

7. The equation of the parabola above is $y = ax^2 + n$ and the vertex of the parabola in the *xy*-plane above is $(0, n)$. Which of the following is true about the parabola with the equation $y = -a(x+m)^2 - n$?

(A) The vertex is (m, n) and the graph opens downward.

(B) The vertex is (m, n) and the graph opens upward.

(C) The vertex is $(-m, -n)$ and the graph opens downward.

(D) The vertex is $(-m, -n)$ and the graph opens upward.

$$h(t) = -38t^2 + 25t + 6$$

8. The function above models the height of a rock h, in feet, t seconds after the rock is tossed straight up into the air. What does the number 6 represent in the function?

(A) The number of seconds before the rock was tossed.

(B) The initial height in feet that the rock was tossed from.

(C) The time in seconds until the rock hits the ground.

(D) The initial speed, in feet per second, that the rock was tossed at.

9. If $y = 10x^2 - 49x + 25$ is graphed in the *xy*-plane, which of the following characteristics of the graph is displayed as a constant or coefficient in the equation?

(A) *y*-intercept

(B) *x*-intercept(s)

(C) *y*-coordinate of the vertex

(D) *x*-coordinate of the vertex

$$y = n(x+7)(x-3)$$

10. In the quadratic equation above, n is a nonzero constant. The graph of the equation in the xy-plane is a parabola with vertex (f, g). Which of the following is equal to g?

(A) $-25n$

(B) $-21n$

(C) $4n$

(D) $10n$

12. FREE RESPONSE: In the xy-plane, a line that has the equation $y = b$ for some constant b intersects the parabola at exactly one point. If the parabola has the equation $y = x^2 + 2x - 8$, what is the value of $-b$?

$$y = 2x^2 - 10x - 12$$

11. The equation above represents a parabola in the xy-plane. Which of the following equivalent forms of the equation displays the x-intercepts of the parabola as constants or coefficients?

(A) $y + 12 = 2x^2 - 10x$

(B) $y = 2(x^2 - 5x) - 12$

(C) $y = 2(x-3)^2 + 2x - 30$

(D) $y = 2(x+1)(x-6)$

$$f(x) = (x+2)(x-5)$$

13. (CALCULATOR) Which of the following is an equivalent form of the function f above in which the minimum value of f appears as a constant or coefficient?

(A) $f(x) = x^2 - 10$

(B) $f(x) = x^2 - 3x - 10$

(C) $f(x) = (x-1.5)^2 - 12.25$

(D) $f(x) = (x+1.5)^2 - 12.25$

14. The scatterplot below shows the number of motorcycles on the road, in thousands, over a 10-year period.

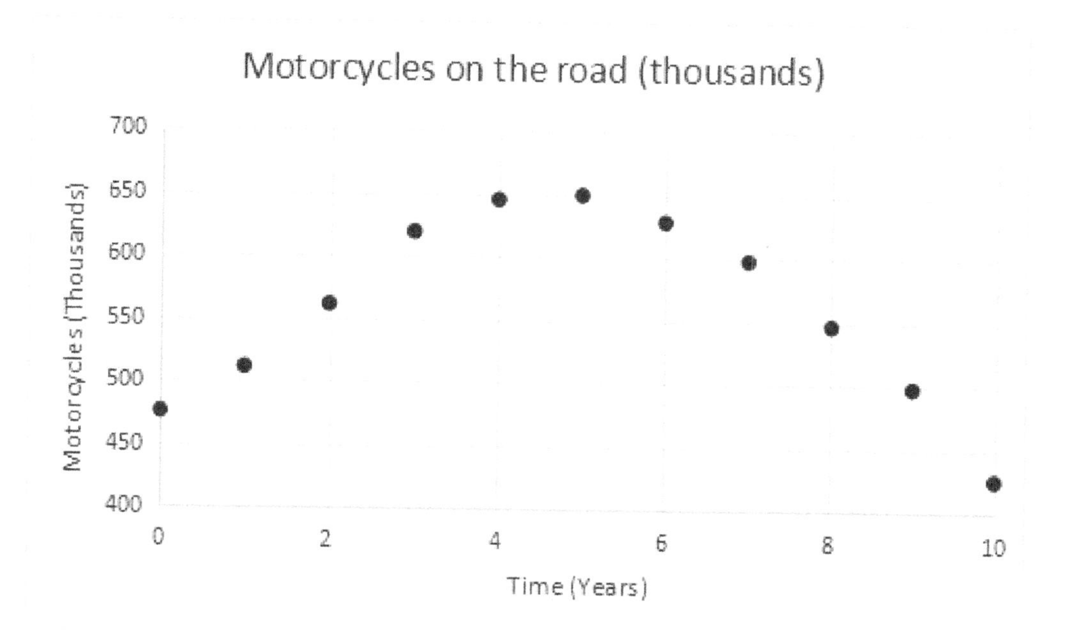

Of the following equations, which best models the data in the scatterplot?

(A) $y = -7.927x^2 + 75.53x + 460.17$

(B) $y = 7.927x^2 + 75.53x + 460.17$

(C) $y = -7.927x^2 + 75.53x - 460.17$

(D) $y = 7.927x^2 + 75.53x - 460.17$

Algebra 2 Parabolas Answers

1. A
2. B
3. B
4. D
5. C
6. C
7. D
8. B
9. A
10. A
11. D
12. 9
13. C
14. A

Algebra 2 Parabolas Explanations

1. **A.** This question explicitly asks for the *y*-intercept of a given parabola equation (we can also tell because the *x*-coordinate of the desired point is at $x = 0$, which is always where the *y*-intercept is located).

It's quite simple to calculate. Simply plug the value $x = 0$ into the equation and calculate the corresponding y:

$$y = 3x^2 + 12x - 15$$
$$y = 3(0)^2 + 12(0) - 15$$
$$y = 0 + 0 - 15$$
$$y = -15$$

The answer is $y = 0$, or in the language of this question, $t = 0$. We also could have read the *y*-intercept straight from the equation - it's the constant "-15" at the tail end.

2. **B.** This question essentially asks us to find the *x*-coordinate of the vertex. We've already learned a good way to do this. First, plug in $y = 0$:

$$0 = (x+6)(x+2)$$

Now, we'd usually factor, but the equation has already been factored for us. We can easily read off the solutions to x (which are the *x*-intercepts) as $x = -6$ and $x = -2$.

Remember that the vertex is always in the dead-center of the parabola, which will be the *x*-midpoint between $x = -6$ and $x = -2$. Add them together and divide by 2:

$$x = \frac{(-6) + (-2)}{2}$$
$$x = \frac{-8}{2}$$
$$x = -4$$

The *x*-midpoint of the parabola, and therefore the *x*-coordinate of the vertex, will be found at $x = -4$.

Go through the list of Answer Choices and eliminate any intervals that do *not* contain the value $x = -4$. That quickly eliminates Choices A, C, and D (note especially in Choices A and C that the inequality signs are "less than," *not* including "less than or *equal* to).

We're left with only Choice B, which *does* contain the value $x = -4$. And that's the answer.

3. **B.** This question asks us to relate the equation of a parabola to a real-life situation of a tennis ball being launched vertically into the air.

Since the *x*-axis for this graph is based on time t, then the *x*-intercept (where the parabola crosses the *x*-axis) will also represent a time value. At the exact moment we hit the *x*-axis, the *y*-value or height will be "0 feet".

Put it together clearly in your mind: if the height above ground is 0 feet, and we have the time at that height, then what would that time value represent?

That's right - if the height above ground is 0, then this is the time when the ball hits the ground - Choice B.

4. **D.** As low-tech as it may seem, the best way to solve this SAT-style question is to pick a point from the graph and start plugging the *x*-coordinate into the given equations. Eliminate any equations that do not return the correct *y*-value.

In cases like this, I strongly recommend *not* picking the vertex as your point. The SAT often anticipates this choice and makes your life more difficult if you do.

Instead, I'll pick the coordinate $(-2, 7)$. My *x*-value is "-2". I will test it by plugging into the given equations. If the equation does *not* return the expected *y*-value of "7", I will eliminate that equation.

First, I'll test Choice A:

$$y = 2(x-3)^2 + 5$$
$$7 = 2(-2-3)^2 + 5$$
$$7 = 2(-5)^2 + 5$$
$$7 = 2(25) + 5$$
$$7 = 50 + 5$$
$$7 \neq 55$$

It clearly returns a false equation, so I can eliminate Choice A.

Now onto Choice B, again testing $(-2,7)$:

$$y = 2(x+3)^2 - 5$$
$$7 = 2(-2+3)^2 - 5$$
$$7 = 2(1)^2 - 5$$
$$7 = 2(1) - 5$$
$$7 = 2 - 5$$
$$7 \neq -3$$

Choice B also returns a false equation, so I can eliminate it. Onto Choice C, again testing $(-2,7)$:

$$y = (x+3)^2 + 5$$
$$7 = (-2+3)^2 + 5$$
$$7 = (1)^2 + 5$$
$$7 = 1 + 5$$
$$7 \neq 6$$

Choice C is another false equation. I have high hopes for Choice D!

$$y = 2(x+3)^2 + 5$$
$$7 = 2(-2+3)^2 + 5$$
$$7 = 2(1)^2 + 5$$
$$7 = 2(1) + 5$$
$$7 = 2 + 5$$
$$7 = 7$$

Nice! Only Choice D returned a true equality when we plugged in a point given on the graph.

Although this may seem like it takes a while, it's actually a relatively efficient way of solving the problem. Plus, you may not have to completely finish each equation if you realize in advance that it's definitely not headed towards a true equality, such as when Choice A was way outside the range of possibility midway through the equation.

5. **C.** This question is an example of Vertex Form, so if you have that form memorized, you immediately can tell that the vertex is at $(4,-a)$ just by reading it off the equation.

If you're like me and you *don't* have the Vertex Form memorized, you should at least remember that this parabola will open *downwards*, because the coefficient of the x^2 term will be negative. (How do I know? Because if you FOIL out $(x-4)^2$ and then distribute the negative sign in front of it, you're guaranteed to get a $-x^2$ term.)

If the parabola opens *downward*, then it will have a *maximum* value at the vertex, not a *minimum*. That quickly eliminates Choices A and B.

The only other issue is figuring out whether the vertex occurs at $x = 4$ or $x = -4$.

Here's how I'll do it. Remember, the *x*-coordinate of the vertex is at the *x*-midpoint. I'll set y equal to 0 first:

$$0 = -(x-4)^2 - a$$

Now watch what I do as I solve for the two *x*-intercepts:

$$0 = -(x-4)^2 - a$$
$$+(x-4)^2 \quad +(x-4)^2$$
$$(x-4)^2 = -a$$
$$\sqrt{(x-4)^2} = \sqrt{-a}$$
$$(x-4) = \pm(-a)$$
$$x - 4 = \pm a$$
$$+4 \quad +4$$
$$x = \pm a + 4$$
$$x = a+4, -a+4$$

I have my two *x*-intercepts as $x = a+4$ and $x = -a+4$. Now find their midpoint by adding them together and dividing by 2, just as we've done throughout this lesson:

$$x = \frac{(a+4)+(-a+4)}{2}$$
$$x = \frac{a-a+4+4}{2}$$
$$x = \frac{8}{2}$$
$$x = 4$$

And now I have the *x*-midpoint of the parabola, which is also the *x*-coordinate of the vertex. Now we know that the parabola opens downwards and has its vertex at *x*-coordinate 4, which only leaves Choice C as an option.

This question is a classic example of the type I mentioned in the section on "Other Parabola Equation Forms." While it's true that a student who knows Vertex Form could answer this question more quickly than I can, I can still use what I know to get there in a relatively short amount of time - and I'm not burdened by remembering yet another formula or equation that can get rusty or forgotten at a critical moment.

At times like this would it be handy to know Vertex Form? Sure it would. It's not necessary, but it makes life easier. It's up to you if you want to memorize it. Personally, I kind of enjoy the challenge of going through the test without it. But, I'll admit it would have been useful on this question.

6. **C.** This is an interesting question. The horizontal line $y = 36$ is intersecting the parabola defined by the equation $y = (x-7)^2$.

Remember from **Linear Equations (Algebraic)** that the intersection point(s) of any two lines (or curves) can be found by setting their equations equal to each other. We can easily do that by plugging $y = 36$ into the parabola equation:

$$36 = (x-7)^2$$

Now FOIL the right side:

$$36 = x^2 - 14x + 49$$

And then set the equation equal to 0:

$$\begin{aligned} 36 &= x^2 - 14x + 49 \\ -36 & \qquad\qquad\quad -36 \\ 0 &= x^2 - 14x + 13 \end{aligned}$$

Now factor the right side:

$$0 = (x-13)(x-1)$$

And we can read off the solutions to this equation as $x = 13$ and $x = 1$. These are the two *x*-coordinates where the horizontal line $y = 36$ intersects the parabola $y = (x-7)^2$.

What is the distance between $x = 13$ and $x = 1$? As we learned in **Algebra 1 Word Problems**, "distance" is the absolute value of one point subtracted from the other point. Subtract them, then take the absolute value:

$$\begin{aligned} & |13-1| \\ &= |12| \\ &= 12 \end{aligned}$$

The distance between the two points of intersection is 12 units, or Choice C.

7. **D.** In this question, we're given a graph of a parabola to start with. Then there are some Parabola Transformations applied to the equation, and we must identify some key characteristics of the transformed graph.

First of all, the easy part: if the graph *originally* opens downwards, and then we make the original coefficient of x negative, then the direction of the graph will flip to upwards. That eliminates Choices A and C.

I understand this can be a little confusing to some, because you might think that $-ax^2$ means the graph *must* open downwards. Sure, it normally seems that way - but wasn't the graph *already* opening downwards? Therefore, the original value of *a* must *already* have been negative. If we flip the sign of a, then the direction of the parabola must also flip.

So, the new graph opens upward. But what about the coordinates of the vertex? Originally the vertex was at $(0, n)$ back when the equation was $y = ax^2 + n$. Now that we've changed to $y = -a(x+m)^2 - n$, what do you think has happened to the *y*-coordinate of the vertex? It used to be positive n, and now it's $-n$. It's a safe assumption that the sign of n in the vertex has also flipped. That leaves us with Choice D.

Notice that we didn't even pay attention to the *x*-coordinate of the new vertex, since it wasn't information necessary to answer the question.

8. **B.** Another question, like Question 3, that asks us to relate the equation of a parabola to a real-life situation. The SAT seems to love launching objects into the air!

Think what would happen if we plugged in 0 for the time value *t*. This would represent a time of 0 seconds before the rock is thrown - in other words, the starting height. And if we plug in 0 for *t*, we get:

$$\begin{aligned} h(0) &= -38(0)^2 + 25(0) + 6 \\ h(0) &= 0 + 0 + 6 \\ h(0) &= 6 \end{aligned}$$

This tells us that at a time of 0 - a moment before tossing the rock - it's already at a height of 6 feet. Only one answer choice makes sense - Choice B, the initial height of the rock immediately before it was tossed.

9. **A.** This is the type of question I explored in the section on "Displayed as Constants or Coefficients," as well as Pretest Question 2. Remember that "constants or coefficients" are just *numbers* that appear in the equation.

Think about what we've done throughout this lesson. Whenever we started with an equation in $y = ax^2 + bx + c$, like the one given in this question, didn't we have to do quite a bit of work before we could find anything about the *x*-intercepts - much less the exact coordinates of the vertex?

But, the *y*-intercept has always been easy to find: just plug $x = 0$ into the equation and see what comes out for *y*. And if we do that right now, what do we get?

$$y = 10x^2 - 49x + 25$$
$$y = 10(0)^2 - 49(0) + 25$$
$$y = 0 + 0 + 25$$
$$y = 25$$

That's right - the *y*-intercept of this equation is "25". And it's sitting right there, in the equation, as the constant (which just means "number") $+25$. That's why our answer to this question is Choice A. The *y*-intercept is already out in the open, on full display as a constant (or "number") in the given equation.

10. **A.** OK, to get this question going, we're shooting for the *vertex* coordinate of the parabola. We've done this many times throughout the lesson.

First, set the *y*-value to 0:

$$0 = n(x+7)(x-3)$$

Now read off the solutions to *x*, which are $x = -7$ and $x = 3$. These are the *x*-intercepts of our graph.

The vertex of the parabola will be located directly in the middle of them. Add the two intercepts together and divide by 2 to find their midpoint:

$$x = \frac{(-7)+(3)}{2}$$
$$x = \frac{-4}{2}$$
$$x = -2$$

The middle of the parabola, and therefore the *x*-coordinate of the vertex, is at $x = -2$.

However, the question is asking for the value of g, which is the *y*-coordinate of the vertex. Go ahead and plug back in $x = -2$ into the original equation to calculate the corresponding *y*-coordinate and get our answer:

$$y = n(x+7)(x-3)$$
$$y = n(-2+7)(-2-3)$$
$$y = n(5)(-5)$$
$$y = -25n$$

And there we have it. This *y*-coordinate may be expressed in terms of n rather than as an exact numerical value, but that fits the answer choices just fine. It looks like our final answer is Choice A.

11. **D.** Another question of the type we explored in the section on "Displayed as Constants or Coefficients," as well as Pretest Question 2 and Practice Question 9. Remember that "constants or coefficients" are just *numbers* that appear directly in the given equation.

And, *don't* fall for the trap of only looking for "equivalent forms" of the equation. More than one of these answer choices are equivalent to the original equation, but only *one* of them gives the *x*-intercepts within the equation.

In this case, we want to see the *x*-intercepts of the given equation displayed as numbers in one of our answer choices. So, it makes sense to first *calculate* the *x*-intercepts of the given equation.

We've done this plenty of times by now. First, set the *y*-value equal to 0:

$$0 = 2x^2 - 10x - 12$$

Now factor. I'll go ahead and factor out a multiple of 2 before factoring the rest of the equation, just to make the numbers a little lower and easier:

$$0 = 2(x^2 - 5x - 6)$$
$$0 = 2(x-6)(x+1)$$

The two solutions to this equation are $x = 6$ and $x = -1$, and these will also be the *x*-intercepts of the parabola. We've seen this many times before.

Not only that, but notice that Choice D gives the values "6" and "1" in it. Sure, the signs are reversed, but don't worry about that now. We know the *x*-intercepts we want to display in our equation are $x = 6$ and $x = -1$, and only Choice D has both of those numbers represented in its equation.

Plus, we've realized by now that the factored form of a Quadratic Equation is the easiest way to read off the x-intercepts of the parabola. All these familiar elements add up to make Choice D a sure bet.

12. **9.** A line with equation $y = b$, where b is a constant, will be a horizontal line, as we learned in **Linear Equations**. If a horizontal line intersects a parabola at exactly (and only) one point, then this horizontal line must go through the vertex of the parabola. Otherwise, it would either cross the parabola twice, or never cross it at all.

In other words, this question is just asking for the y-coordinate of the parabola's vertex, where the horizontal line $y = b$ passes through the vertex. It's just asking that familiar question in an extra-fancy manner.

We've gotten very good at calculating the coordinates of a vertex by now. First, set the Quadratic Equation equal to 0, factor, and solve for x:

$$0 = x^2 + 2x - 8$$
$$0 = (x+4)(x-2)$$

The two solutions for x are $x = -4$ and $x = 2$. Find their x-midpoint and you'll know the x-coordinate of the center of the parabola, which we know is also the x-coordinate of the vertex. Add them together and divide by 2:

$$x = \frac{(-4)+(2)}{2}$$
$$x = \frac{-2}{2}$$
$$x = -1$$

Now we know the x-coordinate of the vertex is "-1". As we've done many times before, plug this back into the original equation to find the corresponding y-coordinate of the vertex:

$$y = x^2 + 2x - 8$$
$$y = (-1)^2 + 2(-1) - 8$$
$$y = 1 - 2 - 8$$
$$y = -9$$

If the vertex is at $y = -9$, and the line $y = b$ touches this parabola at exactly one point, then $b = -9$.

Do Note that the question asked for the value of "$-b$". Therefore, our final answer should be $-(-9)$ or just "9".

13. **C.** Another question of the type we explored in the section on "Displayed as Constants or Coefficients," as well as Pretest Question 2 and Practice Question 9 and 11. Remember that "constants or coefficients" are just *numbers* that appear directly in the given equation.

And, *don't* fall for the trap of only looking for "equivalent forms" of the equation. More than one of these answer choices are equivalent to the original equation, but only *one* of them gives the minimum value within the equation.

This question asks for the *minimum* value of the parabola to be displayed within the equation we select. If the parabola has a *minimum* value, then it must be opening upwards. Basically, we need to calculate the y-coordinate of the vertex of the parabola.

We've done this a dozen times before. First set equal to 0 and find the x-intercepts. The equation is already in factored form, so it's particularly easy this time:

$$0 = (x+2)(x-5)$$

The values of x that satisfy this equation are $x = -2$ and $x = 5$. The vertex will be located directly between these two points, so add them together and divide by 2 to find the x-midpoint:

$$x = \frac{-2+5}{2}$$
$$x = \frac{3}{2}$$
$$x = 1.5$$

The x-coordinate of the vertex is "1.5". Plug $x = 1.5$ back into the original equation to find the corresponding y-coordinate at the vertex:

$$f(x) = (1.5+2)(1.5-5)$$
$$f(x) = (3.5)(-3.5)$$
$$f(x) = -12.25$$

So, we know the minimum value of f (the y-coordinate of the vertex) is "-12.25". The only problem is, *two* of the answer choices contain this value.

We'll have to think of a way to tell the difference between Choices C and D.

What about this - why don't we just FOIL out both equations and make sure they're equal to the original equation?

Our starting equation was $f(x) = (x+2)(x-5)$. FOIL this out:

$$f(x) = x^2 - 3x - 10$$

Now FOIL out Choice C and clean it up. Is it equivalent to the original equation?

$$f(x) = (x-1.5)^2 - 12.25$$
$$f(x) = x^2 - 3x + 2.25 - 12.25$$
$$f(x) = x^2 - 3x - 10$$

Yes, Choice C is equivalent to the original equation. Let's check Choice D and hope that it's *not* the same equation, so we can eliminate it:

$$f(x) = (x+1.5)^2 - 12.25$$
$$f(x) = x^2 + 3x + 2.25 - 12.25$$
$$f(x) = x^2 + 3x - 10$$

Excellent - Choice D is *not* the same as the original equation (notice the difference between "$-3x$" in the original equation vs. "$+3x$" in Choice D.) That eliminates Choice D.

Only Choice C is an *equivalent* form of the original equation that *also* displays the minimum value of "-12.25" within the equation, as the question asked. It's a perfect example of the "Displayed as constants or coefficients"-type of SAT parabola question.

14. **A**. This question is actually crazy-easy compared to some of the others. It's *different* than the rest, but still easy. We're looking for a quadratic equation that represents the scatterplot. Notice that the points resemble a rough parabola opening downwards. That means we must have a *negative* coefficient for our x^2 term (so our parabola can open downwards), and bam - there go Choices B and D, both of which have *positive* coefficients of x^2 and would therefore open the parabola in the wrong direction for this scatterplot.

Next, what's the difference between the remaining Choices A and C? The *only* difference is that one ends with "$+460.17$" and the other ends with "-460.17". You've probably learned by now that this term of the Quadratic Equation gives the *y*-intercept of the parabola - which is clearly a positive number on the scatterplot (the *y*-intercepts appears to be somewhere around 475, it looks like - although it's impossible to be exact from the chart - but it's *definitely* a positive number, not a negative).

Alternately - or just to be sure - you could just enter $x = 0$ into both Choice A and C and see what comes out for y. Again, the chart clearly shows a *positive* *y*-intercept, so no matter how you look at it, Choice C and its *negative* *y*-intercept won't work. We're left only with Choice A, a parabola equation that opens *downward* and has a *positive* *y*-intercept. And that's all we need for this question!

Volume 1: Posttest 1

19 Questions

Answers & Explanations follow the Posttest

Volume 1: Posttest 1

DO NOT USE A CALCULATOR ON ANY OF THE FOLLOWING QUESTIONS UNLESS INDICATED.

$$\frac{3(2+x)}{2} = 2(x-1) - 4(3-2x)$$

1. FREE RESPONSE. What is the value of $5x$ in the equation above?

2. The acceleration A of a certain experimental motorcycle in meters per second can be calculated by the following equation, where r is the radius of the rear tire in centimeters, t is the tension of the chain in newtons, c is the engine size in cubic centimeters, and n is the coefficient of drag:

$$A = \frac{\sqrt{c - 2tn}}{r^2}$$

Which of the following equations can be used to calculate the tension of the chain?

(A) $t = \dfrac{r^2 A - c}{2n}$

(B) $t = \dfrac{c - r^3 A^2}{2n}$

(C) $t = \dfrac{r^4 A^2 - c}{2n}$

(D) $t = \dfrac{c - r^4 A^2}{2n}$

$$5.5 = \left|\frac{1}{2}s - 3\right|$$

3. FREE RESPONSE: If a and b are solutions to the equation above, what is the value of $|-(b-a)|$?

4. (CALCULATOR) Christian orders an average of 14 candles per month. Each of these candles burns for an average of 5 hours before it must be replaced. Christian would like to reduce his monthly expenditures on candles by $70. Assuming each candle costs $12, which equation can Christian use to determine how many fewer hours , h, he should burn candles for each month?

(A) $\dfrac{12h}{5} = 98$

(B) $\dfrac{5h}{12} = 98$

(C) $\dfrac{12h}{5} = 70$

(D) $\dfrac{5h}{12} = 70$

5. (CALCULATOR) FREE RESPONSE: A certain airplane can travel 700 miles in 4 hours at top speed. How many hours would it take the airplane to travel 1,662.5 miles at top speed?

6. FREE RESPONSE: Two soccer teams, the Amazing Animals and the Dastardly Destroyers, are competing in the playoffs. The Amazing Animals have 20 players and their average points per player is 2.5. The Dastardly Destroyers have n players and their average points per player is 4. If the average points per player for both teams combined is 3, what is the value of n?

7. (CALCULATOR) FREE RESPONSE: If an artist is mixing red, orange, pink, and silver paints in a ratio of $1:2:3:5$ by volume respectively, what is the total volume of paint in milliliters if she uses 2.4 milliliters of pink paint?

8. (CALCULATOR) FREE RESPONSE: An obsolete unit of weight called the grzywna was subdivided into 4 wiarduneks. One wiardunek was equivalent to 6 skojecs and 12 grains were equal to 3 skojecs. If one grain is equivalent to 5 obols, how many grzywna were equivalent to 1,440 obols?

9. Line m passes through the points $(4,3)$ and $(-6,-1)$. Line n has a slope of $-\frac{1}{2}$ and intercepts the x-axis at $x=10$. If Lines m and n intersect at the point (a,b), what is the value of $3a-4b$?

(A) -7

(B) 0

(C) 1

(D) 7

Month	October	November	December
Account Value	$6,479	$6,501	$6,523

10. (CALCULATOR) FREE RESPONSE: Henrietta opened a savings account with a certain initial deposit and then made regular deposits each month into her savings account for exactly 20 years. The table above shows the value of her savings account for each of the last three months of these 20 years. What was the amount of Henrietta's initial deposit, in dollars?

	Favorite Toy					
Animal	Tennis Ball	Stuffed Fox	Blanket	Squeak Toy	Stick	Total
Cat	1	1	3	2	1	8
Dog	4	1	2	3	5	15
Turtle	0	0	1	0	1	2
Horse	2	1	3	2	1	9
Parakeet	0	0	2	2	3	7
Total	7	3	11	9	11	41

11. (CALCULATOR) An animal trainer collected information about the favorite toys of the animals he worked with. The table above shows those results. If an animal is selected at random, what is the probability that its favorite toy is a stuffed fox or a stick, given that the animal is <u>not</u> a parakeet or a turtle?

(A) $\frac{9}{32}$

(B) $\frac{5}{16}$

(C) $\frac{14}{41}$

(D) $\frac{7}{16}$

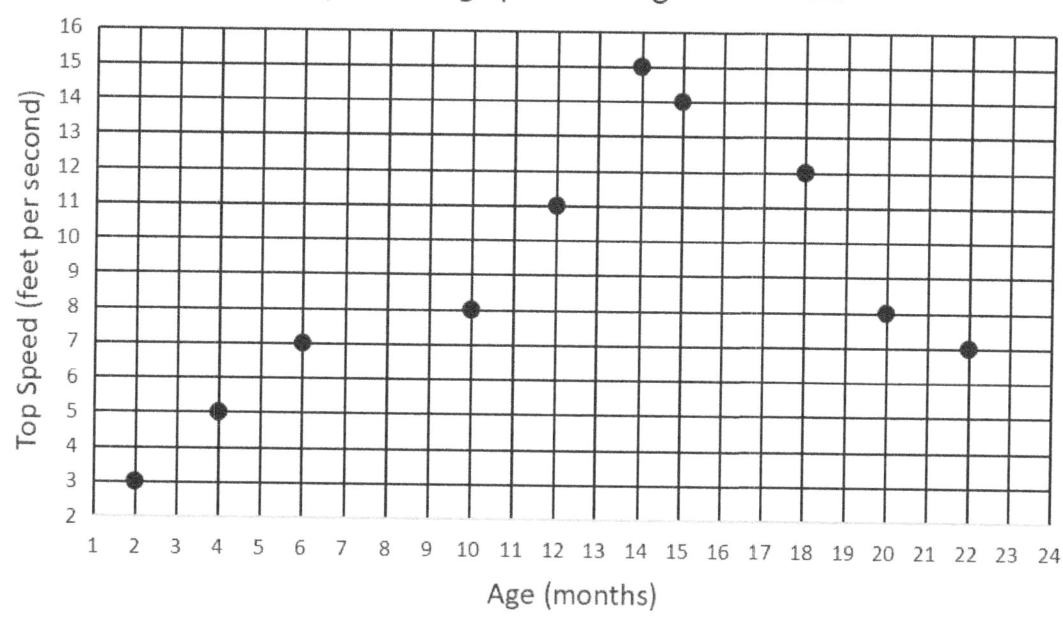

12. A researcher is studying the top running speed, in feet per second, of 10 mice of different ages. What is the difference in the top speed, in feet per second, of the oldest mouse and mouse with the highest top speed?

(A) 4

(B) 7

(C) 8

(D) 12

13. (CALCULATOR) FREE RESPONSE: A certain retro-styled motorcycle jacket was on sale for 15% off sticker price. A sales tax of 10% was applied to this discounted price. If the price paid for the jacket was \$112.20, what was the sticker price of the jacket, rounded to the nearest dollar?

14. If $6x - 3y = 2$, what is the value of $\dfrac{(3^{2x})^{-3}}{(27)^{-y}}$?

(A) $\dfrac{1}{9}$

(B) $\dfrac{1}{\sqrt{3}}$

(C) 9

(D) The value cannot be determined from the information given.

15. (CALCULATOR) FREE RESPONSE: Ian made an initial investment of d dollars on January 1st, 1995. The value of this investment increased by 10% every half of a year for two years until Ian's investment was worth \$9,370.24 on January 1st, 1997. What was the value of Ian's initial investment, in dollars?

$$(x+3)(x-2)+2x=2((x-4)^2-x)$$

16. FREE RESPONSE: What is the sum of all solutions for the equation above?

$$x^2-n=\frac{2m}{3}x$$

17. In the quadratic equation above, m and n are constants. What are the solutions for x?

(A) $\dfrac{m\pm\sqrt{m^2+9n}}{3}$

(B) $\dfrac{2m\pm\sqrt{4m^2+36n}}{3}$

(C) $\dfrac{m\pm\sqrt{m^2-36n}}{6}$

(D) $\dfrac{2m\pm\sqrt{4m^2-9n}}{6}$

18. FREE RESPONSE: What is the sum of all non-negative real solutions for the equation $x^5+36x=13x^3$?

19. Which of the following graphs shows the parabola $y=-x^2+4x-3$ in the xy-plane?

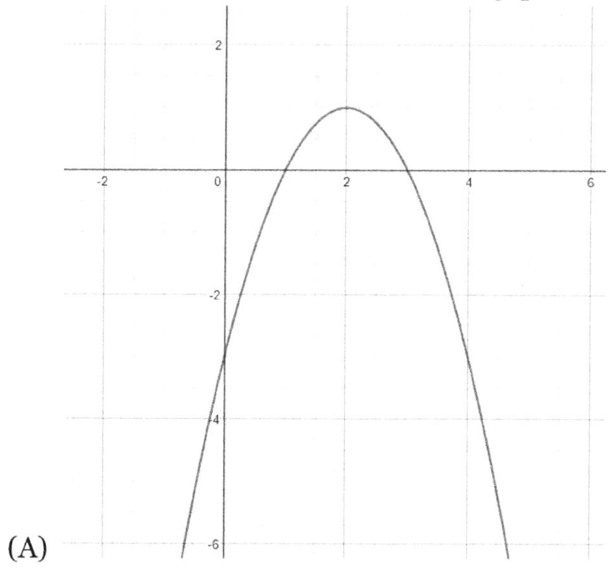

(A)

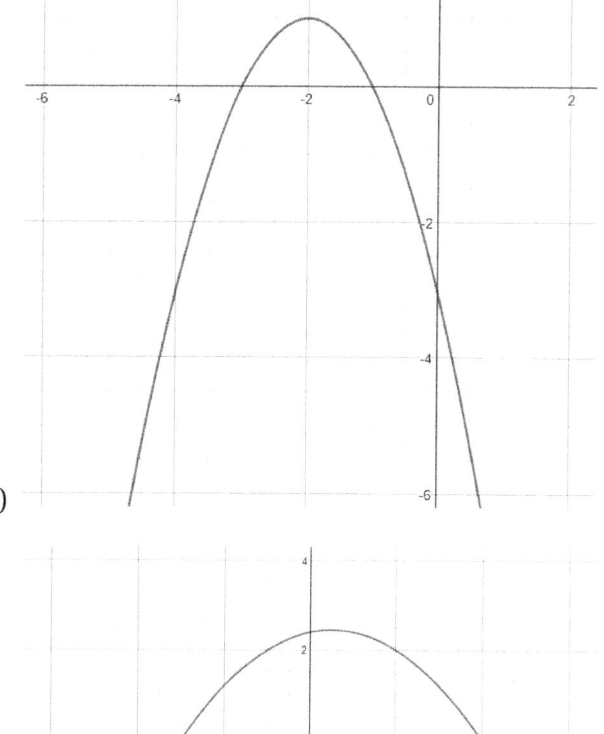

(C)

(B)

(D)

Volume 1: Posttest 1 Answers

1. 10 (Lesson 1 - Basic Algebra 1)
2. D (Lesson 2 - Advanced Algebra 2)
3. 22 (Lesson 3 - Absolute Value)
4. C (Lesson 4 - Algebra 1 Word Problems)
5. 9.5 (Lesson 5 - d=rt)
6. 10 (Lesson 6 - Averages with Algebra)
7. 8.8 (Lesson 7 - Ratios & Proportions)
8. 3 (Lesson 8 - Unit Conversions)
9. B (Lesson 9 - Linear Equations (Algebraic))
10. 1,243 (Lesson 10 - Linear Equations (Words & Tables))
11. B (Lesson 11 - Probability)
12. C (Lesson 12 - Charts & Tables)
13. 120 (Lesson 13 - Percents)
14. A (Lesson 14 - Exponents & Roots)
15. 6,400 (Lesson 15 - Exponential Growth & Decay)
16. 21 (Lesson 16 - Basic Algebra 2)
17. A (Lesson 17 - The Quadratic Formula)
18. 5 (Lesson 18 - Advanced Algebra 2)
19. A (Lesson 19 - Algebra 2 Parabolas)

Volume 1: Posttest 1 Explanations

1. **10.** This question is a **Basic Algebra 1** question. My first move would be to distribute the parentheses multiplications on both sides. Watch your negatives:

$$\frac{3(2+x)}{2} = 2(x-1) - 4(3-2x)$$
$$\frac{6+3x}{2} = 2x - 2 - 12 + 8x$$

Then Combine Like Terms on the right side:

$$\frac{6+3x}{2} = 2x - 2 - 12 + 8x$$
$$\frac{6+3x}{2} = 10x - 14$$

Now multiply both sides by 2 to get rid of the fraction on the left side:

$$(2)\frac{6+3x}{2} = (10x-14)(2)$$
$$6 + 3x = 20x - 28$$

Now get all your x terms together on one side:

$$6 + 3x = 20x - 28$$
$$-3x \quad -3x$$
$$6 = 17x - 28$$

Now get the x terms by themselves:

$$6 = 17x - 28$$
$$+28 \qquad +28$$
$$34 = 17x$$

And finally, divide both sides by 17 to isolate a single x value:

$$\frac{34}{17} = \frac{17x}{17}$$
$$2 = x$$

We've got the value of x, which is 2, but one last step before entering our final answer: make sure the question actually *asks* for the value of x.

Oh wait! The question asks for the value of $5x$, not just x! I'm glad we checked and avoided this Switcheroo **Careless Mistake**.

Let's plug in 2 for x and find that $5x = 5(2) = 10$. So our final answer is 10.

2. **D.** This question is based on Algebra 1 - it's a type of question that we explored in the lesson on **Advanced Algebra 1**. Still, it's not hard (nothing in Algebra 1 really is).

In this type of question, which I called "Word Problem with Rearrangement", we're given a giant word problem that *doesn't really matter* to solving the algebra problem.

Consider the equation we start with:

$$A = \frac{\sqrt{c-2tn}}{r^2}$$

Now consider all of the answer choices we have. They *all* have something in common - namely, the t term has been isolated on one side.

In other words, the word problem doesn't matter at all - our only job is to take the starting equation and use Algebra 1 to isolate the t term by itself.

Let's get to it. First, multiply both sides by r^2 to get rid of the fraction:

$$(r^2)A = \frac{\sqrt{c-2tn}}{r^2}(r^2)$$
$$Ar^2 = \sqrt{c-2tn}$$

Now square both sides to get rid of the square root (we're working towards getting the t term by itself. Be sure to distribute the square correctly on the right side of the equation:

$$(Ar^2)^2 = (\sqrt{c-2tn})^2$$
$$A^2r^4 = c - 2tn$$

Now keep working towards getting t by itself:

$$A^2r^4 = c - 2tn$$
$$-c \quad -c$$
$$A^2r^4 - c = -2tn$$

Almost done:

$$A^2r^4 - c = -2tn$$
$$\frac{A^2r^4 - c}{-2n} = \frac{-2tn}{-2n}$$
$$\frac{A^2r^4 - c}{-2n} = t$$

When we check our equation against the answer choices, we realize that we're almost finished, but the negative sign on the bottom of our equation has been transferred to the top of the answer equation:

$$(\frac{-1}{-1})\frac{A^2r^4 - c}{-2n} = t$$

$$\frac{c - A^2r^4}{2n} = t$$

So now we have our final answer - Choice D.

3. **22.** This question makes use of Algebra with **Absolute Values**. Remember, as explored in that lesson, we must set up *two* "solution branches" - a "positive branch" and a "negative branch" - like so:

$$-5.5 = \frac{1}{2}s - 3 \qquad 5.5 = \frac{1}{2}s - 3$$

One solution branch uses -5.5 on the left side, and the other branch uses positive 5.5. This setup allows us to drop the Absolute Value bars from both equations.

Now solve *both* branches independently:

$$-5.5 = \frac{1}{2}s - 3 \qquad 5.5 = \frac{1}{2}s - 3$$

$$+3 \quad +3 \qquad +3 \quad +3$$

$$-2.5 = \frac{1}{2}s \qquad 8.5 = \frac{1}{2}s$$

$$(2) - 2.5 = \frac{1}{2}s(2) \qquad (2)8.5 = \frac{1}{2}s(2)$$

$$-5 = s \qquad 17 = s$$

We get *two* possible values for s: $s = -5$ and $s = 17$. Both of these are true.

The question asks for the value of $|-(b-a)|$, where a and b are the solutions that we've just found. It doesn't matter which solution is a and which is b (you can test that for yourself if you don't believe me!)

So, just plug in and evaluate for our final answer. Watch your negative signs carefully, though:

$$|-(b-a)|$$
$$=|-((-5)-(17))|$$
$$=|-(-22)|$$
$$=|22|$$
$$=22$$

So, our final answer is 22. If you missed this question, it's important to review the lesson on **Absolute Value**.

4. **C.** This question comes from the lesson on **Algebra 1 Word Problems**, and this is certainly one of the tougher ones in that section. One thing that makes this question relatively tough is that the answer choices aren't actual *values*, they are *equations*.

There are two possible ways to navigate this situation: one is to solve the word problem for a final value for h, then plug your solution into each of the answer choices and eliminate any that create false equations (this is how I will solve it).

The other approach would be to translate the word problem into an Algebra setup and stop short of actually solving the final equation, instead putting it into a final form that matches one of the answer choices.

Both of these approaches have advantages and disadvantages, so I would suggest going with whichever method seems to make more sense to you personally.

My approach will be to calculate the actual solution to the word problem, then test the answer choices.

Christian wants to *save* \$70 on candles per month. The real question is how many *hours* will it cost him to save those \$70?

We could approach this by first calculating the "dollars per hour" it costs to burn a candle, each of which costs \$12 and burns for 5 hours:

$$\frac{\$12 \text{ per candle}}{5 \text{ hours per candle}} = 2.4 \,{}^{\$}\!/_{\text{hr}}$$

So, each hour of candle-burning costs \$2.40. We could divide the desired savings of \$70 by \$2.40 per hour to find how many hours equal \$70 in savings:

$$\frac{\$70}{\$2.40} = 29.166... \text{ or } 29\frac{1}{6} \text{ hours}$$

So, the "answer" to this question is that Christian must cut down on his candle burn-time by $29\frac{1}{6}$ hours per month.

We could plug in $29\frac{1}{6}$ for h in each of the answer choices and eliminate any questions that turn out false.

I'll let you test this yourself with your calculator, but you'll find that only Choice C returns a true equation.

5. **9.5.** This question comes from the lesson on **d=rt** or "Distance equals Rate times Time." These questions can be quite easy if we just recognize them by their focus on distances, times, and speeds, then set up the $d = rt$ equation. Let's do so now:

$$d = rt$$

We know the plane can travel a distance of 700 miles in 4 hours, so $d = 700$ and $t = 4$. (This is also a good time to double-check your units of measurement, but luckily we're already working in miles, hours, and miles per hour throughout the entire word problem, so no additional changes are required):

$$700 = r(4)$$

Now solve for r:

$$\frac{700}{4} = \frac{4r}{4}$$
$$175 = r$$

So, this plane's top speed is 175 miles per hour.

Now, set up a second $d = rt$ equation. This time, plug in 175 for r and the target distance of $1{,}662.5$ for d:

$$d = rt$$
$$1662.5 - 175t$$

Now solve for t:

$$\frac{1662.5}{175} = \frac{175t}{175}$$
$$9.5 = t$$

So, the plane will take 9.5 hours to travel the target distance at top speed. This question shows the ease with which we can handle distance, rate, and time questions using the $d = rt$ equation when appropriate.

Note: You can also solve this using **Ratios & Proportions** with the following setup:

$$\frac{\text{distance}}{\text{time}} = \frac{700 \text{ miles}}{4 \text{ hours}} = \frac{1662.5 \text{ miles}}{x \text{ hours}}$$

6. **10.** This question is based on **Averages with Algebra**. Recall that an Average is calculated with this formula:

$$\text{Avg} = \frac{\text{sum}}{\#}$$

We'll need to set up several average formulas to solve this question - one for the Amazing Animals team, another for the Dastardly Destroyers team, and a final *combined average* for both teams together.

First, let's set up an average for the Amazing Animals team (AA):

$$\text{Avg points per player}_{\text{AA}} = \frac{\text{sum of points}_{\text{AA}}}{\#\text{ of players}_{\text{AA}}}$$
$$2.5 \text{ Avg per player}_{\text{AA}} = \frac{\text{sum of points}_{\text{AA}}}{20 \text{ players}_{\text{AA}}}$$

We can (and should) solve this equation for the total sum of points scored by the players on the Amazing Animals team:

$$(20)2.5 \text{ Avg per player}_{\text{AA}} = \frac{\text{sum of points}_{\text{AA}}}{20 \text{ players}_{\text{AA}}}(20)$$
$$50 = \text{sum of points}_{\text{AA}}$$

So now we know that the Amazing Animals team scored a total of 50 points from all players combined.

Now let's repeat the process for the Dastardly Destroyers team (DD):

$$\text{Avg points per player}_{\text{DD}} = \frac{\text{sum of points}_{\text{DD}}}{\#\text{ of players}_{\text{DD}}}$$
$$4 \text{ Avg per player}_{\text{DD}} = \frac{\text{sum of points}_{\text{DD}}}{n \text{ players}_{\text{DD}}}$$

This equation can also be solved for the total sum of points scored by all players on the Dastardly Destroyers:

$$(n)4 \text{ Avg per player}_{\text{DD}} = \frac{\text{sum of points}_{\text{DD}}}{n \text{ players}_{\text{DD}}}(n)$$
$$4n = \text{sum of points}_{\text{DD}}$$

So, the Dastardly Destroyers scored a total of $4n$ total points for all their players combined. Notice that we are working in terms of n for the total points scored by the Dastardly Destroyers. That is perfectly OK.

Now let's set up a Combined Average equation for the two teams.

But, understand something very important first: the combined average *cannot* be calculated simply by adding the average points per player from the two teams, then dividing by 2. This won't work because the two teams don't have the same number of players.

Instead, we must add the *total points* for both teams and divide by the *total number of players* for both teams combined. So, our setup will look something like this:

$$\text{Avg points per player}_{\text{Both Teams}} = \frac{\text{sum of points}_{\text{Both Teams}}}{\#\text{ of players}_{\text{Both Teams}}}$$

Now plug in all the values we can, using both the information given in the original word problem and the new values we've found from our previous work:

$$3\text{ Avg per player}_{\text{Both Teams}} = \frac{50+4n}{20+n}$$

This equation can be solved using **Basic Algebra 1**. We're done setting up with our Average Equations, and now can just focus on using algebra to solve for the value of n:

$$(20+n)3 = \frac{50+4n}{20+n}(20+n)$$
$$60+3n = 50+4n$$
$$-50 \qquad -50$$
$$10+3n = 4n$$
$$-3n \quad -3n$$
$$10 = n$$

So, now we know that the value of n, the number of players on the Dastardly Destroyers, is 10. It took us a whole three Average setups to get here, but now we're done!

7. **8.8.** This question comes from the lesson on **Ratios & Proportions**. In that lesson, we learned about "Multiple Ratios" such as the ratio of $1:2:3:5$ given in this question - and how to solve them.

The key was to first *add* all of the components of the ratio, giving a total in this case of "11":

$$1+2+3+5=11$$

Then, to think of this mixture of paints as being divided into 11 equal parts.

The red paint is $\frac{1}{11}$ of the total mixture, the orange is $\frac{2}{11}$, the pink is $\frac{3}{11}$, and the silver is $\frac{5}{11}$. Notice that if you combine all four paint colors by adding their fractions, you will return to $\frac{11}{11}$ or "1", representing the entirety of all the paints in the mixture.

Now, if the pink represents $\frac{3}{11}$ of the mixture, and there are 2.4 milliliters of pink paint (as given in the question), then we could set up the following equation, where t is a new variable I'm using for the total volume of paint in the mixture:

$$\frac{2.4\text{ mL pink}}{t\text{ mL all colors}} = \frac{3\text{ parts pink}}{11\text{ parts all colors}}$$

Now we can easily solve this for the value of t. Your calculator will help:

$$\frac{2.4}{t} = \frac{3}{11}$$
$$(2.4)(11) = 3t$$
$$26.4 = 3t$$
$$\frac{26.4}{3} = \frac{3t}{3}$$
$$8.8 = t$$

And we find that $t = 8.8$, so the total volume of paint is 8.8 milliliters - our final answer.

8. **3.** This question is based on **Unit Conversions**. We'll use the Fencepost Method to set up our conversion between grzywna and obols. I'll use abbreviations:

$$(1{,}440\text{ obols})(\frac{1\text{ gr}}{5\text{ ob}})(\frac{3\text{ sko}}{12\text{ gr}})(\frac{1\text{ wiar}}{6\text{ sko}})(\frac{1\text{ grzy}}{4\text{ wiar}})$$

Notice that I start with the 1,440 obols and then use the Fencepost Method to set up diagonal cancellations of units from left to right (the obols are canceled by obols, the grains are canceled by grains, and so forth).

The only unit that will not be canceled out is the grzywna, which is precisely the final unit we're trying to convert into.

Now you can use your knowledge of **Fractions** to work out this multiplication:

$$\frac{(1440)(1)(3)(1)(1)}{(5)(12)(6)(4)} \text{grzywna}$$
$$= \frac{4{,}320}{1{,}440} \text{grzywna}$$
$$= 3 \text{ grzywna}$$

And, after the fractions are multiplied and simplified, we are left with 3 grzywna. Now we just need to take a language course to figure out how to pronounce any of these obsolete units of measurement!

9. **B.** This question comes from **Linear Equations (Algebraic)**. To answer the question, we'll need to find the intersection point of the two lines. And to do that, we'll need the *equations* of the two lines (we'll use the $y = mx + b$ format, since we're most familiar with it).

You must be crystal-clear on the $y = mx + b$ equation before attempting this question. If you have any doubts, be sure to review the entire lesson on Linear Equations (Algebraic) before proceeding.

There are two lines to create equations for. Let's start with the first one, line m.

The question gives us two points on line m: $(4,3)$ and $(-6,-1)$, which we can use to calculate the slope:

$$\frac{y_2 - y_1}{x_2 - x_1}$$
$$= \frac{(-1)-(3)}{(-6)-(4)}$$
$$= \frac{-4}{-10}$$
$$= \frac{2}{5}$$

So, the slope of line m is $\frac{2}{5}$, which we can enter into the $y = mx + b$ equation:

$$y_m = \frac{2}{5}x_m + b_m$$

Next, let's take one of the given (x, y) points on line m and plug it in for x and y in this equation. This will enable us to easily find the b-value or *y*-intercept.

I'll use point $(4,3)$ but you could also use $(-6,-1)$:

$$y_m = \frac{2}{5}x_m + b_m$$
$$3 = \frac{2}{5}(4) + b_m$$

Now evaluate and solve for the *y*-intercept b_m:

$$3 = \frac{2}{5}(4) + b_m$$
$$3 = \frac{8}{5} + b_m$$
$$-\frac{8}{5} \quad -\frac{8}{5}$$
$$\frac{7}{5} = b_m$$

Now we know the b-value or *y*-intercept of line m and we can complete its Linear Equation:

$$y_m = \frac{2}{5}x_m + \frac{7}{5}$$

Now it's time to repeat a similar process for line n to create its $y = mx + b$ equation. We already are given the slope of $-\frac{1}{2}$ in the question:

$$y_n = -\frac{1}{2}x_n + b_n$$

We also know that line n intercepts the x-axis at $x = 10$ - in other words, at the point $(10,0)$. We can plug these coordinates into the equation for x and y, like we did with the previous line:

$$y_n = -\frac{1}{2}x_n + b_n$$
$$0 = -\frac{1}{2}(10) + b_n$$

This enables us to evaluate and solve for the value of b_n:

$$0 = -\frac{1}{2}(10) + b_n$$
$$0 = -5 + b_n$$
$$+5 \quad +5$$
$$5 = b_n$$

And now we know the b-value or *y*-intercept of line n and we can complete its Linear Equation:

$$y_n = -\frac{1}{2}x_n + 5$$

Now that we have our two completed Linear Equations for lines m and n, we can set them equal and find their intersection point:

Our two lines are $y_m = \frac{2}{5}x_m + \frac{7}{5}$ and $y_n = -\frac{1}{2}x_n + 5$.

Set them equal:

$$\frac{2}{5}x + \frac{7}{5} = -\frac{1}{2}x + 5$$

Now solve for the value of x, which will give the x-coordinate at their intersection point. To help manage the **Fractions** I will multiply everything by 10:

$$(10)(\frac{2}{5}x + \frac{7}{5}) = (-\frac{1}{2}x + 5)(10)$$
$$4x + 14 = -5x + 50$$
$$+5x \qquad +5x$$
$$9x + 14 = 50$$
$$-14 \quad -14$$
$$9x = 36$$
$$\frac{9x}{9} = \frac{36}{9}$$
$$x = 4$$

If the x-coordinate the intersect point of these lines is 4, we can plug that in for x into *either* of the two linear equations to find the *y*-coordinate of their intersection point. I'll use our second equation, just because it contains fewer fractions:

$$y_n = -\frac{1}{2}x_n + 5$$
$$y_n = -\frac{1}{2}(4) + 5$$
$$y_n = -2 + 5$$
$$y = 3$$

Now we know the intersection point of these two lines is at $(4, 3)$. To finish the question, we need to calculate $3a - 4b$, where (a, b) is the intersection point. Therefore, $a = 4$ and $b = 3$.

Plug in and evaluate to finish the question:

$$3a - 4b$$
$$= 3(4) - 4(3)$$
$$= 12 - 12$$
$$= 0$$

We end up with a value of 0. I bet you didn't expect that! Nevertheless, it gives us our final answer of Choice B.

This question is certainly among the most difficult and time-consuming of all Algebraic Linear Equations questions, but if you feel that you understand it and could solve it on your own, then you are ready for virtually any Algebraic Linear Equations questions the SAT Math test can give you.

10. **1,243.** This question is based on **Linear Equations (Words & Tables)**. We can tell because the word problem describes a "certain initial deposit" (*y*-intercept) followed by "regular deposits each month" (a slope).

And, like all Linear Equation questions, it's heavily based in the $y = mx + b$ equation.

Could we create such an equation for this table?

Yes, we certainly could - starting with a Slope. We can take any two months from the table and find the difference between them.

For example, the difference between December and November is $\$6,523 - \$6,501$ or \$22. We would find the same difference between November and October. Apparently, Henrietta is depositing \$22 each month, and she has been for the past 20 years.

How many months are in 20 years? Well, each year is 12 months. Therefore, 20 years contain 20×12 or 240 months. So, for 240 months, Henrietta has deposited \$22 each month.

We can therefore calculate the total amount of her deposits after exactly 20 years: \$22 per month, for 240 months, would be $\$22 \times 240$ or \$5,280.

To find the amount of her initial deposit, we could take the final amount in December - the final amount of her 20 years of deposits - and subtract the $5,280 that she deposited for those 240 months:

$$\$6{,}523 - \$5{,}280 = \$1{,}243$$

And so we find that Henrietta's initial deposit must have been 1,243 dollars, which she then supplemented with $22 each month for 240 months, ending at a total of $6,523 in December of the final year.

11. **B.** This question is based on **Probability**. The essential concept of Probability on the SAT Math test is "Desired over Total":

$$\frac{\text{desired}}{\text{total}}$$

We have learned to use "Word Fractions" to clarify our setups before we start reaching for numbers, particularly when the Word Problem and/or Table make the correct values more confusing and less obvious. Let's do that now.

We want an animal selected at random - except this animal can *not* be a parakeet or turtle:

$$\frac{\text{desired}}{\text{total animals (except parakeets or turtles)}}$$

We "desire" to select an animal whose favorite toy is a stuffed fox or a stick - but again, this must *NOT* include any data from animals that are either parakeets or turtles:

$$\frac{\text{stuffed fox/stick (NOT parakeets or turtles)}}{\text{total animals (except parakeets or turtles)}}$$

Now we can find our data in the table and get the numbers to plug into this fraction. Let's start with the bottom of the fraction, where we will total all animals *except* parakeets and turtles (in other words, we can use the total number of 41 animals *minus* the 7 parakeets and 2 turtles):

$$\frac{\text{stuffed fox/stick (NOT parakeets or turtles)}}{\text{total animals (NOT parakeets or turtles)}}$$
$$= \frac{\text{stuffed fox/stick (NOT parakeets or turtles)}}{41 - (7+2)}$$
$$= \frac{\text{stuffed fox/stick (NOT parakeets or turtles)}}{41 - (9)}$$
$$= \frac{\text{stuffed fox/stick (NOT parakeets or turtles)}}{32}$$

Now let's figure out how many animals have the favorite toy of a stuffed fox or a stick, but make sure *not* to include any parakeets or turtles in this:

$$\frac{\text{stuffed fox/stick (NOT parakeets/turtles)}}{32}$$
$$= \frac{(1+1+1)+(1+5+1)}{32}$$

In the top of the fraction above, I've grouped the "stuffed fox" and "stick" favorite toys into parentheses, but made sure to exclude any parakeet and turtles.

This fraction can be cleaned up and reduced to our final answer:

$$= \frac{(1+1+1)+(1+5+1)}{32}$$
$$= \frac{(3)+(7)}{32}$$
$$= \frac{10}{32}$$
$$= \frac{5}{16}$$

And so we find that the probability of this random selection is $\frac{5}{16}$ or Choice B. Probability questions on the SAT Math test are usually simple, since they never move past "Desired over Total" fractions and small variations. The only thing to watch out for is misreading a Word Problem or Table, which is why a step-by-step setup based on "Word Fractions" *before* inputting numbers is so valuable for cutting down on careless mistakes.

12. **C.** This question revisits the topic of **Charts, Tables, and Graphs**. As many of these questions do, it mixes a chart or scatterplot with a word problem. Part of the challenge is simply figuring out what is required of us and how to translate the needed information from the chart or table.

In this question, we need to find the difference in top speed between the *oldest* mouse and the mouse with the *highest top speed*. The *x*-axis of the graph gives the age (in months) of the mice. The *y*-axis gives the top speed (in feet per second) of the mice.

It's really not a very hard question. Since the *x*-axis gives the age of the mice, the oldest mouse will be the data point that is farthest to the right of the graph. It is found at 22 months, and we can tell from the corresponding *y*-value that this mouse has a top speed of 7 feet per second.

Now, to find the mouse with the highest top speed, we just find the data point that is *highest* towards the top of the graph. It is found at 14 months and a top speed of 15 feet per second.

So, now we just use subtraction to calculate the difference in speed between 15 feet per second and 7 feet per second:

$$15-7=8$$

And we find that the difference in speeds is 8 feet per second, or Choice C.

13. **120.** This question draws on our understanding of **Percents**. If you have any trouble following my work in this explanation or understanding the concepts I'm applying, you must make sure to review that lesson in detail.

First, make sure to keep track of the difference between "sticker price" and "price paid." I always find it helpful to imagine I am actually shopping and paying for the item myself. We know the amount *paid* at the cash register, which is $112.20. It's the unknown "starting price" or "sticker price" that we need to solve for.

I will use x to represent the unknown "sticker price" or "starting price" of the jacket *before* the sale discount and the tax.

Next, I must apply a 15% discount to this sticker price. The percent multiplier will be $.85$, representing a decrease of 15% from "sea level" (as we explored in the lesson on Percents):

$$(.85)x$$

Now, we still must apply the sales tax to this amount, which will increase the cost by 10%. The percent multiplier will be 1.1, representing an increase of 10% from "sea level." Remember that we can simply multiply more than one percent change with another (for example, a discount *and* a sales tax):

$$(.85)x(1.1)$$

Now I should include the information that the jacket cost $112.20 *after* the discount and sales tax in my equation. Here's my completed setup:

$$(.85)x(1.1)=\$112.20$$

I can evaluate and solve for x, the sticker or starting price:

$$(.85)x(1.1)=\$112.20$$
$$0.935x=\$112.20$$
$$\frac{0.935x}{0.935}=\frac{\$112.20}{0.935}$$
$$x=\$120$$

And so we find that x, the sticker price of the jacket, must have been $120, and our final answer is 120.

14. **A.** This question uses the lessons we learned in **Exponents & Roots**. It also incorporates topics from Algebra and **Systems of Equations**. However, it is primarily based on Exponent Rules.

The first step is to look at how we could simplify the expression $\frac{(3^{2x})^{-3}}{(27)^{-y}}$. There are several "moves" we can make to improve and simplify it.

First of all, we can rewrite 27 in terms of 3. After all, 27 is simply 3^3:

$$\frac{(3^{2x})^{-3}}{(3^3)^{-y}}$$

This has the significant advantage of giving us a fraction with the same base (3) on top and bottom, which usually comes in handy when working with Exponents.

The next simplification is that when an exponent is raised to another exponent, we can multiply the two exponents together. This can be done on both top and bottom of our fraction:

$$\frac{(3^{2x})^{-3}}{(3^3)^{-y}}$$
$$=\frac{3^{-6x}}{3^{-3y}}$$

The next move is that in division, when the same base (3, in this case) is on top and bottom, we can *subtract* the bottom exponents from the top exponents:

$$\frac{3^{-6x}}{3^{-3y}}$$
$$=3^{(-6x)-(-3y)}$$

Watch your negative signs carefully! This exponent can be cleaned up:

$$3^{(-6x)-(-3y)}$$
$$=3^{-6x+3y}$$

Now, we've done a lot to clean up the original fraction. Compared to the $\frac{(3^{2x})^{-3}}{(27)^{-y}}$ that we started with, our new version of 3^{-6x+3y} is looking significantly cleaner.

Perhaps it's time to turn our attention to the other piece of the puzzle in this question: $6x-3y=2$

In fact, do you notice how similar this is to our exponent in 3^{-6x+3y}?

The $-6x+3y$ and the $6x-3y$ are almost the same. In fact, you could turn them into the same thing by multiplying either equation by -1:

$$(-1)(6x-3y)=2(-1)$$
$$-6x+3y=-2$$

And so now we can find the connection between the two equations/expressions. The exponent of 3^{-6x+3y} can be replaced with the value -2 from the algebra equation:

$$3^{-6x+3y}$$
$$=3^{-2}$$

Almost done with this question. Do you remember how negative exponents work? A negative exponent means to *invert* the value:

$$3^{-2}$$
$$=\frac{1}{3^2}$$
$$=\frac{1}{9}$$

And we're done - finally! Our end result is $\frac{1}{9}$, or Choice A. Successfully solving a question like this one requires a confident mastery of the rules of Exponents. Make sure you lock them down before SAT test day!

15. **6,400.** This question makes use of lessons learned in **Exponential Growth & Decay**, a topic that also uses **Percents**. If any of the following steps don't make sense to you, be sure to review both of those chapters, starting with Exponential Growth & Decay.

If an investment of d dollars increases by 10%, we can write this as multiplying by 1.1:

$$d(1.1)$$

This investment lasts for 2 years, and the 10% increase happens *twice* per year (read carefully! The 10% increases happens every *half* of a year). That means that this increase happens a total of 4 times in two years, so raise the Percent Multiplier to a power of 4:

$$d(1.1)^4$$

Getting this far in the setup is the most important part of the problem, so be sure you understand how everything works so far. Again, if it's unclear, review the lesson on Exponential Growth & Decay.

We know that the value of this investment is worth $9,370.24 *after* the two years, so set our expression equal to $9{,}370.24$:

$$d(1.1)^4=9{,}370.24$$

This equation can be evaluated and solved for the value of d. First, use your calculator to calculate $(1.1)^4$, which equals 1.4641:

$$d(1.4641)=9{,}370.24$$

Now just divide both sides to find d:

$$\frac{d(1.4641)}{1.4641}=\frac{9{,}370.24}{1.4641}$$
$$d=6{,}400$$

And now we know the value of d, or the initial investment, was $6,400.

16. **21.** This question comes from the lesson on **Basic Algebra 2**, which means we might expect to use some techniques like FOILing, Factoring, and solving Quadratic Equations of the form $ax^2+bx+c=0$.

Our first move should be to FOIL the left side:

$$(x+3)(x-2)+2x=2((x-4)^2-x)$$
$$x^2-2x+3x-6+2x=2((x-4)^2-x)$$

Let's clean up the left side by combining like terms:

$$x^2+3x-6=2((x-4)^2-x)$$

Now we need to work on the right side of the equation - first, by FOILing the $(x-4)^2$. Be sure to FOIL this; a common mistake is to think it just equals x^2-16 or x^2+16, but both of these are wrong. Remember that "squared" means "times itself", so we must FOIL out:

$$(x-4)^2$$
$$=(x-4)(x-4)$$
$$=x^2-4x-4x+16$$
$$=x^2-8x+16$$

Now plug this chunk back into our main equation, and combine like terms:

$$x^2+3x-6=2((x-4)^2-x)$$
$$x^2+3x-6=2(x^2-8x+16-x)$$
$$x^2+3x-6=2(x^2-9x+16)$$

On the right side of the equation, we need to distribute the 2 to the parentheses:

$$x^2+3x-6=2x^2-18x+32$$

Now combine all the terms on both sides. Our goal is to reach the basic Quadratic form of $ax^2+bx+c=0$. Remember that this format is the most common way to solve basic Algebra 2 equations:

$$x^2+3x-6=2x^2-18x+32$$
$$-x^2 \qquad\qquad -x^2$$
$$3x-6=x^2-18x+32$$
$$+6 \qquad\qquad +6$$
$$3x=x^2-18x+38$$
$$-3x \qquad -3x$$
$$0=x^2-21x+38$$

The next step is to Factor the right side of this equation. For a complete review of Factoring, review the lesson on **Basic Algebra 2** if needed. Remember that we need two numbers that *multiply* to $+38$ and *add* to -21. The only two numbers that accomplish this are -2 and -19:

$$0=x^2-21x+38$$
$$0=(x-2)(x-19)$$

Almost done. What are the two values of x that will make this equation true? They are the zeros (or "roots" or "solutions") of $x=2$ and $x=19$. (This step should be perfectly clear if you've studied the Basic Algebra 2 lesson).

The final question is "what is the *sum* of all solutions to the equation," so we finish it off by adding $x=2$ and $x=19$ to a final sum of 21.

There is absolutely nothing in this question that departs from the normal steps of a **Basic Algebra 2** question, so be sure you review that crucial lesson if you're having trouble with this one!

17. **A.** This question is based on **Basic Algebra 2**, but the key to solving it comes from **The Quadratic Formula**.

However, we might not realize this at first. How can we tell? There are two major clues: first, the original equation is clearly taken from Algebra 2, with a polynomial format involving both x^2 and x terms. Second, the answer choices can be recognized as Quadratic Formulas.

Remember that the Quadratic Formula is:

$$X=\frac{-b\pm\sqrt{b^2-4ac}}{2a}$$

Comparing this general formula to the answer choices, you should be able to instantly recognize the similarities.

Before we can use the Quadratic Formula, the original equation has to be arranged in the basic Quadratic Equation format of $ax^2+bx+c=0$. Let's do that first:

$$x^2-n=\frac{2m}{3}x$$
$$-\frac{2m}{3}x \quad -\frac{2m}{3}x$$
$$x^2-\frac{2m}{3}x-n=0$$

Although we've got a lot of "extra" unknowns like the constants m and n, this rearranged equation still follows the $ax^2+bx+c=0$ format.

Now, I strongly suggest you get rid of the fraction in this equation *before* entering your values into the Quadratic Formula. This is not a mandatory step, but it will greatly simplify the upcoming work, so let's do it now by multiplying both sides by 3:

$$(3)(x^2-\frac{2m}{3}x-n)=0(3)$$
$$3x^2-2mx-3n=0$$

We can now read our a, b, and c values as $a=3$, $b=-2m$, and $c=-3n$. It's critical to notice that in this Quadratic Equation, our b and c values still contain the unknown constants m and n. It's true that this is a bit unusual, but not a deal-breaker.

Go ahead and plug each of these into their places in the Quadratic Formula:

$$X=\frac{-b\pm\sqrt{b^2-4ac}}{2a}$$
$$X=\frac{-(-2m)\pm\sqrt{(-2m)^2-4(3)(-3n)}}{2(3)}$$

Now evaluate and clean up:

$$X=\frac{-(-2m)\pm\sqrt{(-2m)^2-4(3)(-3n)}}{2(3)}$$
$$X=\frac{2m\pm\sqrt{4m^2+36n}}{6}$$

We can simplify even further by factoring out a 4 from the expression under the square root sign:

$$X=\frac{2m\pm\sqrt{4m^2+36n}}{6}$$
$$X=\frac{2m\pm\sqrt{4(m^2+9n)}}{6}$$

The 4 can be "pulled out" from under the square root (which calls back to the lesson on **Exponents & Roots**) The square root of 4 is 2:

$$X=\frac{2m\pm\sqrt{4(m^2+9n)}}{6}$$
$$X=\frac{2m\pm2\sqrt{m^2+9n}}{6}$$

Almost done - but this fraction can be simplified even further, by reducing the 2 s on top and the 6 on bottom:

$$X=\frac{2m\pm2\sqrt{m^2+9n}}{6}$$
$$X=\frac{m\pm\sqrt{m^2+9n}}{3}$$

And now we have our final solution - which matches Choice A. If you're comfortable with the Quadratic Formula, you'll understand that this final format is an acceptable solution to the equation - a bit ugly-looking perhaps, but a valid solution nonetheless (plus, we can't go any further without knowing the exact values of m and n, which is why the answer choices also stop here!)

18. **5.** This type of question is explored in the lesson on **Advanced Algebra 2**. It is based on Factoring - just as covered in **Basic Algebra 2** - with the only difference that this question focuses on the higher-order equation $x^5+36x=13x^3$, which has an x^5 term, instead of our basic 2nd-order polynomials, which are limited to x^2 terms.

If we want to find the solutions for a polynomial, we should set the equation equal to 0 and try to factor the equation as much as possible.

First, set the equation equal to 0:

$$x^5+36x=13x^3$$
$$-13x^3-13x^3$$
$$x^5-13x^3+36x=0$$

Then, factor out an x from all three terms:

$$x^5-13x^3+36x=0$$
$$x(x^4-13x^2+36)=0$$

We can continue to factor using two numbers that *multiply* to 36 and *add* to -13. The only pair of numbers that satisfy this requirement are -4 and -9:

$$x(x^4-13x^2+36)=0$$
$$x(x^2-4)(x^2-9)=0$$

Notice that there are two Difference of Squares and we can continue to factor each one even further:

$$x(x^2-4)(x^2-9)=0$$
$$x(x-2)(x+2)(x-3)(x+3)=0$$

Now we can read the *five* possible values (zeros, roots, or solutions) for x that would make this equation true. From left to right in our factored form, these are $x=0$, $x=2$, $x=-2$, $x=3$, and $x=-3$.

The question asks for the sum of all non-negative solutions, which are 0, 2, and 3. Since $0+2+3=5$, our final answer is 5.

19. **A.** This question comes from the lesson on **Algebra 2 Parabolas**. We should be very familiar with the basic shape and essential qualities of a Parabola (e.g. Vertex, x-intercepts or "zeros," and whether it opens upwards or downwards).

The given equation $y=-x^2+4x-3$ has a *negative* coefficient for the x^2 term, which means the graph must open downwards. This eliminates Choice B right away.

Our next move should be to set the equation equal to 0 and Factor it, as we practiced in **Basic Algebra 2**. This will enable us to find the exact x-intercepts of the graph.

It's easier to factor if we first get rid of that negative sign on the x^2 term. We can factor out a -1 from the entire equation. Watch out for your negative signs:

$$0=-x^2+4x-3$$
$$0=-(x^2-4x+3)$$

Now we need two numbers that *multiply* to 3 and *add* to -4. The only pair of numbers that fit this description are -3 and -1:

$$0=-(x^2-4x+3)$$
$$0=-(x-3)(x-1)$$

Now we can read our two solutions for x that make this equation true, which are $x=3$ and $x=1$. Remember that these solutions can also be called zeros, roots, or x-intercepts.

In other words, the graph of this parabola must cross or intercept the x-axis at coordinates $x=3$ and $x=1$. The only graph choice that both opens downwards and has the correct x-intercepts at $+3$ and $+1$ is Choice A.

Volume 1: Posttest 2

19 Questions

Answers & Explanations follow the Posttest

Volume 1: Posttest 2

DO NOT USE A CALCULATOR ON ANY OF THE FOLLOWING QUESTIONS UNLESS INDICATED.

$$4-2n\leq\frac{4}{5}$$

1. Which of the following gives all solutions for n in the inequality above?

(A) $n\leq\frac{8}{5}$

(B) $n\geq0$

(C) $n\geq\frac{8}{5}$

(D) $n\leq\frac{12}{5}$

$$2\sqrt{x+6}=2x$$

2. What is the set of solutions to the equation above?

(A) $x=\{-2,3\}$

(B) $x=\{3\}$

(C) $x=\{-2\}$

(D) There are no solutions to the given equation.

3. A number line contains two different points N and M. These points are both 9 units from the point with coordinate -4. The solutions to which of the following equations gives the coordinates of both points?

(A) $|x+4|=9$

(B) $|4-x|=9$

(C) $|9-x|=4$

(D) $|x+9|=4$

4. (CALCULATOR) FREE RESPONSE: When designing a radio tower, a construction firm uses the tower dimensions formula $4n+b=40$, where n is the height in feet of a single level of the tower, and b is the perimeter in feet of the tower's base. A construction firm wants to use the tower dimensions formula to design a tower with a total height of 108 feet, a base perimeter of at least 24 feet but no more than 25 feet, and an odd number of levels. Within these constraints, which of the following must be the perimeter, in feet, of the tower base? (Note: Round your answer to the nearest tenth of a foot.)

5. An experimental rocket is powered by a new chemical reaction. After starting from rest on the launch pad, the rocket travels m meters in s seconds, where $m = 18s\sqrt[3]{s^2 - 4s}$. Which of the following gives the average speed of the rocket, in meters per second, over the first s seconds after it launches?

(A) $18\sqrt[3]{s-4}$

(B) $18\sqrt[3]{s^2-4s}$

(C) $18s^2\sqrt[3]{s^2-4s}$

(D) $18s\sqrt[3]{s^3-4s^2}$

6. (CALCULATOR) FREE RESPONSE: A new movie receives 30 reviews between 0 and 5 stars, inclusive. Decimal ratings are allowed to the nearest tenth of a star. In the first 20 reviews, the average of the ratings was 4.4 stars. What is the lowest rating, in stars, that the movie can receive for the 25th rating and still be able to have a rating of at least 4.5 stars for the first 30 ratings?

7. (CALCULATOR) FREE RESPONSE: Building A is 240 feet tall, and Building B is 80 feet tall. The ratio of heights of Building B to Building A is equal to the ratio of the heights of Building C to Building D. If Building C is 380 feet tall, what is the height of Building D in feet?

8. (CALCULATOR) FREE RESPONSE: An angula is an obsolete unit of length measurement equal to eight barley-corns. There are 1.5 barley-corns per vitasti and 2 vitasti per hasta. A dhanusha is equivalent to 4.5 hasta. How many angula were equivalent to 48 dhanusha?

$$2y - 5x = 4$$
$$bx = 4y - 2$$

9. FREE RESPONSE: If the System of Equations above has no real solutions, what is the value of b?

$$S(n) = 12n + 15$$

10. A certain model rocket is launched from a catapult just prior to firing the engine of the rocket. The velocity S in meters per second of the rocket is a linear function of the length of time in seconds n that the rocket's engine has been firing. The equation above can be used to calculate S. Which of the following is the best interpretation of the meaning of the number 12 in the equation?

(A) The total distance traveled by the rocket, in meters.

(B) The initial velocity of the rocket before the engine is fired.

(C) The current velocity of the rocket, in meters per second.

(D) The increase in velocity per second (acceleration) of the rocket, in meters per second.

Favorite Place	Number of Students
Mall	4
Beach	x
Movies	5
Restaurant	4
House	9

11. FREE RESPONSE: The table above shows the results of a survey that asked a group of students their favorite places to hang out. If the probability that a student chosen at random has the favorite place of either the beach or a restaurant is $\frac{1}{4}$, what is the value of x?

Time (Minutes)	Concentration (mg/mL)
0	2.5
4	2.1
8	1.8
12	1.6
16	1.5
20	1.4

12. (CALCULATOR) FREE RESPONSE: The table above shows the concentration, in milligrams per liter, of a certain medicine in a patient's bloodstream of a patient. According to the data, how many fewer milligram of medicine are present in 6 milliliters of the patient's blood at 16 minutes than in 14 milliliters of the patient's blood at 4 minutes?

Year	Albums Sold
1987	5,000
1988	5,100

13. (CALCULATOR) FREE RESPONSE: The album sales for a certain band in the years 1987 and 1988 are shown in the table above. If the percent increase in albums sold by this band from 1988 to 1989 was twice the percent increase in albums they sold from 1987 to 1988, how many albums were sold in 1989?

14. The expression $\frac{(x^{\frac{2}{3}}y^{-\frac{5}{4}})^{-4}}{\sqrt[4]{x^{-\frac{8}{3}}y^{8}}}$, where $x \neq 0$ and $y \neq 0$, is equivalent to which of the following?

(A) 1

(B) $\frac{y^3}{x^2}$

(C) $\frac{y^{\frac{13}{4}}}{x^{\frac{4}{3}}}$

(D) $\frac{x^8}{y^{27}}$

15. A researcher is studying the effect of a new pesticide on a mosquito colony. The colony begins with 20,000 mosquitoes. If the effect of the pesticide over 8 weeks can be modeled by a 3% decrease in mosquito population per week, which of the following is closest to the mosquito population at the end of 8 weeks?

(A) 1,153

(B) 8,800

(C) 14,767

(D) 15,675

16. The expression $\frac{5}{6}x^2 - 15$ can be rewritten as $\frac{5}{6}(x-b)(x+b)$, where b is a positive constant. What is the value of b?

(A) $3\sqrt{2}$

(B) $\sqrt{15}$

(C) 18

(D) 225

$$y = \frac{5}{2}x^2 - 4x + 3$$
$$6x = y + 7$$

17. How many solutions exist for the system of equations above?

(A) There are no real solutions.

(B) Exactly 1 real solution.

(C) Exactly 2 real solutions.

(D) Exactly 4 real solutions.

$$(4x+3)(ax^2+2bx-1) = 12x^3 - 7x^2 - 16x - 3$$

18. FREE RESPONSE: The equation above is true for all x, where a and b are constants. What is the value of $a+b$?

19. FREE RESPONSE: The equation of a certain parabola is given by $y - x^2 = 16 - 10x$. The vertex of this parabolas is at (n, m) and the parabola has a y-intercept at $y = t$. What is the value of $n + m + t$?

Volume 1: Posttest 2 Answers

1. C (Lesson 1 - Basic Algebra 1)
2. B (Lesson 2 - Advanced Algebra 2)
3. A (Lesson 3 - Absolute Value)
4. 24 (Lesson 4 - Algebra 1 Word Problems)
5. B (Lesson 5 - d=rt)
6. 2 (Lesson 6 - Averages with Algebra)
7. 1,140 (Lesson 7 - Ratios & Proportions)
8. 81 (Lesson 8 - Unit Conversions)
9. 10 (Lesson 9 - Linear Equations (Algebraic))
10. D (Lesson 10 - Linear Equations (Words & Tables))
11. 2 (Lesson 11 - Probability)
12. 20.4 (Lesson 12 - Charts & Tables)
13. 5,304 (Lesson 13 - Percents)
14. B (Lesson 14 - Exponents & Roots)
15. D (Lesson 15 - Exponential Growth & Decay)
16. A (Lesson 16 - Basic Algebra 2)
17. B (Lesson 17 - The Quadratic Formula)
18. 1 (Lesson 18 - Advanced Algebra 2)
19. 12 (Lesson 19 - Algebra 2 Parabolas)

Volume 1: Posttest 2 Explanations

1. **C.** This question is a **Basic Algebra 1** question that also uses inequalities. Here's how I'd do it.

My first move would be to get the n terms by themselves on one side:

$$4-2n \leq \frac{4}{5}$$
$$-4 \qquad\qquad -4$$
$$-2n \leq \frac{4}{5}-4$$

Then I would use my knowledge of **Fractions** to clean up the subtraction on the right side of the inequality. First I convert "4" into $\frac{20}{5}$ so I have a common denominator of 5. Then I can subtract the top values. Watch your negatives!

$$-2n \leq \frac{4}{5}-4$$
$$-2n \leq \frac{4}{5}-(\frac{20}{5})$$
$$-2n \leq -\frac{16}{5}$$

Now, speaking of negatives, I need to get n by itself. That requires dividing both sides by -2. Remember that any time you divide or multiply an inequality by a negative number, the direction of the inequality sign will flip.

$$\frac{-2n}{-2} \leq \frac{-\frac{16}{5}}{-2}$$
$$n \geq \frac{-\frac{16}{5}}{-2}$$

Notice that the direction of the sign has flipped.

Now, that **Fraction** on the right side of the inequality can seem a little ugly. I find it easier if I rewrite it as multiplication instead of division:

$$n \geq -\frac{16}{5} \times \frac{1}{-2}$$

Now I can finish the fraction multiplication (remember, "top times top and bottom times bottom") and simplify. Be sure you cancel your negatives:

$$n \geq -\frac{16}{5} \times \frac{1}{-2}$$
$$n \geq \frac{-16}{-10}$$
$$n \geq \frac{8}{5}$$

So our final answer is that n can be any value greater than or equal to $\frac{8}{5}$, or Choice C.

2. **B.** This type of question - a "False Roots" question - originally appeared in our lesson on **Advanced Algebra 1**, although it also makes a quick reappearance in **Basic Algebra 2**.

These questions are distinctive because they typically include an algebra equation with a square root in it, and most importantly, a set of answer choices with two answer values (in this case -2 and 3) that are either given together or just one or the other.

The *easiest* way to solve these questions on the SAT is by skipping the algebra and simply plugging in the answer choices. We only have to do this twice, since our only options are -2 and 3.

Let's do this for both answer options, starting with $x=3$:

$$2\sqrt{x+6}=2x$$
$$2\sqrt{(3)+6}=2(3)$$
$$2\sqrt{9}=6$$
$$2(3)=6$$
$$6=6$$

This first attempt of plugging in $x=3$ returns a true equation $6=6$, so it appears that $x=3$ is a valid solution.

Now let's test the other answer choice of $x=-2$:

$$2\sqrt{x+6}=2x$$
$$2\sqrt{(-2)+6}=2(-2)$$
$$2\sqrt{4}=-4$$
$$2(2)=-4$$
$$4\neq-4$$

Plugging in the choice $x=-2$ results in a *false equality* of $4\neq-4$. This tells us that $x=-2$ is *not* a solution, but $x=3$ *is* a solution, leaving us with Choice B as our final answer.

Notice that we never "did any algebra," we just plugged in the two possibilities from the answer choices and evaluated to look for true or false equalities.

If you missed this question, I highly recommend checking this topic out again: go to **Advanced Algebra 1** and carefully study the sub-section on "False Solutions & Square Roots." It's explored again in **Basic Algebra 2** under "Checking for False Roots."

3. **A.** This question asks about a number line, but the topic itself comes from the lesson on **Absolute Value** (you can also tell by the answer choices, which all use Absolute Value bars in them).

One easy way to handle this question would be to simply draw a number line and label the coordinate -4 on it (try this now). Then move to the right by nine points - you can count on your fingers, or you can calculate $-4+9$. Either way, you'll end up at 5.

Then return to the starting coordinate of -4 and move to the *left* by nine points - again you can count on your fingers, or you can calculate $-4-9$. Either way, you'll end up at -13.

So, the two solutions to this question are $x=5$ and $x=-13$.

What you could do now is test the answer choices by plugging in 5 and -13 for x and see which answer choice returns true equations for both values. You'll find that only **Choice A** works.

But here's another way. Remember that *distance* is *difference*, and "difference" is subtraction. You can calculate distance by taking the Absolute Value of the difference between two values.

We could set up an equation:

$$|x-(-4)|=9$$

What this equation "says" is the *distance* (subtraction) between x and a point at -4 is 9.

We can clean it up the negative signs little bit more:

$$|x+4|=9$$

And notice that it's identical to Choice A. So, you can see there are two solid ways of solving this question. The first one is easier if you don't feel confident setting up an equation, and the second one is better and faster if you *do* feel in control of distances and absolute value.

4. **24**. This question comes from **Algebra 1 Word Problems** and is similar to a question given in the practice set. It's a relatively complicated word problem, with several phases. Let's get into it!

First, we know we'll be using the "tower dimensions formula" $4n+b=40$. The question gives us a base perimeter "of at least 24 feet but no more than 25 feet"; however, it doesn't provide us any possible heights for each level of the tower.

So, let's find the minimum and maximum possible heights for each level. We can do this by plugging in *both* the minimum and maximum options of 24 feet and 25 feet for b and solving the two separate equations for the minimum and maximum possible n or height values:

$$\begin{aligned} 4n+b&=40 \\ 4n+(24)&=40 \\ -24\quad&-24 \\ 4n&=16 \\ \frac{4n}{4}&=\frac{16}{4} \\ n&=4 \end{aligned} \qquad\qquad \begin{aligned} 4n+b&=40 \\ 4n+(25)&=40 \\ -25\quad&-25 \\ 4n&=15 \\ \frac{4n}{4}&=\frac{15}{4} \\ n&=3.75 \end{aligned}$$

So, the minimum possible height per level is 3.75 feet, and the maximum is 4 feet. We know from the question that the tower has a total height of 108 feet. We can divide this by *both* 3.75 feet and 4 feet to find the minimum and maximum possible number of levels for this tower:

$$\frac{108\text{ feet}}{3.75\text{ feet per level}}=28.8\text{ levels}$$

$$\frac{108\text{ feet}}{4\text{ feet per level}}=27\text{ levels}$$

So, we know the tower has a minimum of 27 levels and a maximum of 28.8 levels. Of course, there's no such thing as ".8 of a level" (just like there's not ".8 of a floor" in an office building). So, we must round down - in other words, this tower has either 27 levels or 28 levels.

The question tells us that the tower has an "odd number of levels." So there *must* be 27 levels to this tower.

If each tower level is the same height, that means that each level will be $\frac{108}{27}$ or 4 feet high. We can return to the tower dimensions formula and plug in 4 feet for n, the height, then solve for b, the perimeter of the tower base:

$$\begin{aligned} 4n+b&=40 \\ 4(4)+b&=40 \\ 16+b&=40 \\ -16 \quad &-16 \\ b&=24 \end{aligned}$$

So, now we know our final answer - the base perimeter of this tower must be 24 feet.

5. **B.** This is a **d=rt** question. It focuses on distances, times, and speeds. So, we should begin with our $d=rt$ equation:

$$d=rt$$

It's also good to double-check our units of measurement at this point, but luckily we are working in meters, seconds, and meters per second throughout the problem, so no unit conversions are needed.

Now plug in the known values from the word problem. The distance is equal to $m=18s\sqrt[3]{s^2-4s}$, and this is traveled in s seconds. Don't get confused by the new variables the question gives us: m is just our distance and goes in for d, and s is just our time value for t:

$$\begin{aligned} d&=rt \\ 18s\sqrt[3]{s^2-4s}&=rs \end{aligned}$$

Now that we've plugged in the given distance and time, we can easily solve for the speed or rate r by dividing both sides by s:

$$\begin{aligned} \frac{18s\sqrt[3]{s^2-4s}}{s}&=\frac{rs}{s} \\ \frac{18s\sqrt[3]{s^2-4s}}{s}&=r \end{aligned}$$

The resulting equation can be slightly simplified by canceling the s on top of the fraction with the s on the bottom, resulting in our final answer $18\sqrt[3]{s^2-4s}=r$, or Choice B.

Not hard, if you recognize from the clues in the word problem that it's time to use our $d=rt$ equation!

6. **2.** This question is based on **Averages with Algebra**. Recall that an Average is calculated with this formula:

$$\text{Avg}=\frac{\text{sum}}{\#}$$

Let's start by setting up an Average equation for the first 20 reviews:

$$\begin{aligned} \text{Avg}_{1\text{-}20}&=\frac{\text{sum}_{1\text{-}20}}{\#_{1\text{-}20}} \\ 4.4&=\frac{\text{sum}_{1\text{-}20}}{20} \end{aligned}$$

This equation can be solved for the sum total points earned in the first 20 reviews:

$$\begin{aligned} (20)4.4&=\frac{\text{sum}_{1\text{-}20}}{20}(20) \\ 88&=\text{sum}_{1\text{-}20} \end{aligned}$$

So, the first 20 reviews combined add to a total of 88 points.

Now, let's set up an Algebra equation for the first 30 reviews:

$$\begin{aligned} \text{Avg}_{1\text{-}30}&=\frac{\text{sum}_{1\text{-}30}}{\#_{1\text{-}30}} \\ 4.5&=\frac{\text{sum}_{1\text{-}30}}{30} \end{aligned}$$

Note that we already know the sum of the first 20 reviews, which we can include in the equation:

$$4.5=\frac{88+\text{sum}_{21\text{-}30}}{30}$$

Now, if we want to know the *lowest* possible value for one of the remaining reviews, we should plug in the *highest* possible value for the other 9 reviews. (This is a classic math concept: if you want to *minimize* one value, you should *maximize* all other possible values).

The maximum review value is 5, so let's act like 9 of the final 10 reviews were perfect 5's, and our final unknown "lowest" review will be represented by the unknown x:

$$4.5 = \frac{88 + 9(5) + x}{30}$$

This equation can be cleaned up a little bit:

$$4.5 = \frac{88 + 9(5) + x}{30}$$
$$4.5 = \frac{88 + 45 + x}{30}$$
$$4.5 = \frac{133 + x}{30}$$

Now we can solve this setup for x, the value of the lowest-possible review:

$$(30)4.5 = \frac{133 + x}{30}(30)$$
$$135 = 133 + x$$
$$-133 \quad -133$$
$$2 = x$$

And so we see that 2 stars is the lowest possible rating the movie can receive for the 25th rating while still managing to reach a 4.5 star average review for the first 30 ratings.

7. **1,140.** This question is based on **Ratios & Proportions**. We can use the word problem to set up the following proportion of the heights of the buildings:

$$\frac{B}{A} = \frac{C}{D}$$

Then plug in the given heights of Buildings A, B, and C:

$$\frac{80}{240} = \frac{380}{D}$$

Then simply solve for the value of D, starting with cross-multiplication:

$$\frac{80}{240} = \frac{380}{D}$$
$$(80)D = (240)(380)$$
$$80D = 91{,}200$$
$$\frac{80D}{80} = \frac{91{,}200}{80}$$
$$D = 1{,}140$$

And now we have the value of D or the height of Building D as 1,140 feet.

8. **81.** This question is based on **Unit Conversions**. We'll use the Fencepost Method to set up our conversion between dhanusha and angula. I'll use abbreviations.

$$(48 \text{ dha})(\frac{4.5 \text{ has}}{1 \text{ dha}})(\frac{2 \text{ vit}}{1 \text{ has}})(\frac{1.5 \text{ barley}}{1 \text{ vit}})(\frac{1 \text{ angula}}{8 \text{ barley}})$$

Notice that I start with the 48 dhanusha and then use the Fencepost Method to set up diagonal cancellations of units from left to right (the dhanusha are canceled by dhanusha, the hasta are canceled by hasta, and so forth).

The only unit that will not be canceled out is the angula, which is precisely the final unit we're trying to convert into.

Now you can use your knowledge of **Fractions** to work out this multiplication:

$$\frac{(48)(4.5)(2)(1.5)(1)}{(1)(1)(1)(8)} \text{ angula}$$
$$= \frac{648}{8} \text{ angula}$$
$$= 81 \text{ angula}$$

After the fractions are multiplied and simplified, we are left with 81 angula as our final answer. The Fencepost Method comes through again on Unit Conversions! Make sure you understand how to set it up and use it.

9. **10.** This question comes from **Linear Equations (Algebraic)**. We've seen these interesting questions before in that lesson, where a System of Linear Equations has either "no real solutions" or "infinite solutions."

In that lesson, we explore in detail how "solutions" to a pair of Linear Equations are the coordinates where the lines of those equations intersect; therefore, a system with "no solutions" means the lines never intersect, which means the lines must be parallel, and therefore have identical Slope values.

So, if we put these two equations into $y = mx + b$ format, we could directly compare their m-values or Slopes, and ensure that those two slopes are parallel.

Note: the unknown b in the equation from the original question is *not* the same thing as the b in the generic Linear Equation format $y = mx + b$, which represents the *y*-intercept of a line. This question deliberately used the letter b in the problem to add to your possible confusion.

Let's start with the top equation, and maneuver it into $y = mx + b$ format:

$$2y - 5x = 4$$
$$+5x \quad +5x$$
$$2y = 5x + 4$$
$$\frac{2y}{2} = \frac{5x+4}{2}$$
$$y = \frac{5}{2}x + 2$$

Note that the slope of this line must be $\frac{5}{2}$. Now repeat the process for the bottom equation:

$$bx = 4y - 2$$
$$+2 \quad +2$$
$$bx + 2 = 4y$$
$$\frac{bx+2}{4} = \frac{4y}{4}$$
$$\frac{b}{4}x + \frac{1}{2} = y$$
$$y = \frac{b}{4}x + \frac{1}{2}$$

So, the slope of this second line must be $\frac{b}{4}$. And, since the slopes of the two lines must be the same, we can set them equal to each other:

$$\frac{b}{4} = \frac{5}{2}$$

Now solve for the value of b, starting with cross-multiplication:

$$\frac{b}{4} = \frac{5}{2}$$
$$b(2) = (4)(5)$$
$$2b = 20$$
$$\frac{2b}{2} = \frac{20}{2}$$
$$b = 10$$

And we have our final answer - the value of b must be 10.

10. **D.** This question is based on **Linear Equations (Words & Tables)**. In this question we must interpret the meaning of a value taken from a $y = mx + b$ Linear Equation, in the context of a word problem about the velocity of a model rocket.

This linear function gives S, the velocity of the rocket. From our understanding of the generic $y = mx + b$ equation, we know that m is the slope, and b is the *y*-intercept or starting value.

Therefore, in the equation $S(n) = 12n + 15$, the constant 15 is the b value and must represent the starting value of the rocket's velocity; in other words, the rocket starts at 15 meters per second just before the engine fires (this must be the speed of the rocket right as it leaves the catapult, immediately before its engines ignite).

As for the $12n$ term, the 12 must be the slope of the equation. In other words, for every increase in n (which represents seconds that the rocket's engine has been firing), the value of velocity S must increase by 12.

Therefore, we conclude that the meaning of the number 12 in the equation must be the slope of the velocity function - the increase in velocity, in meters per second, of the rocket, for each second of n - or Choice D.

11. **2.** This question is based on **Probability**. The essential concept of Probability on the SAT Math test is "Desired over Total":

$$\frac{\text{desired}}{\text{total}}$$

We have learned to use "Word Fractions" to clarify our setups before we start reaching for numbers, particularly when the Word Problem and/or Table make the correct values more confusing and less obvious. Let's do that now.

$$\frac{\text{desired}}{\text{total}} = \frac{\text{favorite is beach or restaurant}}{\text{total students}}$$

We also know that this probability is equal to $\frac{1}{4}$ from the word problem:

$$\frac{\text{favorite is beach or restaurant}}{\text{total students}} = \frac{1}{4}$$

The "Total" for this question is *all* students from the table. Unfortunately, we are missing one of the data values, for the number of students whose favorite place is the beach.

Still, we can represent this total by adding up all the known values and simply including an x term to represent the missing value:

$$\frac{\text{favorite is beach or restaurant}}{\text{total students}} = \frac{1}{4}$$
$$\frac{\text{favorite is beach or restaurant}}{4+x+5+4+9} = \frac{1}{4}$$
$$\frac{\text{favorite is beach or restaurant}}{22+x} = \frac{1}{4}$$

Now, about the top of this fraction, we can easily add the students from the table whose favorite spot is either the beach (x students) or a restaurant (4 students):

$$\frac{\text{favorite is beach or restaurant}}{22+x} = \frac{1}{4}$$
$$\frac{x+4}{22+x} = \frac{1}{4}$$

OK, we're making excellent progress. This work of translating the Word Problem and Table into a "Desired over Total" probability fraction has created a single-variable Algebra equation that we can solve for x, starting with cross-multiplication:

$$\frac{x+4}{22+x} = \frac{1}{4}$$
$$(x+4)(4) = (22+x)(1)$$
$$4x+16 = 22+x$$
$$-16 \quad -16$$
$$4x = 6+x$$
$$-x \qquad -x$$
$$3x = 6$$
$$\frac{3x}{3} = \frac{6}{3}$$
$$x = 2$$

And so we find that the value of x is 2. Check carefully to make sure that this is the final value that the word problem asked for (it is). So, we are done!

12. **20.4.** This question revisits the topic of **Charts, Tables, and Graphs**. As many of these questions do, it mixes a table of values with a word problem. Part of the challenge is simply figuring out what is required of us.

In this case, we are asked the difference ("how many fewer") in milligrams of medicine in two *different* amounts of blood at two *different* times.

The first value comes from 14 milliliters of blood at 4 minutes. The table tells us that at 4 minutes, the concentration of medicine is 2.1 mg/mL. Therefore, we can find the amount of medicine by multiplying 14 milliliters times 2.1 mg/mL:

$$14 \text{ mL} \times 2.1 \text{ mg of meds per mL} = 29.4 \text{ mg of meds}$$

The second value comes from 6 milliliters of blood at 16 minutes. The table tells us that at 16 minutes, the concentration of medicine is 1.5 mg/mL. We can find the amount of medicine at this time by multiplying 6 milliliters times 1.5 mg/mL:

$$6 \text{ mL} \times 1.5 \text{ mg of meds per mL} = 9.0 \text{ mg of meds}$$

Now we simply use subtraction to find the difference between these two amounts of medicine:

$$29.4 \text{ mg} - 9.0 \text{ mg} = 20.4 \text{ mg}$$

So, the difference is 20.4 mg, and our final answer is 20.4.

13. **5,304.** This question draws on our understanding of **Percents**. If you have any trouble following my work in this explanation or understanding the concepts I'm applying, you must make sure to review that lesson in detail.

It also uses a word problem and table to make our work a bit more confusing. Part of the problem is simply figuring out what we're supposed to be doing. The other part is correctly applying our understanding of Percents.

The key information is that "the percent increase... from 1988 to 1989 was *twice* the percent increase... from 1987 to 1988."

The table only gives data about 1987 and 1988, so we should calculate the percent increase for these two years.

Percent Change is easily calculated using the fraction $\frac{\text{difference}}{\text{original}}$ (although remember that this will give us a *decimal*, which must be multiplied by 100 to convert to a percent).

$$\frac{\text{difference}}{\text{original}} = \frac{5100-5000}{5000}$$

In this situation, the "original" would be the albums sold in the first year of 1987. The "difference" is the subtraction of the old value (5000) from the new value (5100).

We can evaluate and find the percent change from 1987 to 1988:

$$\frac{5100-5000}{5000} = \frac{100}{5000} = .02$$

Remember to multiply by 100 to convert this decimal to percent:

$$.02(100\%) = 2\%$$

So, there was a 2% increase from 1987 to 1988. And, we know that the increase from 1988 to 1989 was *twice* this 2% increase, or a 4% increase.

So, we must calculate the result of a 4% increase from the albums sold in 1988. Our percent multiplier will be 1.04, representing a 4% increase in "sea level" from the albums in 1988 (remember, if any of these concepts are unclear, you must carefully review the lesson on Percents):

$$(1.04)5{,}100 = 5{,}304$$

And so we find that the band must have sold 5,304 albums in 1989, a 4% increase over the number of albums they sold the year before.

14. **B.** This question uses lessons we learned in **Exponents & Roots**. The starting expression of $\frac{(x^{\frac{2}{3}}y^{-\frac{5}{4}})^{-4}}{\sqrt[4]{x^{-\frac{8}{3}}y^{8}}}$ must be greatly simplified, using the rules of Exponents & Roots, to arrive at a "cleaner" version that matches one of the answer choices.

I'll also note that the $x \neq 0$ and $y \neq 0$ is *not* very important or useful. This simply ensures that we don't end up with a fraction that divides by 0, producing an "undefined" result.

Instead, our focus should be on the rules of Exponents.

First of all, whenever an exponent is raised to another exponent, the two exponents can be simplified by multiplying them together. The top of our fraction gives an excellent opportunity to do so. Be sure to distribute the -4 exponent to *both* the x and y terms, and watch your negatives:

$$\frac{(x^{\frac{2}{3}}y^{-\frac{5}{4}})^{-4}}{\sqrt[4]{x^{-\frac{8}{3}}y^{8}}} = \frac{x^{\frac{-8}{3}}y^{\frac{20}{4}}}{\sqrt[4]{x^{-\frac{8}{3}}y^{8}}}$$

The fractional exponent $\frac{20}{4}$ on the y term can also be reduced like any ordinary fraction:

$$\frac{x^{\frac{-8}{3}}y^{5}}{\sqrt[4]{x^{-\frac{8}{3}}y^{8}}}$$

The next step is to rewrite the $\sqrt[4]{\ }$ on the bottom of the fraction as the exponent "$\frac{1}{4}$". Remember that any root can be rewritten as a fractional exponent:

$$\frac{x^{\frac{-8}{3}}y^{5}}{(x^{-\frac{8}{3}}y^{8})^{\frac{1}{4}}}$$

Now on the bottom, apply again the rule that whenever an exponent is raised to another exponent, the two exponents can be simplified by multiplying them together. We get:

$$\frac{x^{\frac{-8}{3}}y^5}{x^{-\frac{8}{12}}y^{\frac{8}{4}}}$$

Both of the fractional exponents on the bottom can be reduced:

$$\frac{x^{\frac{-8}{3}}y^5}{x^{-\frac{2}{3}}y^2}$$

This expression certainly looks a lot prettier than it once did. But we're not done yet. Remember that negative exponents mean "invert", so any negative exponents on top of the fraction should move to the bottom, and any negative exponents on the bottom of the fraction should move to the top:

$$\frac{x^{\frac{-8}{3}}y^5}{x^{-\frac{2}{3}}y^2}$$
$$=\frac{x^{\frac{2}{3}}y^5}{x^{\frac{8}{3}}y^2}$$

This can continue to simplify, because when a fraction has the same bases on top and bottom (such as x on top and bottom), then the exponents of the bottom can be subtracted from the top:

$$\frac{x^{\frac{2}{3}}y^5}{x^{\frac{8}{3}}y^2}$$
$$=(x^{\frac{2}{3}-\frac{8}{3}})(y^{5-2})$$

These fraction subtractions can be cleaned up:

$$(x^{\frac{2}{3}-\frac{8}{3}})(y^{5-2})$$
$$=x^{-\frac{6}{3}}y^3$$

And the fractional exponent on the x term can be reduced:

$$x^{-2}y^3$$

Finally, we can again apply the negative exponent on the x term to invert it to the bottom of a fraction:

$$x^{-2}y^3$$
$$=\frac{y^3}{x^2}$$

And *finally* we're done! Our final, simplified version of the expression is $\frac{y^3}{x^2}$. This question really tests our understanding of the rules of exponents and roots. However, our final answer perfectly matches Choice B.

15. **D.** This question makes use of lessons learned in **Exponential Growth & Decay**, a topic that also uses **Percents**. If any of the following steps don't make sense to you, be sure to review both of those chapters, starting with Exponential Growth & Decay.

If a colony of 20,000 mosquitoes decreases by 3%, we can write this as multiplying the initial population by $.97$:

$$20,000(.97)$$

This decrease happens once per week for 8 weeks, so raise the Percent Multiplier to a power of 8:

$$20,000(.97)^8$$

Getting this far in the setup is the most important part of the problem, so be sure you understand how everything works so far. Again, if it's unclear, review the lesson on Exponential Growth & Decay.

This expression can be evaluated (probably on your calculator) for the final population of the mosquito colony after 8 weeks:

$$20,000(.97) \approx 15,674.867...$$

This final value can be rounded to 15,674, and the closest answer is Choice D.

16. **A.** This question comes from the lesson on **Basic Algebra 2**, so we'd expect that it involves some mix of FOILing, Factoring, and/or the basic Quadratic Equation form of $ax^2 + bx + c = 0$.

Here's I'll handle this question. First, set the two expressions equal to each other, like this:

$$\frac{5}{6}x^2 - 15 = \frac{5}{6}(x-b)(x+b)$$

It's valid to set the equations equal since the question tells us that one side is just the "rewritten" form of the other side.

Next, I would multiply both sides by $\frac{6}{5}$ to eliminate the fractions. Make sure you don't forget to distribute this to the -15 on the left side of the equation:

$$(\frac{6}{5})(\frac{5}{6}x^2 - 15) = (\frac{5}{6}(x-b)(x+b))(\frac{6}{5})$$
$$x^2 - 18 = (x-b)(x+b)$$

Now, FOIL out the right side of the equation. It's good to notice that it is a Difference of Squares:

$$x^2 - 18 = (x-b)(x+b)$$
$$x^2 - 18 = x^2 + xb - xb - b^2$$

Of course, the right side can be cleaned up (if you notice the Difference of Squares, you're able to skip the previous step entirely):

$$x^2 - 18 = x^2 - b^2$$

Now subtract the x^2 from both sides:

$$\begin{aligned} x^2 - 18 &= x^2 - b^2 \\ -x^2 \quad & \quad -x^2 \\ -18 &= -b^2 \end{aligned}$$

We're almost done - let's find the value of b:

$$(-1) - 18 = -b^2(-1)$$
$$18 = b^2$$
$$\sqrt{18} = \sqrt{b^2}$$
$$\sqrt{18} = b$$

Although $\sqrt{18}$ is the correct value of b, it is not any answer choice that we can see. That is because $\sqrt{18}$ can be simplified further, using the techniques we learned in the lesson on **Exponents & Roots:**

$$\begin{aligned} &\sqrt{18} \\ &= \sqrt{9 \times 2} \\ &= \sqrt{9} \times \sqrt{2} \\ &= 3\sqrt{2} \end{aligned}$$

So, now we have our final answer - $3\sqrt{2}$, or Choice A.

17. **B.** This question is based our Algebra 2 lesson on **The Quadratic Formula**. Specifically, it will use the "Discriminant," which we covered in-depth in that lesson.

The Discriminant is the perfect tool for determining *how many solutions exist* for a Quadratic Equation. It doesn't tell you what those solutions *are*, but then again, this question doesn't ask for them - just how *many* there are.

To calculate the Discriminant, we first need to combine these two equations into one. The best way to do that would be to Isolate, then Substitute a variable (this technique is explored deeply in **Systems of Equations**).

The second equation is much simpler. It would make sense to Isolate y, since the top equation has a y term just sitting out in the open on the left side.

$$\begin{aligned} 6x &= y + 7 \\ -7 \quad & \quad -7 \\ 6x - 7 &= y \end{aligned}$$

Now that we know $y = 6x - 7$ from the bottom equation, we can substitute this for y in the top equation:

$$6x - 7 = \frac{5}{2}x^2 - 4x + 3$$

To use the Discriminant, our equation must first be arranged into the standard $ax^2 + bx + c = 0$ Quadratic Equation format. That's our next step:

$$6x - 7 = \frac{5}{2}x^2 - 4x + 3$$

$$+7 \qquad\qquad +7$$

$$6x = \frac{5}{2}x^2 - 4x + 10$$

$$-6x \qquad\qquad -6x$$

$$0 = \frac{5}{2}x^2 - 10x + 10$$

We can now read our a, b, and c values as $a = \frac{5}{2}$, $b = -10$, and $c = 10$.

The Discriminant is calculated by $b^2 - 4ac$. Again, all of this is covered in detail in the lesson on **The Quadratic Formula**. Let's plug our a, b, and c values in and see what it tells us:

$$\begin{aligned}
&b^2 - 4ac \\
&= (-10)^2 - 4(\frac{5}{2})(10) \\
&= 100 - (10)(10) \\
&= 100 - 100 \\
&= 0
\end{aligned}$$

The value of the Discriminant for this System of Equations is 0. Remember that if the Discriminant is *positive*, then the equation will have *two* solutions. If the Discriminant is *negative*, the equation will have *no* real solutions. And if the Discriminant is *zero* the equation will have exactly *one* real solution.

And that gives us our final answer. This System of Equations must have exactly one real solution, or Choice B. Again, the question doesn't require that we find what those solutions *are*, only how *many* there would be - making the Discriminant the perfect tool for the job.

18. **1.** This question is a very "SAT-style" Algebra 2 question that we explored in the lesson on **Advanced Algebra 2**.

The general process to solve this is first to multiply out the left side of the equation, then simplify as much as possible, and then start looking for useful similarities between the left and right sides of the equation.

Let's start by multiplying out the left side. I will ignore the right side for the moment to keep things looking a little cleaner:

$$(4x+3)(ax^2+2bx-1)$$
$$=4ax^3+8bx^2-4x+3ax^2+6bx-3$$
$$=4ax^3+3ax^2+8bx^2-4x+6bx-3$$

Note that in the final line of my work, I've just rearranged the terms to "group" similar terms next to each other.

Now let's bring the right side of the equation back in:

$$4ax^3+3ax^2+8bx^2-4x+6bx-3=12x^3-7x^2-16x-3$$

We could go ahead and get rid of the -3 on both sides:

$$4ax^3+3ax^2+8bx^2-4x+6bx=12x^3-7x^2-16x$$

At this point, though, a lot of students will get stuck. The key is to compare the $4ax^3$ on the left side with the $12x^3$ on the right side. Since this is a balanced equation, the two sides of the equation must be equal. And since the two sides are equal, it's reasonable to assume that the coefficients of the x^3 terms will be equal to each other as well.

In other words, it would make sense if $4ax^3=12x^3$. And if this is true, we can solve for the value of a:

$$\frac{4ax^3}{x^3}=\frac{12x^3}{x^3}$$
$$4a=12$$
$$\frac{4a}{4}=\frac{12}{4}$$
$$a=3$$

If $a=3$, we can plug this value back into all a terms in the equation. We had already reached the setup below:

$$4ax^3+3ax^2+8bx^2-4x+6bx=12x^3-7x^2-16x$$

Remember, $a=3$. Plug it in:

$$4ax^3+3ax^2+8bx^2-4x+6bx=12x^3-7x^2-16x$$
$$4(3)x^3+3(3)x^2+8bx^2-4x+6bx=12x^3-7x^2-16x$$

Clean it up:

$$12x^3+9x^2+8bx^2-4x+6bx=12x^3-7x^2-16x$$

We can cancel out the $12x^3$ terms by subtracting them on both sides:

$$9x^2+8bx^2-4x+6bx=-7x^2-16x$$

Now, just as we compared the coefficients of the x^3 earlier in the explanation, we will compare our x^2 on the right and left sides of the equations.

Remember, this is a balanced equation, so it's reasonable to expect a balanced number of x^2 terms on the left and right side. Therefore:

$$9x^2 + 8bx^2 = -7x^2$$

We can use this setup to solve for the value of b:

$$9x^2 + 8bx^2 = -7x^2$$
$$-9x^2 \qquad\qquad -9x^2$$
$$8bx^2 = -16x^2$$
$$\frac{8bx^2}{x^2} = \frac{-16x^2}{x^2}$$
$$8b = -16$$
$$\frac{8b}{8} = \frac{-16}{8}$$
$$b = -2$$

So now we've found that the value of b must be -2. If you wanted to be extra-careful, you could plug $b = -2$ back into our previous work to make sure it checks out:

$$12x^3 + 9x^2 + 8bx^2 - 4x + 6bx = 12x^3 - 7x^2 - 16x$$
$$12x^3 + 9x^2 + 8(-2)x^2 - 4x + 6(-2)x = 12x^3 - 7x^2 - 16x$$

Now evaluate and clean up the left side to make sure it's equal to the right side:

$$12x^3 + 9x^2 - 16x^2 - 4x - 12x = 12x^3 - 7x^2 - 16x$$
$$12x^3 - 7x^2 - 16x = 12x^3 - 7x^2 - 16x$$

And indeed, we find that the values $a = 3$ and $b = -2$ have produced equal expressions on both the left and right sides of our equation. This is proof that our work is correct and $a = 3$ and $b = -2$.

To finish the question, simply calculate the value of $a + b$, as requested:

$$a + b$$
$$= (3) + (-2)$$
$$= 1$$

And this gives our final answer of 1. It's the kind of Algebra 2 question that I've only ever seen on the SAT Math test - nowhere else! (They deserve points for creativity... I guess?)

19. **12.** This question comes from the lesson on **Algebra 2 Parabolas**. We should be very familiar with the basic shape and essential qualities of a Parabola (e.g. Vertex, x-intercepts or "zeros," and whether it opens upwards or downwards).

This question tests our ability to calculate some of the most important points on the graph of a parabola.

We should start by finding the coordinates of the Vertex. Remember that the Vertex is the highest or lowest point of the parabola (depending on whether the parabola opens upwards or downwards). The Vertex is also found at the exact *middle* of the parabola, because parabolas have horizontal symmetry.

Before we can find the vertex, though, we should put our equation into the standard quadratic equation form of $ax^2+bx+c=0$. Our starting equation is $y-x^2=16-10x$. Let's rearrange it:

$$\begin{aligned} y-x^2 &= 16-10x \\ +x^2 \quad & \quad +x^2 \\ y &= x^2-10x+16 \end{aligned}$$

To find the center of this parabola, and from there the Vertex, we need to calculate the x-intercepts. Set $y=0$ and Factor, as we learned to do in **Basic Algebra 2**:

$$\begin{aligned} y &= x^2-10x+16 \\ 0 &= (x-8)(x-2) \end{aligned}$$

This gives us x-intercepts at $x=8$ and $x=2$. I'm glossing over some of the details because you should already understand out work so far based on the work we did in Basic Algebra 2.

If the x-intercepts are at $x=8$ and $x=2$, we can find the center of the Parabola at the *midpoint* of these coordinates. Add the two coordinates and divide by two to find their midpoint (you can imagine this as the **Average** of the two coordinates).

$$\frac{8+2}{2}=\frac{10}{2}=5$$

The midpoint of the Parabola will be at $x=5$.

However, we still need the y-coordinate of the vertex. It's easily found by plugging $x=5$ back into the equation of the Parabola:

$$\begin{aligned} y &= x^2-10x+16 \\ y &= (5)^2-10(5)+16 \\ y &= 25-50+16 \\ y &= -9 \end{aligned}$$

The y-coordinate of the vertex is at $x=-9$. The complete coordinates of the vertex are $(5,-9)$.

This question also requires that we find the y-intercept of the Parabola. The y-intercept of any graph is found at the point where the graph crosses the y-axis, where $x=0$. Simply plug $x=0$ into the equation of Parabola and evaluate:

$$\begin{aligned} y &= x^2-10x+16 \\ y &= (0)^2-10(0)+16 \\ y &= 0-0+16 \\ y &= 16 \end{aligned}$$

Now we have everything we need to finish the question. The vertex, given as (n,m) in the question, is $(5,-9)$, so $n=5$ and $m=-9$. The y-intercept, given as t in the question, is at 16, so $t=16$.

To finish the question, simply evaluate $n+m+t$:

$$\begin{aligned} & n+m+t \\ &= (5)+(-9)+(16) \\ &= 12 \end{aligned}$$

The sum of $n+m+t$ is 12, which is our final answer.

Review of the SAT Math Test

This brings us to the end of the math lessons in Volume 1 of this book. We've diagnosed our weak spots with the Pretest and Posttest and covered a huge variety of topics from Algebra 1 and Algebra 2.

If you've followed the recommended study plan in my book, you are fully prepared to move onto Volume 2 of *SAT Math Mastery* (available on Amazon at https://amzn.to/2z6hMge), where we'll explore more advanced Algebra 2 topics, as well as the complete set of Geometry and Statistics questions on the SAT Math test.

Remember, Algebra 1 and Algebra 2 compose almost 85% of the SAT Math test. Geometry and Statistics are certainly important, but at only about 15% of the test questions, they are secondary topics when compared to your essential Algebra skills.

Algebra 1 is the core of the test. **Basic Algebra 1**, **Linear Equations**, **Algebra 1 Word Problems**, and **Systems of Equations** (covered in Volume 2) are the beating heart of the test; the foundation that everything else is built on.

Algebra 2 also contains a crucial set of commonly-tested skills. Mastery of **Basic Algebra 2** - factoring, FOILing, Parabolas and Quadratic Equations - should be one of the highest priorities for intermediate-level students.

The SAT also puts a heavy emphasis on Charts, Tables, and Graphs - which can be mixed with a variety of other topics. Be sure you're comfortable with reading key information from these question formats.

Always watch your **Careless Errors**! Remember to write all your steps clearly and double-check your Negative Signs and Distributing, and make sure you've read the question correctly.

Before you put this book away, please read **What to Do Next** and **Final Words** on the following pages. I want to make sure you understand how to continue your studies and keep improving your score on the SAT Math test.

What to Do Next

If you've completed both Pretests, studied the lessons for all the topics you missed, completed both Posttests and carefully reviewed the lessons for any topics you missed, then it's time to move on to Volume 2 of *SAT Math Mastery*, which is available on Amazon.com at https://amzn.to/2z6hMge.

Whenever you have the time, return to this first volume for review. You may not have finished all the Pretest or Posttest questions yet - come back and do them in the near future. Let these diagnostic tests help you uncover your weaknesses in the math topics. Study the lessons you struggle with and complete all of the practice questions for each lesson.

You can always work old questions again on a clean sheet of paper. You will find that doing the same questions again will reveal new connections and build your confidence. After all, the SAT Math test itself just asks variations on these same questions.

There is still a lot to do. Your score increase will directly relate to the time and effort you put in. Order the second volume of *SAT Math Mastery* from Amazon and keep your momentum going!

Final Words

There is nothing to fear from the SAT Math test as long as you practice and prepare.

If there's one pattern I've consistently noticed among my students, it's that almost none of the math questions seem difficult once they're broken down into steps and explained.

The biggest problems you face are either not recognizing the problem type or not remembering what you've learned about it - or simply feeling like you never understood the topic in the first place.

By investing your time and energy into this book, you've taken the necessary steps towards overcoming all three of these challenges.

I believe in you, as I do all my students. I know you want to do better on the SAT Math test, and I know you can, and you will. Never let this test intimidate you. Face it head-on and give it all you've got.

SAT Math is like anything else in life. If you practice it, if you invest time into breaking it down and understanding each step, you'll improve.

Like any other skill, it takes time. It takes patience, and it takes commitment and willpower. But if you diagnose your weak spots with the Pretests and Posttests, master the lessons in Volumes 1 and 2 of this book, then practice in the Official SAT Study Guide, you *will* get a higher score on the SAT Math Test.

And it will make a difference. With your higher scores, you'll have better options for colleges and universities and better chances of winning scholarships. The impact on your life and career can be profound.

Best of all, you'll experience the feeling of being proud of your results and confident in yourself.

Thank you for studying with me! And if you have a moment right now, **please leave a review** and feedback for this book on Amazon.com :)

Finally, let me know however I can help. You can always send an email to Help@LovetheSAT.com and someone from my team will get back to you as soon as possible.

About the Author

Christian Heath has been teaching SAT & ACT Prep to high school students full-time since 2009, making him one of the most experienced SAT & ACT specialists in the world.

In 2005, he scored a 1590 out of 1600 on the SAT as a senior in high school and was a National Merit semifinalist before being accepted into his top-choice school of Pomona College in Claremont, California - one of the most selective liberal arts colleges in the United States. He graduated from Pomona in 2009 with a B.A. in Music and additional concentration in Piano Performance, winning three awards in music before graduating.

Since then, he has achieved perfect scores on both the SAT and ACT tests, including perfect scores on each individual section of both tests.

Over the past 10 years he has focused on teaching, writing, and building the best SAT and ACT prep curriculum in the world. He's worked personally with over 1,500 students - including the younger siblings of many former students - and racked up the astonishing sum of over 12,000 teaching hours in both 1-on-1 and group lessons.

After a decade of tutoring, he's learned to see the universal underlying patterns of SAT and ACT prep, allowing him to predict the help families need and answer key questions that they may never have even thought to ask.

Students and parents consistently describe Christian's positive impact on their SAT and ACT scores in dozens of pages of glowing testimonials. He is the top-rated SAT & ACT Prep tutor in his hometown of Austin, Texas.

He's authored 10 books and courses for high school students and parents. If you liked this book, check out his textbook *SAT & ACT Grammar Mastery, Ed. 2* on Amazon at https://amzn.to/2vpyqpo, which will teach any student to master the SAT Writing & Language test and the ACT English test. Also, don't forget to order Volume 2 of *SAT Math Mastery*, also available on Amazon at https://amzn.to/2z6hMge.

Christian's free blog articles have attracted readers from all around the world, giving him the opportunity to offer consultations to students and parents on every continent (except Antarctica).

He's also been honored with the opportunity to teach SAT Prep in person to students at an international school in Chengdu, China - an invitation he's accepted three times.

In his free time, Christian loves to read sci-fi and fantasy books, ride motorcycles, play guitar and piano, write songs and learn about small business - and of course, to keep developing his SAT and ACT Prep skills!

You can learn more about Christian, including student testimonials, credentials, and even his favorite books, music & movies on our website at www.LovetheSAT.com/about-founder-Christian-Heath.

About Love the SAT Test Prep

Founded in 2011 by Christian Heath, Love the SAT Test Prep is a small, independent SAT & ACT Prep company based in Austin, Texas.

Our mission is "to help high school students significantly improve their SAT and ACT scores and provide an awesome experience for students, parents, and our tutors."

We work exclusively with high school students to raise their SAT & ACT test scores. It's our sole mission - we don't tutor any other topics - not even the SAT 2 Subject Tests or College Counseling - although families ask us for our advice on these topics all the time!

As a result of our specialized focus and dedication to this mission, we've achieved the highest ratings for any SAT & ACT Prep center in the competitive test-prep market of our home city.

Visit our main website and blog at www.LovetheSAT.com, or find us on Instagram or Twitter (both @LovetheSAT) or Facebook at www.facebook.com/lovethesat.

Better yet, check out the next page for a special offer exclusively for readers of this textbook.

We are quite happy to share our expertise and experience with you. Please contact us any time by email for questions at Help@LovetheSAT.com.

A Special Bonus for Readers of this Book

As one of the top SAT prep tutors in the world, I can help you increase your SAT score far beyond the lessons in this math textbook.

I've got some special SAT prep stuff available for you at www.LovetheSAT.com/sat-math-mastery-bonus.

Just follow the link above or enter it into your browser. This bonus content is *only* available to people who have this math book.

If you're trying to get a higher SAT score, I *know* these bonus materials will help you. Follow the link and claim your score-raising SAT bonuses!

Also by Christian Heath

SAT Math Mastery, Vol.2: Advanced Algebra, Geometry and Statistics

Get higher SAT math scores - guaranteed!

The second volume of SAT Math prep adds another 19 critical math lessons that break the math test down into easy topics to master before test day. Over 325 more realistic SAT practice questions exclusive to this textbook. Comprehensive Pretest & Posttest diagnostics to quickly identify your weak spots.

Available on Amazon at https://amzn.to/2z6hMqe.

SAT & ACT Grammar Mastery, Ed.2

Get higher SAT & ACT grammar scores - guaranteed!

A revolutionary nesw grammar textbook for higher SAT & ACT scores. Master the seventeen rules of the SAT Writing and Language and ACT English sections in record time.

17 lessons break the grammar tests down into easy topics to master before test day. Over 320 realistic SAT & ACT practice questions exclusive to this textbook. Comprehensive Pretest & Posttest diagnostics to quickly identify your weak spots.

Available on Amazon at https://amzn.to/36LW9Nl.

Ultimate Time Management for Teens and Students

If there's one thing that unites every high school student, it's that they never have enough time or energy to get everything done. It's time for that to change.

This book contains an arsenal of tips, tricks, and strategies from a veteran SAT & ACT tutor and elite-college graduate that will work for every high school student at any point in their high school career.

Get better grades, have more fun, reduce your anxiety, enjoy life more, win more scholarships, and get into a better college!

Available on Amazon at https://amzn.to/2SxWPj8.

Made in the USA
Coppell, TX
17 February 2022

73694431R00267